Scale in Spatial Information and Analysis

Scale in Spatial Information and Analysis

Jingxiong Zhang, Peter M. Atkinson, and Michael F. Goodchild

CRC Press
Taylor & Francis Group
Boca Raton London New York

CRC Press is an imprint of the
Taylor & Francis Group, an **informa** business

CRC Press
Taylor & Francis Group
6000 Broken Sound Parkway NW, Suite 300
Boca Raton, FL 33487-2742

First issued in paperback 2017

Version Date: 20140306

ISBN 13: 978-1-4398-2937-0 (hbk)
ISBN 13: 978-1-138-07536-8 (pbk)

Library of Congress Cataloging-in-Publication Data

Zhang, Jingxiong, 1964-
    Scale in spatial information and analysis / Jingxiong Zhang, Peter M. Atkinson, and Michael F. Goodchild.
        pages cm
    Includes bibliographical references and index.
    ISBN 978-1-4398-2937-0 (hardcover : alk. paper)
    1. Geographic information systems. 2. Spatial analysis (Statistics) I. Title.

G70.212.Z52 2014
910.285--dc23   2013048103

Visit the Taylor & Francis Web site at
http://www.taylorandfrancis.com

and the CRC Press Web site at
http://www.crcpress.com

Contents

Preface...ix

Acknowledgment xi

Authors... xiii

Chapter 1 Introduction 1

    1.1   Issue of Scale 1

    1.2   Models of Scale 4

    1.3   Scaling Up and Down... 9

    1.4   Book Chapters 12

Chapter 2 Geographic Representations.. 17

    2.1   Geo-Atoms 20

    2.2   Geo-Fields 21

    2.3   Geo-Objects... 25

    2.4   Hierarchical Data Structures 28

         2.4.1   Quadtrees 28

         2.4.2   Hierarchical TINs... 29

    2.5   Discussion... 31

         2.5.1   Geo-Dipoles and Spatial Interactions 31

         2.5.2   Temporal Dimension 32

         2.5.3   Scale Dependence of Representations................ 32

Chapter 3 Geospatial Measurements 35

    3.1   Framework for Spatial Sampling................................... 36

         3.1.1   Spatial Sampling Design 36

         3.1.2   Data Supports 43

         3.1.3   Spatial Resolution in Remotely Sensed Images 44

              3.1.3.1   Rayleigh Criterion.............................. 45

              3.1.3.2   MTF and Its Equivalent Bandwidth 45

              3.1.3.3   Effective Radiometric Resolution Element... 46

    3.2   Optical Remote Sensing and Resolution 47

         3.2.1   Spatial and Spectral Responses.......................... 47

         3.2.2   Geometric Characteristics................................. 50

    3.3   Microwave Remote Sensing and Resolution 55

         3.3.1   Resolution.. 55

         3.3.2   Radar Signals, Imaging, and Distortions 58

         3.3.3   Passive Microwave 62

    3.4   Discussion... 64

Chapter 4   Geostatistical Models of Scale 67

    4.1   Geostatistical Fundamentals and Variograms..................... 68
        4.1.1   Random Fields... 68
        4.1.2   Variograms and Covariance Functions 70
        4.1.3   Indicator Variograms and Transition
               Probability-Based Models... 75
    4.2   Variogram Regularization and Deregularization 77
        4.2.1   Regularization 77
        4.2.2   Deconvolution of Regularized Variograms.............. 80
    4.3   Statistics for Determining Measurement Scales 82
    4.4   Discussion.. 87

Chapter 5   Lattice Data and Scale Models.. 91

    5.1   Lattice Data 93
    5.2   Spatial Autocorrelation and Its Measures 95
    5.3   Local Models... 99
    5.4   Discussion.. 103

Chapter 6   Geostatistical Methods for Scaling 107

    6.1   Kriging 109
    6.2   Indicator Approaches... 116
    6.3   Upscaling by Block Kriging... 118
    6.4   Downscaling by Area-to-Point Kriging 120
    6.5   Geostatistical Inverse Modeling..................................... 124
    6.6   Discussion.. 126

Chapter 7   Methods for Scaling Gridded Data 129

    7.1   Upscaling.. 131
        7.1.1   Statistical Approaches...................................... 131
        7.1.2   Mechanistic Approaches 134
    7.2   Downscaling... 137
        7.2.1   Deconvolution 137
        7.2.2   Super-Resolution Mapping................................ 141
        7.2.3   Subpixel Mapping 144
    7.3   Discussion.. 145

Chapter 8   Multiscale Data Conflation.. 147

    8.1   Multivariate Geostatistics... 149
    8.2   Image Fusion 154
        8.2.1   Conventional Approaches.................................. 154
        8.2.2   Multiresolution Representations Based on Wavelets ... 155
        8.2.3   Wavelet-Based Image Fusion 161

8.3   Other Multiscale Methods 163
8.4   Discussion 166

Chapter 9   Scale in Terrain Analysis 169

9.1   Digital Elevation Data and Their Scales 171
9.2   Terrain Derivatives 176
        9.2.1   Slope and Aspect 176
        9.2.2   Curvature 180
        9.2.3   Secondary Properties 186
9.3   Models of Scale in Topography 187
9.4   Methods for Scaling Terrain Variables 193
        9.4.1   Upscaling 193
        9.4.2   Downscaling 196
9.5   Discussion 198

Chapter 10   Scale in Area-Class Mapping 201

10.1   Area-Class Mapping 204
        10.1.1   Discriminant-Space Model 204
        10.1.2   Methods for Mapping Area Classes 209
10.2   Spatial Scales and Patterns in Area Classes 213
10.3   Scaling Area-Class Information 219
        10.3.1   Upscaling 219
        10.3.2   Downscaling 220
10.4   Discussion 223

Chapter 11   Information Content 227

11.1   Information Theory 230
11.2   Information Content in Remotely Sensed Images 238
        11.2.1   Informational Characteristics of Images 238
        11.2.2   Information Conveyed in Images with
                  Additive Noise 242
        11.2.3   Information in Images with Multiplicative Noise 247
11.3   Image Resolution and Information Content 250
        11.3.1   Effects of Resolution on Information Content 250
        11.3.2   Information Capacity and Super Resolution 253
11.4   Information Content in Map Data 255
11.5   Discussion 261

Chapter 12   Uncertainty Characterization 265

12.1   Accuracy Metrics and Assessment 268
        12.1.1   Accuracy in Interval/Ratio Fields 268
        12.1.2   Accuracy in Area Classes 271

　　　　　　　12.1.3　Positional Accuracy 273
　　　　　　　12.1.4　Confidence Intervals and Hypothesis Testing 274
　　　12.2　Geostatistical Approaches to Validation 278
　　　12.3　Analytical Approaches to Error Propagation 282
　　　12.4　Geostatistical Simulation 287
　　　　　　　12.4.1　Simulation over a Point Support 287
　　　　　　　12.4.2　Point-Support Simulation Conditional to
　　　　　　　　　　　Areal Data 291
　　　　　　　12.4.3　Simulation over Blocks 293
　　　12.5　Discussion 296

Epilogue 299

References 309

Index 347

Preface

Scale can be a frustrating concept. It has multiple meanings even in a single scientific domain, and there is significant variation in meaning across domains. It underlies how we as humans perceive and act in the world, from the scale of our own bodies to that of our immediate surroundings, to the scale of life experiences, and to the scale of the cosmos. It can function as a noun or as a verb, with different sets of meanings in the two cases. It is implicit in the definitions of many properties that we work with as environmental or social scientists. Statements such as "Minnesota has 10,000 lakes" from that state's license plate, "The coastline of Italy measures 7600 km" (CIA 2009), "This is the flattest spot in the United States," or "This area has the world's highest population density" are all scale specific, and yet we frequently make or accept statements such as these without specifying scale or even recognizing the importance of scale in them.

This book attempts to resolve some of these issues and to alleviate some of this frustration by taking a rigorous, scientific approach to scale and its various meanings in relation to the geographic world. It covers methods for measuring aspects of scale and for changing scale through upscaling and downscaling. It discusses scale in relation to the various systems that are now available for acquiring geographic data, for modeling environmental and social processes, and for visualizing phenomena and their distributions over the surface and near the surface of the Earth.

This is not the first book to appear on these issues, and citations of many of the previous ones can be found at various points in this book. It will certainly not be the last, because geographic technologies are advancing rapidly, along with their importance in numerous areas of human activity, and because our thinking and theorizing about scale are also in a constant state of flux. New techniques for downscaling, new methods of analysis, new models of spatial and temporal scale, and new theories of how scale affects and is affected by social and environmental processes are all likely to emerge in the next few years.

Almost two decades ago the Alexandria Digital Library was funded by the US National Science Foundation as an early attempt to build an online system for searching, evaluating, and retrieving geographic datasets—in effect, an Internet-based map library (Smith et al. 1996). One of our first concerns in developing the library was to prioritize search criteria: what would people consider most important in searching for online data? First, clearly, was coverage in the form of a bounding box defined by latitude and longitude; second was theme, or the topic of the data; third was time, or the date of validity of the data. Missing from this list was scale: the degree of detail, since all geographic data must be generalized, sampled, or abstracted in some way, the real world being infinitely complex. Scale was considered too difficult for the average user to conceptualize, too difficult to characterize, and not of sufficient importance. We, the authors, hope this book will move us some distance in the direction of resolving these issues and of promoting scale to the importance it deserves. Level of detail is surely an inescapable property of any digital representation of any aspect of the Earth and is an essential parameter in any modeling, prediction, or decision making based on such representations.

This book is intended for an audience primarily of researchers and instructors of courses at the advanced undergraduate or graduate levels. Parts of the book are heavily mathematical, and, unfortunately, the mathematics of spatial data is inescapably difficult. Other parts of the book are more conceptual, and we hope that readers will be able to adapt the book to their own level of mathematical knowledge and ability. The book draws heavily on work in geographic information science, so readers will be well advised to have at least a rudimentary background in that field (see, for example, Longley et al. 2011) as well as in statistics and mathematics.

Acknowledgment

The research related to this book is supported by the National Natural Science Foundation of China (grant nos. 41071286 and 41171346).

Authors

Jingxiong Zhang earned his BEng and MEng, both in photogrammetry and in remote sensing, from Wuhan Technical University of Surveying and Mapping (now merged into Wuhan University), in 1985 and 1988, respectively. He earned a PhD in geography from the University of Edinburgh in 1997. His research interests include spatial uncertainty characterization, models of scale and methods for scaling, quantitative methods (including information-theoretic perspectives) in remote sensing and GIS, land cover and land cover change analysis, and geomatics (photogrammetry, remote sensing, and surveying). He serves on the editorial board of *Geographical Analysis* and is a member of the Standing Committee of the International Spatial Accuracy Research Association (http://www.spatial-accuracy.org/).

Peter M. Atkinson earned his BSc in geography from the University of Nottingham in 1986 and his PhD (dissertation title: "Optimal Sampling Strategies for Remote Sensing Investigations") from the University of Sheffield in 1990. His research interests focus on the application of spatial statistical, geostatistical, and spatial process modeling approaches to environmental problems and hazards, such as spatial epidemiology, disease transmission modeling, health management information systems, hydrological and geomorphological process modeling, environmental hazards, and spatial downscaling of remote-sensing imagery for various goals including information handling, land-cover classification, and change detection. His main awards include fellow of the Royal Statistical Society (FRSS) (2001–), fellow of the Remote Sensing and Photogrammetry Society (FRSPSoc) (1997–), Life Member IAMG (1992–), and fellow of the Royal Geographical Society (with the IBG) (FRGS) (1989–).

Michael F. Goodchild is now professor emeritus of geography, University of California, Santa Barbara (UCSB). He earned a BA in physics from Cambridge University and a PhD in geography from McMaster University. He joined UCSB in 1988, and he has served as chair of the Department of Geography (1998–2000); chair of the Executive Committee of the National Center for Geographic Information and Analysis since 1997; associate director of the Alexandria Digital Library Project since 1994; chair of the Mapping Science Committee, National Research Council, 1997–1999; and director of NCGIA's Center for Spatially Integrated Social Science since 1999. His research interests include urban and economic geography, geographic information systems, and spatial analysis. Considered the father of GIScience, Prof. Goodchild's many honors include being elected as a member of the National Academy of Sciences and foreign fellow of the Royal Society of Canada in 2002, being awarded the Founder's Medal of the Royal Geographical Society in 2003, being elected as a fellow of the American Academy of Arts and Sciences in 2006, foreign member of the Royal Society, and corresponding fellow of the British Academy (2010). He was awarded the Prix Vautrin Lud, St Dié-des-Vosges, France, 2007, which is regarded by many as geography's equivalent of the Nobel Prize. He was inducted into the GIS Hall of Fame, Urban and Regional Information Systems Association, in 2007. He won a Lifetime Achievement Award from the Geospatial Information and Technology Association the same year. He was editor of Geographical Analysis between 1987 and 1990; the Methods, Models, and Geographic Information Sciences section of the Annals of the Association of American Geographers from 2000 to 2006; and serves on the editorial boards of 10 other journals and book series. According to the flyer for his retirement celebration in Santa Barbara in May 2012, he has published more than 500 papers, made more than 500 presentations, and secured more than US$55 million in research grants after 43 years of academic excellence and research. These numbers will increase, as he is active in GIScience research and education even after his retirement.

1 Introduction

1.1 ISSUE OF SCALE

Spatial information or geo-referenced information has widespread applications in science, the economy, the environment, and spatial problem solving, making it an essential part of the infrastructure of modern society. Spatial information technology, such as the global positioning system (GPS), remote sensing, and geographic information systems (GIS), is becoming increasingly sophisticated and integrated, contributing to the proliferation of spatial data and information products. These have provided important impetuses for fundamental research that drafts the roadmap toward sustainable developments in geospatial science and technology.

Geographic information science (GIScience) has been advocated as the methodological and theoretical foundation underlying the developments and applications of geospatial technology (Goodchild 1992). Some of the theoretical issues in GIScience, such as geographic representation, spatial data acquisition and integration, scale, space–time analysis and modeling, and uncertainty, are on the research agenda of learned societies, such as the University Consortium for Geographic Information Science (http://www.ucgis.org/; McMaster and Usery 2004). Scale is an essential dimension in geospatial research and applications, transcending various facets of spatial information and analysis (Atkinson and Tate 2000).

Widespread applications of spatial information technology have made scale issues more acute. This is because it is easy to combine datasets using GIS functions, such as map overlay, where due consideration about scale differences and their implications is not guaranteed. Also, it is common to compare maps with finer detail with aggregated maps of coarser detail. This may lead to erroneous results in map comparison, as inconsistencies in geometry, attributes, and semantics are bound to be incurred owing to scale differences. The following discussions seek to highlight the relevance and importance of scale in spatial information and analysis.

First, it is necessary to discuss various definitions of scale, such as geographic scale, operational scale, measurement scale, and cartographic scale. As reviewed by Lam and Quattrochi (1992) and Lam et al. (2004), geographic scale (or observational scale) refers to the size or spatial extent of the study. For example, studies of land cover (DeFries and Townshend 1994; Lawrence and Chase 2010), population (Bracken and Martin 1989; Hay et al. 2005), topography (Pike 1988, 2000), and other geographic distributions may be conducted at global, regional, or local scales (implying different geographic scales or spatial extents), depending on specific needs. Operational scale refers to the spatial (or temporal) extent at which certain processes operate in the environment and is also known as process scale or intrinsic scale (e.g., Blöschl and Sivapalan 1995), as described in the context of ecology by Wu and Li (2006).

Determining the process scale of a geographic phenomenon (used interchangeably with geographic distribution or geographic property in this book, unless indicated otherwise) is an important step, because it can help suggest both the geographic scale (i.e., spatial extent as defined above) and the measurement scale (i.e., resolution as defined below) needed to observe the pattern resulting from the process, as reviewed by Lam and Quattrochi (1992) and Lam et al. (2004). Measurement scale, commonly called resolution, refers to the smallest distinguishable parts of an object (Tobler 1988), such as pixels in remote-sensing images, as discussed by Lam et al. (2004). Fine resolution is useful, though at some cost, for getting detailed information about a study area, whereas coarse resolution (implying less storage and computing cost as opposed to the case of fine resolution) may be what is adequate or needed for a certain purpose. Resolution is often referred to as grain in ecology (Turner et al. 1989). For ecologists, grain is the finest resolution (in space or time) of a phenomenon or a dataset within which homogeneity is assumed and may be considered as the pixel size for raster data or the minimum mapping unit for vector data (Wu and Li 2006).

Cartographic scale refers to the ratio between the length measurements on a map and the actual measurements on the ground and is also known as map scale, because cartographic scale is most commonly understood with respect to maps. A large-scale map covers a smaller area with more detailed information and *vice versa*. Unlike the types of definitions of scale discussed earlier (i.e., geographic scale, process scale, and measurement scale), which refer to data characteristics, cartographic scale refers to data representation. Therefore, as commented by Lam et al. (2004), once spatial data are encoded digitally, their resolutions are fixed, so that zooming in or out during display can only increase or decrease cartographic scale, not measurement scale.

Data support is yet another concept related to scale. Spatial data are often acquired by sampling over the problem domain, highlighting the role of sampling unit, as will be described in Chapter 3 where spatial sampling is one of the key issues concerned. Although certain properties can be measured or evaluated at point-like locations (e.g., spot heights at survey monuments and rainfall measurements at rain gauges), for many others, the sampling units have positive finite geometric dimensions (e.g., sample plots in forest inventory and pixels in remote-sensing images). In geostatistics, sample support generally refers to the length, area, or volume associated with a measurement or observation (Gotway Crawford 2006). Finite supports are common in geographic studies, as spatial aggregation is often necessary for analysis over areal units (e.g., forest stands and land parcels) or for inference about an area of interest (e.g., the severity of drought or crop yields in an agricultural region). Changing the support of a variable creates a new one, which is related to the original one, but has different statistical and spatial properties (Gotway and Young 2002). This is a central idea in quantification of scale effects and will provide the conceptual basis for change of scale. Also, data support should be factored in for representations of geospatial entities and phenomena, as follows.

Geographic information representation or data modeling is a central issue for studies concerning phenomena on the Earth's surface and, of course, is an important component in GIScience (Goodchild et al. 2007). A scheme for representing data is required and is closely linked with spatial analysis and environmental modeling. There are two complementary models for geospatial conceptualization, which refer to whether the real world is conceived of as populated by discrete entities or objects $ID(x, attr)$ that are geo-referenced by position x (although it is conventional to denote a position in space and time using a vector notation **x**, we make an exception for it in this book, denoting a position using an italic lower case x) and associated with some qualitative and/or quantitative attributes *attr*, or consisting of layers of single-valued functions defined everywhere over each problem domain. In object models, attributes are usually assumed to be the responsibility of specialists collecting application-specific data, such as census data and utility maintenance data, implying that it is often positional data that are dealt with by geospatial information specialists (Zhang and Goodchild 2002). For fields, there are continuous variables $Z(x)$, measured at interval and ratio scales (and readily extended to the case of vector fields), and discrete variables $C(x)$, measured at nominal and ordinal scales. In both object- and field-based representations, x should be viewed as data support rather than a mere notion of position, thus indicating the multiplicity of scales, as will be elaborated in Chapter 2.

As the representations of land-surface properties and processes via objects and fields are linked inherently to the scales of analyses and applications, scale needs to be properly incorporated in data modeling. This is because changing the scale of data without understanding its effects may result in biases with respect to the representations and geo-processing intended. As will be elaborated in later chapters, in order to characterize a pattern or process at a scale other than the scale of observation, some knowledge of how that pattern or process changes with scale is needed to adjust representations and data analyses accordingly. Multiscale representations are valuable for extension of geospatial data modeling, whereas knowledge about scale behaviors is used to construct the hierarchy of data representations with minimal loss of information (for related discussions, see Lam et al. 2004).

Spatial analysis encompasses a wide range of techniques for analyzing, visualizing, and interpreting spatial data. Spatial analysis well reflects the width of research on scale not only in data, but also in their analyses. Spatial analysis needs to consider scale effects, which usually refer to the changes in the result of a study due to a change in the scale at which the study is conducted. Effects of changing scale on sampling and experimental design, statistical analyses, and modeling have been well documented in ecology and geography (e.g., Openshaw 1983; Turner et al. 1989; Levin 1992; Quattrochi and Goodchild 1997; Marceau 1999; Gotway and Young 2002; Wu 2004; Wu and Li 2006). For example, to characterize urban land use or urban land use change using census-type data and remotely sensed images (Montgomery 2008; Patino and Duque 2013), different results can be obtained using data based on different zoning systems, as they are affected by the size, shape, and position of the spatial units for data aggregation and analysis, as cautioned by Openshaw (1983).

Research has suggested that the scale of analysis be commensurate with the intrinsic scale of the phenomenon under study (Blöschl and Sivapalan 1995; Legendre et al. 2002), as discussed by Wu and Li (2006). Despite the requirements for scale match between data and analyses, data irregularity in terms of scale, and complex scale dependencies of spatial patterns, processes, and their relationships, suggests that scale change will be necessary. Scaling is often used as a synonym for scale change, which implies the transfer of information between scales. Scaling is necessary whenever predictions need to be made at a scale that is different from the scale at which data are acquired (Wu and Li 2006). For example, resource inventory may have been conducted at a local scale, but management and policy making often have to be dealt with at a regional scale, meaning that scaling is required to bridge such scale gaps. Moreover, spatial phenomena often occur over a wide range of scales, with hierarchical linkages existing among them (Wu and Li 2006), echoing the importance of relating information across scales. For these, we need to identify data scales and scale thresholds in spatial processes in deciding upon appropriate scales of analysis. Scale sensitivity analysis and robust methods for scaling will help to develop geographic models that allow users to adjust or compensate for the effects of data scale on the analysis results, as explained in the research agenda of UCGIS (McMaster and Usery 2004).

Spatial information technology has continued to improve, well reflected in advances in data techniques, such as diversification in the means of data acquisition, and spatial data proliferation. Newly available remote-sensing satellites are providing unprecedented amounts of data on the Earth environment, and new sources of demographic, social, and economic data are becoming available at finer spatial detail. These signal vigorous developments in geospatial technology.

However, despite the growing numbers of multisource spatial data, our ability to maximize information potentials from massive data lags behind. It remains nontrivial to analyze spatial datasets acquired from different sources, for different purposes, at different resolutions, or using different classification schemes. For instance, data conflation poses a serious impediment to map overlay (Saalfeld 1988; Lam et al. 2004; Chen et al. 2008). Techniques for data conflation must be developed that can handle scale dependencies and multiplicity effectively and efficiently. Although image fusion and map conflation have been addressed in remote sensing and GIS, respectively, it is evident that a coherent framework and related techniques remain underdeveloped (Zhang et al. 2007a; Zhang 2008). More sophisticated methods need to be explored for effectively handling scale mismatch and, thus, truly integrated processing of massive geospatial data.

In summary, we need to investigate models for describing the scale of data, to explore methods for characterizing scale dependencies of patterns and processes (e.g., Bian and Walsh 1993), to gain an improved understanding of scale effects, to develop methods for changing scale, and to devise methods for multiscale data conflation, as also discussed in the research agenda of UCGIS. Ultimately, we should be able to process and present geographic information up and down from local and regional to global scales. Information and remote-sensing systems of the future can inform users about the implications of scale dependencies and provide tools to deal with scale matters (Lam et al. 2004). These will be discussed further in the next two sections where models of scale and methods for scaling are reviewed, respectively, before we outline the contents of individual chapters.

1.2 MODELS OF SCALE

There exist various techniques to measure scale in a given dataset and to model scale in the underlying processes that generated the data. These include fractals, geographical variance, local variance, wavelets, and spatial statistics. Scale descriptors and models will allow the determination of appropriate scales for data collection, information extraction, and spatial analysis.

Mandelbrot (1967) found that lengths of geographic curves with complex detail (e.g., coastlines and other natural geographic borders) are often infinite; however, many such complex curves are statistically *self-similar*, meaning that each portion can be considered a reduced-scale image of the whole. In that case, the degree of complexity can be described by a quantity $D_{fractal}$ that has many properties of a dimension, known as the Hausdorff dimension, which is between 1 and 2 for curves. According to Mandelbrot (1967), the measured lengths $L(\Delta)$ of complex curves are related to the measurement scale Δ via a function of the form $L(\Delta) = k\Delta^{1-D_{fractal}}$. Thus, a plot of the quantity $L(\Delta)$ vs. scale Δ on a log–log plot gives a straight line whose slope indicates the fractal dimension $D_{fractal}$. Mandelbrot interpreted this result as showing that coastlines and other geographic features can have a property of statistical self-similarity (so they are called fractal features), with the exponent $D_{fractal}$ measuring the Hausdorff dimension of the curve (e.g., 1.25 for the west coast of Britain) (Mandelbrot 1967; Shelberg et al. 1982). Fractal models have been used widely by researchers (Goodchild 1980a; Pentland 1984; Lam 2004). For instance, the fractal dimension provides a means of characterizing the effects of cartographic generalization and the scale dependence of estimates derived from sampled data with an inherent scale of measurement, as discussed by Goodchild (1980a) and Lam and Quattrochi (1992). Fractal surfaces with self-similarity are useful as initial or null hypothesis landscapes in the study of geomorphic processes (Goodchild and Mark 1987).

The geographic variance method proposed by Moellering and Tobler (1972) can be used to analyze scales in hierarchical geographic regions, as analyzing the data at different levels of the hierarchy is equivalent to analyzing the data at different geographic scales. This method, which resembles variance analysis in statistics, can determine the relative variability of each level in a nested hierarchy where lower-level units are contained entirely in the higher-level units (so that the values at a higher level are aggregated from lower-level units). According to this method, the total variability can be divided using the sum of squares at each level, whereas the underlying process may occur at the level where the percent sum of squares is the largest. In other words, the process scale of a phenomenon is believed to coincide with the scale of maximum variability in the data. Many types of spatial data can be constructed as a nested hierarchy through aggregation, and then the geographic variance method can be applied (Wu et al. 2000; Wu and Li 2009). However, it would be hard to generalize this method because of the variation among different datasets and their level-wise percent sum of squares (and thus the largest percent sum of squares that indicates the underlying process scale). Another limitation to this method is the requirement for a nested hierarchy, precluding its applicability for data that are not organized in this nested hierarchical structure.

For analysis of scales in remote-sensing images, Woodcock and Strahler (1987) proposed the local variance method for quantifying the relationship between the size of objects in a scene and the spatial resolution of sensors. The local variance is the mean value of the variance of a moving window (say 3 × 3 pixels) over the entire image. By plotting the local variance as a function of scale, the spatial structure of an image can be quantified and measurement scale suitable for the image can be determined (Atkinson and Aplin 2004; Ming et al. 2011). This is based on the reasoning below. When the spatial resolution is considerably finer than the size of objects in the scene, most pixel values in the image will be highly correlated with their neighbors (i.e., pixel values in the neighborhood are similar), suggesting small local variance. As the size of pixels increases up to the point where it approximates that of objects, the neighbors become less likely to be similar, causing the local variance to increase. If the size of pixels increases further to be greater than that of individual objects, it becomes more likely that one single pixel contains multiple objects, with the local variance gradually decreasing because of the increased likelihood that individual pixels become mixels of multiple objects and hence exhibit reduced variation due to averaging or convolution (Woodcock and Strahler 1987; Drăguţ et al. 2011). Therefore, generally speaking, a peak appears when the pixel size (i.e., spatial resolution) approximately matches the size of the objects. Multiple local-variance peaks may indicate multiple scales of variation in the scene (e.g., in a nested scene model where one level of elements is aggregated to another one at a higher level or larger elements) (Woodcock and Strahler 1987; Wu and Li 2009). However, local variance-based intercomparison between images may not be straightforward (i.e., the values of local variance for one image cannot be compared directly with those from other images), as interpretability of scale in the underlying scene based on local variance is dependent on the image's global variance. Thus, measurements of local variance derived from an image have meaning only when compared with those derived from the same image degraded to different resolutions (Woodcock and Strahler 1987; Chen 1998). Despite this, the local variance method has been used widely for multiscale analysis in various applications, such as geomorphometry (Drăguţ et al. 2011).

The local variance method is one of several texture-analysis techniques that seek to measure the spatial variability of image data. This permits the heterogeneity of the scene being studied to be quantified using a variety of indices, such as variance and entropy (a key concept in information theory to be described in Chapter 11), as discussed in the context of image-based urban land cover classification by Su et al. (2008). The texture indices for images of different spatial resolution can be computed and compared, with the largest index indicating the scale of the underlying process, as discussed in the context of fractal-based image texture analysis by Emerson et al. (1999). It can be seen from previous descriptions that geographic variance, local variance, and texture analysis are all based on the assumption that the scale with the maximum variability indicates the scale at which most processes operate (Woodcock and Strahler 1987; Emerson et al. 1999).

There are also methods in the field of image analysis that may be explored for analysis of scale. For instance, Fourier decomposition may be used so that a signal can be decomposed into weighted sums of oscillating functions with different periods. The resulting Fourier spectrum gives the relative magnitude of the different frequencies in a signal. As described by Pike and Rozema (1975), we can derive the spectral density function via the discrete Fourier transform of an autocovariance function ascribed for a terrain profile. The resulting spectrum gives the variance associated with different-sized terrain undulations as the function of a range of topographic wavelengths. Pike and Rozema (1975) adapted spectral analysis to topographic profiles obtained from 1:24,000 contour maps for five contrasting areas in the conterminous United States (two fluvial, one aeolian, and two glacial areas, approximately bracketing topographic variation therein). The spectra (one-dimensional) derived from elevation profile data can be used to describe four independent aspects of topography: magnitude of landform periodicities, spacing of periodicities, overall roughness, and contrast between slopes of large and small features. The last three numerically expressed attributes can be incorporated for landform classification and other geomorphic problem solving. However, as remarked by Pike and Rozema (1975), the technique, which was limited to a resolution of between 240 and 3700 m, does not exhaust the information content of a topographic map or profile: such characteristics as the arrangement of slopes and the degree and direction of lineation were omitted; even the two-dimensional spectrum cannot describe certain geometric aspects, such as skewness and dispersion of elevation and terrain properties. Nevertheless, spectral analysis is valuable for scale description in topography and landscape analysis (Perron et al. 2008), as will be shown in Chapter 9.

Unlike the Fourier transform using sine or cosine functions to form an orthonormal basis, the wavelet transform method (Mallat 1989) uses a localized function in time or space so that the wavelet's size can be adjusted and shifted to analyze a dataset, allowing features of interest in the dataset to be studied at an appropriate scale (Graps 1995). Multiresolution analysis can also be performed on the basis of the wavelet decomposition of a dataset into a linear combination of simple orthonormal functions parameterized by a scale factor (Mallat 1989). The wavelet variance (Percival 1995) can be calculated by the wavelet transform method and is thus a natural tool to investigate the spatial scales in remote-sensing data (Lindsay et al. 1996). It can be used as an indicator to quantify the spatial scale and as a suitable tool to study the multiscale relationships of spatial pattern and heterogeneity. For example, in investigating the scales of variability of the surface temperature and albedo of spring pack ice in the Beaufort Sea using both the reflective (channel 3, 30 m of resolution) and the thermal (channel 6, 120 m of resolution) band data of Landsat TM images, Lindsay et al. (1996) performed wavelet variance analysis and covariance estimation and found that more than half of the variance was accounted for by scales of less than 800 m. Scale dependence of cross-variable correlation analysis can also be performed using wavelet transforms (Milne et al. 2011).

Theoretically, scale is related to the structure, extent, and strength of spatial dependence inherent to many spatial processes and patterns. Spatial dependence is the propensity for nearby locations to influence each other and to possess similar attributes (Goodchild 1992). In other words, although everything is related to everything else, things that are close together tend to be more related than things that are far apart (Tobler 1970). Terrain elevations, vegetation types, and land surface temperatures, for instance, are more likely to be similar at points a few meters apart than at points kilometers apart. As spatial dependence means that conditions at nearby points are not independently and randomly distributed, but instead show remarkable levels of interdependence, it is thus possible to divide the surface into regions that exhibit substantial internal similarity, such as arid zones, because of the lack of rainfall, and tropical rainforests, where rainfall is abundant and vegetation is dense (de Smith et al. 2009). Also, spatial dependence suggests that interpolation can be performed to predict a spatial property's values at unsampled locations based on nearby samples (because sampled locations contain information about the unsampled location, thanks to spatial dependence) and that it is possible to represent continuous geographic phenomena in digital form (i.e., to discretize them to facilitate computer geo-processing without significant loss of information).

A statistical measure of the similarity of properties over space is called spatial autocorrelation (Cliff and Ord 1973). Traditional statistics usually assumes independence between sampled entities and will not be suitable for modeling spatial distributions and the data that depict them. Although past research on scale descriptions has resulted in a decent basis for further developments, scale metrics should be reassessed in the light of spatial dependence so that previous methods and their extensions can be approached from a general framework. Also, we can incorporate scale in spatial analysis by building upon spatial statistics (Haining 2003; de Smith et al. 2009; Fortin et al. 2012).

As mentioned previously, spatial dependence is a fundamental concern in the geosciences, since both spatially distributed phenomena and their representations exhibit spatial dependence. The magnitude of spatial dependence can be measured using several statistical methods based on spatial autocorrelation (Cliff and Ord 1981; Goodchild 1986). Geostatistics is well established and provides a solid basis for quantifying and utilizing spatial dependence in spatial analysis based on the theory of regionalized variables (Journel and Huijbregts 1978). For example, geostatistics can be used to explore and describe spatial variation in remotely sensed and other spatial data, to design optimum sampling strategies, and to increase the accuracy with which remotely sensed data can be used to classify land cover or estimate continuous variables (Curran and Atkinson 1998; Collins and Woodcock 1999; Garrigues et al. 2006; Zhang 2008; Jin et al. 2012).

In geostatistics, covariance, correlograms, or variograms are used to quantify spatial dependence in regionalized variables (a key geostatistical concept to be described in Chapter 4) with both deterministic and stochastic characteristics. Covariance functions or correlograms show how spatial autocorrelation decreases with increasing lag (the distance and direction of separation), whereas variograms measure the dissimilarity in values of geographic properties (e.g., terrain slopes and soil erosion susceptibility) at two sample locations as a function of the lag between them (i.e., the spatial dissimilarity behavior) (Journel and Huijbregts 1978). There are three named parameters in a typical variogram model: nugget, sill, and range. These parameters can be used to characterize and quantify spatial dependence. The nugget is the discontinuity of the variogram at the ordinate. It can be used to judge whether uncorrelated noise (measurement error) exists or spatial structures are smaller than the spacing of observations. The sill (if one exists) is the value that the variogram may reach when the lag distance heads toward infinity. It indicates the maximum spatial variability of the data. The range (if one exists) is the lag distance at which the variogram reaches a sill, indicating the scale of spatial variation over the domain. We will describe these variogram parameters in more detail in Chapter 4 where geostatistical fundamentals are reviewed. Clearly, variograms can be explored as scale descriptors. For example, Tarnavsky et al. (2008) applied variogram modeling to evaluate the differences in spatial variability at different scales and to characterize the impact of scale.

On the other hand, the Earth's surface is characterized by variety, from the Tibetan plateau to the highly urbanized Yangtze River Delta, from the Scottish glens to metropolitan London. It is difficult to imagine any subset of the Earth's surface being a representative sample of the whole. The results of any analysis over a particular area are likely to be different from those obtained over another area or the results that would be obtained for the surface of the Earth as a whole. These concepts are collectively described as *spatial heterogeneity*, and they tend to affect almost any kind of spatial analysis conducted on geographic data (de Smith et al. 2009). Many techniques such as geographically weighted regression (Fotheringham et al. 2002) focus on providing results that are specific to each area (i.e., local or place-based).

Spatial heterogeneity implies nonstationarity in terms of variogram models across the problem domain. Scale parameters should thus be conceived of as spatially varied. This suggests local models of scale. In fact, geostatistics can be adapted for local scale quantification via localized variogram models and local implementation of scale change (Sampson and Guttorp 1992; Corstanje et al. 2008). This will contribute to statistically and spatially coherent strategies for cross-scale data analyses.

Clearly, although spatial dependence is inherent to spatial forms and processes, spatial heterogeneity is the other side of the coin. In other words, spatial dependence and spatial heterogeneity can be seen as interacting with each other, with the former as a general pattern and the latter carving out local variation. Thus, both the global and the local aspects of scalability should be discussed, built upon a sound understanding of the interaction between spatial dependence and spatial heterogeneity. This helps to accommodate spatial variation in measurement and process scales, and to quantify scales and operationalize scale changes, as elaborated in the following section.

1.3 SCALING UP AND DOWN

In addition to scale quantification, methods for change of scale are required, which aim for developing theory and exploring methods for information communication across multiple scales. This is because discrepancies in scales among data, analysis, and applications are common and need to be handled properly, as developed below.

Spatial models with many different data sources are often used in practice, and such models differ according to the spatial and temporal scales at which they operate. Also, the scale at which available data are collected is usually different from the scales required by the models. Moreover, the scale of the model output is rarely in tune with the scale at which information is required or management measures are implemented, as mentioned in the introductory section. To overcome scale discrepancy, different methods have been developed to transfer data and model outputs from a finer to a coarser scale (upscaling or aggregation) or from a coarser to a finer scale (downscaling or disaggregation) (Wu and Li 2006).

Consider change of support as a special case for scaling. The change-of-support problem is concerned with inference about the values of a variable at locations (points or blocks) that are different from those where it has been observed, as discussed by Gelfand et al. (2001). For the case of areal data, which will be studied with respect to scale in Chapter 5, this kind of problem is often referred to as the modifiable areal unit problem (Openshaw 1983).

There is a large literature on scaling (O'Neill and Rust 1979; Band 1989; Wiens 1989; Van Gardingen et al. 1997; Bian and Butler 1999; Marceau and Hay 1999; Simic et al. 2004; Hong et al. 2011). Consider measurements, retrievals, and products in remote sensing. There exist various scaling methods for remote sensing, many of which seek to characterize the land-surface heterogeneity and to compensate for scaling effects. As discussed by Wu and Li (2009), area-based weighting, based on the law of conservation of energy or matter, is often applied for scale changes and may be applicable in homogeneous regions. For remote measurements of radiance emitted or reflected by rugged terrain, such a method should be modified by taking the influence of slopes into account (Liu et al. 2006). Also, as retrieval models derived from one scale cannot be guaranteed for use at another scale, corrections need to be made to the area-weighted measurements in order to scale products correctly (Bierkens et al. 2000). Various methods have been developed to reduce the scaling effects due to the nonlinearity of a retrieval model at a local scale (Raffy 1992). Tian et al. (2002) discuss scaling based on radiative transfer. Corrections based on the quantification of the intrapixel heterogeneity have also been explored (Garrigues et al. 2006), as reviewed by Wu and Li (2009).

Other scaling methods may also be explored, such as regression methods and fractal methods, as follows. Regression methods can be used to relate the products at different scales. Fernandes et al. (2004) used empirical relationships between spectral vegetation indices and *in situ* measurements of the leaf-area index (LAI) to estimate coarse-scale LAI. The regression relationship is usually site, time, model, and scale dependent. If any one factor is changed, the relationship may need to be recalibrated. Fractal-based methods are based on power-law scaling characteristics, as mentioned previously. Based on the log–log linearity relationships between the statistical moment and the scale factor, and the nonlinear dependence of scaling exponents with orders of statistical moments (i.e., scaling exponent $\beta(q)$ in a spatially scaling property G that satisfies $E\{[G(\lambda\Delta)]^q\} \propto \lambda^{\beta(q)} E\{[G(\Delta)]^q\}$ is a nonlinear function of moment order q; Jones and Brunsell 2009), it is possible to extend the statistical properties (moments) of a variable from one scale to another (Dubayah et al. 1997; Peters-Lidard et al. 2001).

Whether simple area-based weighting or more complex algorithms are used for scaling, there seems to be a lack of coherence in the approaches to scaling in remote sensing. Scale descriptors reviewed in the preceding section also appear rather limited in terms of commonality. With these in mind, we believe that the key to research on scale issues is to promote systematic strategies for establishing scale metrics and developing methods for scaling. It will be seen that the way forward lies in the integration of geostatistics and process-model-based methods, in which spatial dependence and spatial heterogeneity are characterized and incorporated in the description and modeling concerning the underlying processes that control the information dynamics. This integration also provides coherence in the organization of the book's chapters, as will be overviewed in the next section.

In geostatistics, punctual (point) and block kriging techniques are well developed and may be adapted for scale change. In the context of block data, which can be sensibly viewed as averaging over point data, a unified geostatistical framework can be proposed that provides the foundation for prediction from points to points, points to blocks, blocks to points, and blocks to blocks (Gelfand et al. 2001). For a variable whose value for a block can be viewed as a sum of sub-block values, as in population or disease counts, areal weighting offers a simple imputation strategy. Since areal allocation does not capture tendencies toward spatial clustering, model-based approaches using covariates other than area have emerged (Goodchild et al. 1993; Gelfand et al. 2001).

As for scale modeling, spatial heterogeneity can also complicate the process of scaling. For example, extrapolation of plot-scale data to the landscape scale is a trivial matter in a spatially homogeneous (uniform or random) environment. However, simple averaging of local conditions usually leads to poor estimates at the landscape scale in a heterogeneous landscape, as ecological relationships often become distorted when translated across scales in heterogeneous environments (Wu and Li 2006). Also, new patterns and processes may emerge as a result of scale changes, because their controlling factors tend to shift across scales. Thus, fine-scale observations may miss important processes operating on broader scales, whereas broad-scale observations may not have enough detail to characterize fine-scale processes, as discussed by Wu and Li (2006). Models used in remote sensing are based mostly on relationships between process variables and reflectance measurements. If these models, which assume homogeneity over the problem domain, are applied to reflectance measured over heterogeneous surfaces without the corrections necessary to compensate for the change of scale due to local variation with regard to the relationships being modeled, there will be inaccuracy in the modeling results. Thus, we should also develop locally adaptive methods of scaling, similar to local models of scale.

We can also draw upon developments in ecology and environmental science concerning scaling. Methods for scaling may be classified into similarity based and dynamic model based, with the former being simpler and the latter focusing on state variables and modeling aspects (Blöschl and Sivapalan 1995; Wu and Li 2006; Wu et al. 2006). As our goal in scaling is not only to describe and predict across scales, but also to understand patterns and processes at multiple scales (e.g., eco-regionalization for delineation of ecosystem regions; Bailey 1996), dynamic model-based methods are better suited for scaling. This is because they can explicitly consider dynamic processes and their interactions, and can couple patterns and processes in a spatially explicit manner (e.g., Bailey 1996). Finally, scientists are often interested in quantifying how spatial heterogeneity interacts with spatial processes in their efforts to scale across space. This requires an approach that deals with spatial processes directly and explicitly. For this, dynamic model-based strategies hold advantages over empirical ones.

Existing methods for scaling can be seen as lying along a spectrum, with statistical and mechanistic being the two extremes. Commonalities in different scale descriptors and models should be pursued, as this will shed light on the complementarity of alternative models. Geostatistical methods can be integrated usefully with process models to gain further understanding about the interactions between spatial dependence and spatial heterogeneity, nonlinear scale effects due to spatial heterogeneity, and controlling factors underlying scale multiplicity and scale transfers. Most significantly, geostatistics and process models provide a unified framework for scale modeling and scale change in the spatial information flow from measurement, through analyses, to applications, as we work with both objects and fields, both continuous and discrete variables, both geostatistical data and lattice data, and both static patterns and dynamic processes. For instance, it is important to identify and model the underlying processes that have generated the data observed (i.e., through the imaging mechanisms of remote sensing). Equally important is the process view about how terrain properties are derived from elevation data, how land-cover information is extracted from remote measurements and auxiliary data, and how objects are delineated from fields.

Spatial information gathering, processing, and communication hinge on how the dimension of scale is encoded and scale changes are performed. Information-theoretic perspectives are valuable as a unifying framework for quantifying scales and understanding scale changes. Last but not least, uncertainty is closely related with scale and must be treated in association with it rather than separately. The next section will outline the roadmap by which a synthesis of past work on scale issues is accomplished and new directions are highlighted.

1.4   BOOK CHAPTERS

Chapter 1 sets the scene for this book. Some of the scale issues concerning spatial information and analysis are first brought into focus, with the key concepts concerning scale and scaling introduced. This is followed by a review of past research on scale descriptors and methods for scale change, including how to effectively measure and characterize scale, and how scale and scale change affect spatial analysis and applications. This review is not meant to be comprehensive, but is hopefully informative. A coherent framework for scale modeling and scale change is provided by geostatistics integrated with process-mechanistic perspectives about spatial information, in which both spatial dependence and spatial heterogeneity are central to scale description and transfers are taken care of effectively. The chapter concludes with this introduction to the book's chapters.

Scale is known to constrain the detail with which the real world can be represented, observed, and analyzed. Chapter 2 discusses spatial data modeling. As briefly described earlier, fields and objects are the two complementary conceptualizations for representing spatially distributed phenomena and entities. Geo-fields and geo-objects will be described as aggregates from geo-atoms, the atomic form of geographic representations. Different model variants for both geo-fields and geo-objects will be elaborated on in light of scale dependence. Scale-adaptive models, which can accommodate scale multiplicity, are important extensions to data models. For this, hierarchical data structures will be described, including quadtrees for regular tessellation and hierarchical triangulation for irregular tessellation, which can adapt to the spatially varied resolution in geographic representations. Given the growing importance of spatial interaction and the temporal dimension in spatial information, geo-dipoles (in parallel to geo-atoms) and spatio-temporal data models are reviewed in the discussion section. Finally, raster and vector data structures are re-evaluated with respect to scale dependence and provision for scale-dependent information representation and processing.

With geospatial conceptualization described in Chapter 2, the emphasis of Chapter 3 is on measurement. A large quantity of spatial data is collected from various sources, often at different scales. Before these data can be integrated for spatial problem solving, fundamental issues must be addressed. The process of data acquisition and the associated measurement-scale issues will be described through the processes of spatial sampling and remote sensing, both optical and microwave, so that the mechanisms by which scale is inherently imprinted in the measurement process will be formulated quantitatively. The process of measurement needs to consider sampling design in terms of sample size and density. Thus, scale-related issues in the sampling are necessarily discussed in order to link sampling and data support. For remote sensing, key concepts, such as point-spread functions, convolution, and transfer functions, will be introduced. This will be important for understanding scale not only in measurement, but also in the underlying processes that have generated the data, either *in situ* or remotely measured.

Geostatistical models of scale are the focus in Chapter 4. Variogram models provide the basis for determining the scale of the underlying phenomena under study and, thus, the commensurate scale for measurement. The notion of point variograms is described as the basis for scale change and for exploring scale effects. For upscaling and downscaling, regularization (convolution) and de-regularization (deconvolution) can be performed to estimate upscaled and downscaled variograms, respectively. Scaled variograms can be used to quantify the scale dependence of variograms themselves. Lastly, nonstationary variogram models, which provide the basis for locally adaptive spatial prediction, are described to deal with spatial heterogeneity intrinsic to patterns and processes and their representations. The interactions between spatial dependence and spatial heterogeneity are seen to control the balance between global and local stationarity in the modeled process and will lay the foundation for scale-dependent spatial prediction and analysis.

Geostatistical data are those involving the measure of a variable of interest over a region where the measurement points can vary continuously, for which Chapter 4 has described models of scale based on variograms. There is another type of spatial data known as lattice data, for which there is no space between lattice units where additional sample points could be obtained. Chapter 5 will describe models of scale for lattice data, based on measures, both global and local, of autocorrelation. Regularization of variograms may be applied usefully for quantifying spatial dependence in the resultant lattice data. However, the transformation from lattice to geostatistical data is not straightforward. Correlograms are the equivalents of variograms for lattice data. Neighborhood relationships should be used for lattice data analysis, such as prediction of quantities across distances. Thus, the definitions of neighborhood play a central role in lattice data analysis, in particular with respect to scale determination. This chapter will discuss the quantification of spatial autocorrelation in lattice data, with Moran's I and Geary's c introduced. Local indicators of spatial autocorrelation are then described, which can accommodate spatial heterogeneity in the strength of spatial autocorrelation in a region.

Chapter 6 describes geostatistical approaches to prediction of regionalized variables at unsampled locations, given samples at nearby locations. Variants of kriging techniques are discussed, ranging from simple to ordinary and universal methods for upscaling. Geostatistical upscaling is then described and formulated as an approach for spatial prediction of random fields over unsampled locations given samples in the neighborhood. The emphasis is placed upon developing kriging equations for predicting the unknowns over block data supports, based on point support data. In contrast to upscaling, downscaling refers to situations where prediction is performed over a data support smaller than that of the existing data. As in upscaling kriging, explicit reference to data support has to be made in the development of the algorithms. Geostatistical inverse modeling is also described as an alternative geostatistical method for downscaling. This strategy is advantageous over conventional kriging-based downscaling methods, especially those assuming uniform data supports. This is because it can provide parameter estimation for spatial covariance models as well as fine-resolution predictions of the underlying distributions based on coarse-resolution data. Moreover, geostatistical inverse modeling is able to incorporate convolution kernels in downscaling problem solving, although kriging-based downscaling may be adapted for accommodating nonuniform sampling kernels implied in data of larger supports.

The main objective of Chapter 7 is to describe methods for upscaling and downscaling of gridded lattice data. They originate in remote sensing and image analysis but complement the geostatistical methods described in Chapter 6. For upscaling, both statistical and mechanistic approaches will be reviewed. The former is based on such statistical operators as mean and median, assigning suitable values to the cells of larger support from cell values of smaller supports that fall within the bounds of the larger support, and can be usefully formulated as a convolution process (for the mean operator, for instance). For the latter, image degradation based on convolution will be described as an example, in which point-spread functions and related quantities are employed for simulating images at multiple resolutions. For downscaling, there is an impressive literature on image deconvolution in image analysis, which is a type of downscaling technique well suited for remote-sensing images and can be seen as the inversion of the process of image degradation. Super-resolution mapping will also be described by examining the ways to restore bandlimited images beyond their cutoff frequencies. This can be done through extrapolation in the frequency space or statistical methods using multiple images for the same scenes. Spectral unmixing and subpixel classification are also reviewed as part of the super-resolution mapping methods. Finally, a prospective view is given of the integration of sensor modeling, statistical image analysis, and geostatistics for scaling gridded data.

The existence of multisource data implies that multiscale properties are intrinsic to the process of information flow. Thus, integration of multiple datasets and fusion of information have to deal with scale multiplicity. Chapter 8 synthesizes the methods for conflating spatial data, both maps and images. This chapter first addresses the topic of multivariate geostatistics, whereby multiple variables operating at different scales are often involved. Both auto- and cross-variogram models need to be quantified, before cokriging procedures may be pursued. Then, attention is given to image fusion, which has drawn increased research effort and represents a group of data conflation techniques designed for raster images. Alternative methods for image fusion will be described, including the classical intensity-hue-saturation method and wavelet-based method. Other multiscale methods, such as spatially extended Kalman filters and tree models, are described to furnish an overview of process-model-based strategies for fusion of multiscale data.

As important components in geographic information science and environmental modeling, digital elevation models (DEMs) and terrain analysis have drawn impressive research efforts toward scale issues. Therefore, Chapter 9 focuses on scale in digital terrain analysis so that some of the key concepts and methods related to scale and scaling can be reviewed with specific reference to terrain data. Characteristics of commonly available DEM datasets will be reviewed. Methods for computing terrain derivatives from DEMs, such as slope, aspect, and curvature, will be reviewed, hinting at the scale dependence of these terrain derivatives. This is followed by a section comparing the alternative scale descriptors for terrain data. Methods for scaling in elevation data and terrain derivatives will also be described. An important message from this chapter is that scale descriptions and models should be established from a process-based perspective in the sense that scale characterization of a variable is better approached from its underlying controlling variables. This is also true for developing methods for scaling.

Provision and analysis of land-cover information across local, regional, and global scales are becoming increasingly important for resources management, environmental modeling, and global-change research. Land cover and many other nominal fields are often depicted as area-class maps. In line with the process-model-based strategy for modeling scale and devising methods of scaling, Chapter 10 will describe a discriminant-space-based strategy for area-class mapping, scale description, and scaling, which is built upon the understanding that area classes are controlled by certain process variables (i.e., discriminant variables), such as the control of vegetation by temperature and precipitation. Therefore, area-class scale description and scaling are performed in the discriminant space and later projected to geographic space. Alternative scale descriptors and methods for scaling will also be reviewed.

Information theory, originally developed in communication, has been a topic of discussion in geography for quite some time. Intuitively, finer-resolution (spatial) images tend to contain more information than coarser-resolution images, although the relationships between information content and resolution may not be linear or straightforward. The same can be said of maps, as larger-scale maps provide the media to represent finer-grained spatial entities and distributions (thus, providing more information content). Chapter 11 will take on this topic in association with scale issues and seek to provide a review of some of the fundamentals on, and research efforts toward, informational analysis of geospatial datasets (e.g., remotely sensed images and maps), complementing the spatial statistical, fractal, and pattern index types of scale models that the majority of the book content has elucidated. Information theory will first be introduced. This is followed by a description of the characteristics of images in terms of information theory. Methods for quantifying information content in images contaminated with additive and multiplicative noise will then be discussed. As mentioned above, since images with finer spatial, radiometric, and spectral resolution tend to contain more information than those with coarser resolution, it is interesting to examine the interaction between information content and resolution. Clearly, spatial dependence needs to be quantified and accommodated in the informational analysis of images and maps and for exploring scale–information relationships. Lastly, methods for quantifying information content in maps will be reviewed.

Uncertainty is ubiquitous in spatial information and analysis, as it enters and creates weak links in spatial information chains. Scale is known to have complex interactions with uncertainty, with the former being part of the descriptors for the latter, whereas the latter is usually affected by the former. Chapter 12 will explore the interactions between scale and uncertainty, and seek to incorporate scale dependencies in uncertainty characterization. In this chapter, accuracy metrics and assessment methods will first be reviewed, with scale dependencies highlighted. Validation methods that can deal with scale mismatch will be described. Fine-resolution reference data are often used to evaluate map products at coarse scales. Thus, methods are required for aggregating reference data from fine to coarse scales, and for assessing the errors introduced into the reference data by scaling. The other important aspect of uncertainty characterization in spatial information is to quantify the variance of a variable. Traditionally, uncertainty is approached from error analysis methods that deal with error description and propagation by assuming spatial independence in error distributions. As spatial dependence is characteristic of spatial data and their error occurrences, a better strategy for error modeling is based on stochastic simulation, which can generate large numbers of dense and equally probable realizations whereby spatial structure inherent to the field being examined is reproduced and sample data are honored at their locations if conditional simulation is required. Geostatistical simulation will be discussed, catering for both point-to-area and area-to-point scenarios.

2 Geographic Representations

Scale is a fundamental concept in the geosciences and provides key information on geographic representations, measurements, analysis, and modeling. Thus, studies about representing and integrating spatial data at different scales are important for spatial problem solving (Amrhein and Wong 1996; Longley and Batty 1996; Quattrochi and Goodchild 1997; Atkinson and Tate 2000; Goodchild 2011). As outlined in Chapter 1, geographic conceptualization concerns how spatial entities and distributions should be represented (i.e., in which data models and at what scales) to facilitate discretization, data acquisition and integration, geo-processing, and visualization in a computerized environment, and is fundamental for research on scale in spatial information and analysis. This chapter will describe geographic representations and associated scale issues (e.g., scale dependence of these representation models and representations that are adaptive to scale), based on the conceptual framework laid out by Goodchild et al. (2007).

Spatial information, interchangeably called geographic information in this book, is geo-referenced information, associating locations in the geographic domain with the properties of those locations. The geographic domain includes the Earth's surface and near surface, with the latter including the domains of geology, oceanography, and atmospheric science. In the temporal dimension, retrospective studies billions of years into the past may be conducted by geologists, whereas decadal forecasting into the future may be the interest of environmental modelers, highlighting the enormous discrepancy in temporal spans involved in geospatial studies (Goodchild et al. 2007).

For the geographic domain, spatial and temporal resolution are important metrics of geographic information and are also application specific. Spatial resolution will usually be no finer than 1 cm, and temporal resolution no finer than 1 s, in most cases, as asserted by Goodchild et al. (2007). The diversity of spatial and temporal resolution of geospatial information is well reflected by developments in Earth observation, as remotely sensed images provide sources of geographic information at a wide range of spatial and temporal resolutions. For example, the National Aerial Photography Program (NAPP), operational from 1987 to 2007, was coordinated by the USGS as an inter-agency project for aerial photography over the United States (http://eros.usgs.gov/#/Find_Data/Products_and_Data_Available/NAPP). The project obtained more than 1.3 million cloud-free black-and-white and color-infrared aerial photographs. These aerial photographs were taken with a 6-inch focal length lens at a scale of 1:40,000 and were made available as medium or fine spatial resolution digital images. With spaceborne sensors, such as NASA's Moderate Imaging Resolution Spectroradiometer (MODIS) with spatial resolutions of 250 m to 1 km, surface processes can be remotely monitored at a range of spatially and temporally balanced resolutions to generate information products of global coverage (Zhan et al. 2000; Justice et al. 2002; Friedl et al. 2010). Spaceborne remotely sensed data are also available at increasingly fine spatial resolution (1 m and sub-meter), such as from the IKONOS (1999), QuickBird (2001), OrbView (2003), and Worldview-1 (2007) sensors, which are valuable sources of information for applications requiring fine spatial detail, as reviewed by Blaschke (2010).

Within the limits mentioned previously on spatial and temporal resolution, a vast and potentially infinite number of properties might be found at a geographic location at any point in time, although in practice they are often limited for purposes of standardization and prioritization by authorities such as the US Federal Geographic Data Committee (http://www.fgdc.gov), as noted by Goodchild et al. (2007). For example, agriculturalists might be concerned with acreage and yields of different crop types; cadastral surveyors and real-estate marketing analysts may need information about parcel boundaries and ownerships along with other properties of interest; and an ecologist might record land cover and vegetation types. In all such cases, the semantic meanings of the properties are clearly important for information communication and sharing and must be understood by both information producers and users.

As described by Goodchild et al. (2007), "geographic location" may refer to a point (e.g., a survey monument), a line (e.g., a property boundary), an area (e.g., a salt flat), a volume (e.g., a reservoir or rock), or some indeterminate region with fuzzy boundaries (Burrough and Frank 1996), and in the temporal dimension, the property may be associated with an instant interval, an extended interval, or an indeterminate interval (Peuquet 2001). In statistics, the geometric basis for a statement about a property is referred to as its support. This is also true of geostatistics where data support is a key concept (Gotway and Young 2002).

The concepts of discrete objects and continuous fields were introduced into the GIScience literature in the late 1980s and early 1990s, and have since become the dominant conceptual models of information embedded in geographic space (Goodchild 1991, 1992; Goodchild et al. 2007). Discrete objects are better suited for describing well-defined entities and, hence, are widely employed in cartography, whereas continuous fields are more useful for modeling various forms and processes in geography, as reviewed by Longley et al. (2011). There are numerous examples of object- and field-based representations of spatial entities and distributions: discrete objects are clearly well suited to representing well-defined entities such as the Forbidden City in Beijing and Wall Street in New York, whereas fields are inherently more effective in depicting spatial distributions such as terrain elevation, vegetation biomass, and vegetation composition and abundance. Nevertheless, objects and fields are often employed in combination for accurate representation of spatial distributions. For instance, object-based representations of ridges, streams, shorelines, and other features may well be used in terrain modeling that is essentially field based. Cropland units may be better represented as polygonal objects for easier integration with historical and other agricultural data that are based on areal units, whereas image classification and yield prediction may well be based on field-based representations.

Many applications of GIS address phenomena distributed over one-dimensional structures embedded in two or three spatial dimensions, commonly represented as networks of roads or rivers as in transportation and hydrology (Goodchild et al. 2007). Such networks are particularly useful for applications where routing, identification of nearest facilities, and other network analyses are required and can be performed using GIS (Miller and Shaw 2001; Biba et al. 2010; Longley et al. 2011). As discussed by Goodchild et al. (2007), networks can be regarded as special cases of field-based and object-based conceptualizations in which the set of possible locations x is limited to the one-dimensional network structure, rather than being assigned to a third case.

One of the common grounds for fields and objects seems to lie in the individual points defining the geographic locations and their properties. The concept of the geo-atom was introduced by Goodchild et al. (2007) as a kind of unified geographic conceptualization based on point sets that provides the foundation for both objects and fields. A geo-atom is defined as a tuple $\langle x, Z, z(x) \rangle$, where x is a location in space–time, Z is a property, and $z(x)$ is the value of the property at that location. Both objects and fields aggregate the locations, which are the first element of the geo-atom tuple. Geo-fields are formed by aggregating the geo-atoms for a single property, the second element of the geo-atom tuple; and geo-objects are formed by aggregating geo-atoms according to rules defined on the geo-atom's value, the tuple's third element (Goodchild et al. 2007). The concept of a field object could be defined as a geo-field whose domain is a geo-object with internal heterogeneity represented through fields, as described in the context of storm system characterization by Yuan (1999).

This chapter will first describe geo-atoms as the building bricks for field-based and object-based representations of the geographic world, to which all geographic information can be reduced. Field models and object models are then elucidated, with emphasis being placed upon their scale dependence in representing spatial distributions and entities. Fields and objects are called geo-fields and geo-objects, respectively, to underline the fact that they are aggregated from geo-atoms. Nevertheless, the notions of fields and objects are used interchangeably with those of geo-fields and geo-objects, respectively, without causing any confusion.

The amount of information shown on a map varies enormously from area to area, depending on the local variability: e.g., dense streets and residences in urban areas vs. those in sparsely populated remote areas. It would make sense then to use rasters of different sizes depending on the density of information. Thus, large cells are used in smooth areas, and small cells in rapidly varying areas. Unfortunately, unequal-sized squares will not fit together except under unusual circumstances, for example, when small squares nest within large ones, as explained by Goodchild and Kemp (1990). There are, however, some methods for compressing raster data that allow for varying information densities. Quadtrees for regular tessellation and hierarchical triangulation for irregular tessellation can adapt to the spatially varying resolution at which the underlying spatial phenomena can be represented (Samet 1984; Bóo and Amor 2009). Hierarchical data structures are useful for GISs because of their ability to focus on the interesting subsets of the data (Samet 1984). Hierarchical data structures will be described in this chapter, after the descriptions of field-based and object-based geographic conceptualizations.

In the concluding section of this chapter, we will first discuss representations for spatial interactions. The concept of a geo-dipole, introduced in parallel to geo-atoms by Goodchild et al. (2007), provides a basis for dealing with such properties as distance, direction, interaction, and flow, which are useful for understanding the driving processes of physical and social landscapes. Then, we describe the temporal extensions of fields and objects, which should be included in geographic representations to accommodate the dynamics of spatial processes (Pultar et al. 2010). Lastly, we will provide a brief overview of scale dependence in field and object models. We highlight the fact that, although fields and objects are theoretically independent of scale, all representations implemented in practice are scale dependent and, moreover, that rasters are more informative for indicating explicitly the underlying resolution of representation and better suited for scale-aware geo-processing, as also discussed by Goodchild (2011).

2.1 GEO-ATOMS

As noted earlier, a geo-atom is defined as an association between a point location in space–time and a property, and is an atomic form of geographic information (Goodchild et al. 1999, 2007). For example, a geo-atom might indicate that at 114.9° E, 30.5° N, at about 50 m above mean sea level, and at local noon time on April 11, 2011 (a four-dimensional definition of x), the air temperature (the property Z) was 23°C (the value $z(x)$). Many similar examples can be listed, from both natural and built environments.

All geographic information can be reduced to the geo-atoms of different instances (Goodchild et al. 1999). For example, a polygon defining the outline of Hubei, China can be regarded as a set of atomic statements about the points constituting (or filling) the province, a polyline defining the centerline of US Freeway 101 can be regarded as a set of atomic statements about points along the centerline, and the extents of croplands affected by a prolonged hot and dry period can be decomposed to a set of geo-atoms representing the location and severity of drought-stricken croplands.

With geo-atoms, a point location x can be four-dimensional in space and time. In practice, however, static and planar implementations are common, as the temporal and the third (vertical) spatial dimension are often ignored. Also, it is assumed that, even though there can be many properties ascribed to the same location x, any single property Z can take only one value at any instant and location in four-dimensional space–time, although the issue of single values becomes more complex when dimensionality is reduced, as in the case of cliffs and intricate three-dimensional structures (Goodchild et al. 2007).

Although point-like observations are often made in such disciplines as meteorology (e.g., rain gauge data), many other forms of geographic observations concern larger geographic aggregates (e.g., tree cover density and the steppe of central Kazakhstan), and it is much easier to store them in aggregate form. Even point-like observations must place some lower limits on their spatial footprints (Goodchild et al. 2007). Thus, a geo-atom may be denoted as a tuple $\langle v_x, Z, z(v_x) \rangle$, where v_x is a generalization of the point x, which may be a point (point-like to be more accurate) or a block (volume) of definite size. However, with proper understanding about its implications, we can still use x to denote data support of zero or finite dimension in the remainder of this chapter.

In principle, geographic features of more than zero dimensions contain an infinite number of points and therefore an infinite number of geo-atoms of point support (Goodchild et al. 2007). In practice, however, the number of statistically independent observations that can be made in space–time is limited, because of spatial dependencies, widely known as Tobler's first law of geography (TFL) (Tobler 1970), although there are exceptions with certain phenomena (Longley et al. 2011). TFL ensures the existence and representational efficiency of geo-objects of measurable size and finite positional precision, which would otherwise be nonexistent or uncountable (Goodchild et al. 2007). For instance, terraced rice fields near Ubud, Bali, Indonesia, which follow the natural contours of the hills, may be neatly represented as a mosaic of polygons rather than as an infinite set of geo-atoms recording locations planted with rice. With TFL, rules applied to the values of geo-atoms, such as using "mean annual temperature" and "mean annual precipitation," can be expected to identify connected areas of measurable size and relative categorical homogeneity, a logic for mapping area classes (Goodchild et al. 2007, 2009). There are various examples of area-class mapping in regional vegetation classification and ecological regionalization (Peet 1978; Bailey 1983; Franklin et al. 2012). Area-class mapping will be discussed further in Chapter 10.

Applications of geographic information always impose a limit on spatial and temporal resolution, as shown above. Thus, data support v_x will often be of finite size within a problem domain. Moreover, it is impossible to measure location on the Earth's surface exactly. Thus, the location-mapped x' is seldom the same as x, but is better conceived as a distribution around x. Similar constraints apply to time (Goodchild et al. 2007). The net results of finite data supports and limited positional precision are that geographic information is usually represented over measurable aggregates of geographic locations and that representation and geoprocessing efficiency are better achieved through matching between data resolution and information granularity. We will discuss related issues in the context of spatial sampling in Chapter 3.

2.2 GEO-FIELDS

As described by Goodchild et al. (2007), a geo-field defines the variation of one or more properties over a space–time domain D. As such, it constitutes an aggregation of geo-atoms, where each geo-atom defines the same set of properties and aggregation is over a set of data supports $\{v_x\}$, which may be spatially varied. Thus, a geo-field aggregates geo-atoms over space by property Z, irrespective of its value. A geo-field for a single property such as elevation is termed a scalar geo-field, whereas a vector geo-field might describe the spatial (and temporal) variation of magnitude and direction of a phenomenon such as wind or terrain aspect. The terms "coverage" and "surface" often convey the sense of geo-field.

Many properties can be conceptualized as geo-fields (Longley et al. 2011). Elevation and land surface temperature are examples of properties measured on interval scales, whereas land cover and soil type are examples of properties measured on nominal scales. For this latter type of property measured on nominal or ordinal scales, we can use C rather than Z to denote the field variable. Thus, a field for property Z can be considered as a single-valued function defined over a problem domain D, that is, $\{z(x), x \in D\}$. We can represent distributions of area classes (Goodchild et al. 2009), such as land cover C, as $\{c(x), x \in D\}$, where c denotes a discrete class label (i.e., a value for C). To confirm their probabilistic distributions, we can view the cs as outcomes of a stochastic class model $\mathbf{p}(x) = \{p_1(x), ..., p_K(x)\}$, so that $c(x) = \arg \max_k (p_1(x), p_2(x), ..., p_K(x))$. As shown above, geo-fields act as functions of geographic locations, since they map locations in space–time within the domain D to the values of one or more properties, although not to more than one value per property (Goodchild et al. 2007).

In principle, any geo-field over a finite domain D aggregates an infinite number of geo-atoms. However, unless it can be represented accurately by a mathematical function, a geo-field is necessarily sampled, selected, or approximated to be represented in a store of finite size, leading to field representations of limited spatial and/or temporal resolution (Goodchild et al. 2007). One of the rationales underlying the finite spatial and/or temporal resolution implied in field-based representations is due to spatial dependence, which means increasing spatial redundancy at closer spatio-temporal proximity. According to Shannon's sampling theorem, it is possible to achieve an accurate reconstruction of original signals (of limited bandwidth) given data sampled at the Nyquist rate (Shannon 1949; Jerri 1977; we will take a closer look at sampling in the next chapter). Spatial dependence and Shannon's sampling theorem suggest that accuracy in geographic representations can be maintained with finite sets of geo-atoms, as shown in the context of spatial sampling design by Modis et al. (2008). Furthermore, in practice and as noted earlier, the definitions of many properties require convolution over a neighborhood around the point (e.g., fields representing density properties, such as population density). Also, convolution may be inherent in the process of measurement, as is the case with remote sensing (Goodchild et al. 2007). We will discuss the issue of convolution in depth in Chapter 3.

For two-dimensional geographic domains, six models for fields are in common use in geographic information systems, although many more are encountered in other domains, such as finite element meshes, Fourier transforms, and wavelets (we will describe wavelets and wavelet-based multiresolution representations of image data in Chapter 8), as discussed by Goodchild et al. (2007). The six include polygons, triangulated irregular networks (TINs), regular grids, irregularly spaced sample points, regularly spaced sample points, and contours (Goodchild 1993; Goodchild et al. 2007). Although polygons and grids imply finite data supports for the variables being mapped, all of the other four variants employ point or quasi-point data supports, suggesting apparent variations of spatial resolution in these representations. Grids and regularly spaced points are commonly termed as "raster" and the remainder as "vector." To maintain the regularity of grid cells and point spacing when working with distributions over regional and global extents, we should take the Earth's curvature into account (Dutton 1991; Sahr et al. 2003). We will discuss this issue in the context of spatial sampling design in Chapter 3.

To elaborate further, the six commonly used ways of representing geo-fields (Goodchild 1993) are termed discretizations (Goodchild et al. 2007) and are described further below:

- Polygons: piecewise constant, such that the variable is constant within each of a set of nonoverlapping, space-exhausting polygons (the choropleth map, and the area-class map in the nominal case).
- TINs: piecewise linear with continuity of gradient within triangular elements and continuity of value across triangle edges. In an extension to the conventional TIN model, variation within each triangle is modeled as a quintic polynomial rather than as a linear function (Akima 1978).
- Grids: piecewise constant in a regular grid of cells (normally rectangular), a special case of polygons.
- Irregular points: sampled at a set of irregularly spaced points.
- Regular points: sampled at a set of points in a regular array (normally rectangular).
- Contours: sampled along a set of isolines of the property Z.

Thus, discretization is a necessity for the digital representation of any geo-field, along with a domain and a property, which also implies an inherent limitation in the spatial resolution of the underlying represented distribution. The two point-based discretizations of a two-dimensional geo-field, irregular and regular points, constitute samples of geo-atoms. Grids and regular points, the two raster options, are duals of each other, and the distinction between them should not be overlooked (Goodchild et al. 2007). Similarly, we may also consider Thiessen polygons (also known as Voronoi or Dirichlet polygons), which are named after a climatologist who used them to perform a transformation from point climate stations to watersheds, as duals of irregularly spaced points. The former imply homogeneity within each polygonal area (hence resulting in the typical appearance of a choropleth map), whereas the latter record the original values of the property Z without specifying the values in-between the sample points.

Thiessen polygons can be used to describe the area of influence of a point in a set of points. Points contained in each Thiessen polygon are closer to the point on which the polygon is built than to any other point in the dataset. Given a set of points, we can construct a TIN by computing the Delaunay triangles. By definition, Delaunay triangulation is a triangulation such that no point is inside the circumcircle of any triangle in the triangulation. Delaunay triangles are the dual of Thiessen polygons or Voronoi tessellations, and are formed by connecting every pair of points that indicates the existence of an edge in the Thiessen boundary network. On the other hand, if one bisects each edge of the Delaunay triangles perpendicularly and creates closed polygons with the perpendicular bisectors, the result will be a set of Thiessen polygons. There are efficient algorithms developed for computing Thiessen polygons and their dual, the Delaunay triangulation, in the literature (Bowyer 1981; de Berg et al. 2008).

As shown by Goodchild et al. (2007), in the first three types of discretization (i.e., polygons, TINs, and grids), the user is able to query the value of the field at any point—for polygons and grids, the value returned will be the value associated with the enclosing polygon or rectangle, whereas for TINs, it will be a value obtained by linear interpolation within the enclosing triangle. In the remaining three cases, however, some (unspecified) form of interpolation procedure that is not inherent in the discretization is needed to obtain an estimate, unless the query point coincides with a sample point or isoline. For this reason, Goodchild (1993) has termed the first three models complete and the last three models incomplete representations of fields on two dimensions, respectively.

The variable of interest can be sampled at a number of locations and subsequently analyzed to estimate its statistical characteristics across the landscape that are deemed relevant in a particular context. The concept of a sampling scale triplet of extent, spacing, and support is briefly mentioned here (Skøien and Blöschl 2006), to be further described in Chapter 3. Extent is the overall size of the domain sampled. It may range from centimeters to kilometers, depending on the applications. Extent is often selected as the area of interest, but sometimes it can be smaller for the convenience of survey and analysis. Spacing is the characteristic distance of two sample locations, that is, it is small if the samples are narrowly spaced and large if only a few samples are taken over a given domain. In many cases, the number of samples over a domain is a function of the budget available rather than based on sampling design for mapping the variable of interest accurately. A sample's support is synonymous with grain and footprint, indicating its area or volume (Skøien and Blöschl 2006).

For a two-dimensional domain D, we can make the distinction of geostatistical vs. lattice types for data sampled from geo-fields. The former type (geostatistical) of data refers to data indexed continuously in D, whereas the latter type (lattice) includes those sampled over finite areal units. Clearly, the two point-based discretizations of geo-fields (regular and irregular points) fall into the former type, whereas polygonal and gridded data belong to the latter type. The major content of this book will be on scale and scaling with respect to geo-fields, although models and methods oriented to geo-fields may be extensible to geo-objects, given the advantage of geo-fields over geo-objects in addressing scale issues. Chapter 5 will discuss how scale may be modeled for lattice data and under what circumstances geostatistical models may be applicable to lattice data. Chapter 7 will describe some of the methods for scaling in the case of lattice data of regular support (e.g., gridded data).

The concept of a sampling scale triplet of extent, spacing, and support can be linked usefully with the field representation models described previously. For instance, discretizations via relatively sparse points in a smaller survey extent may generate data that are more informative than those based on cheap gridded samples, such as remotely sensed imagery. However, their differentiation should not be taken as a reason for using one over the other. Rather, they should be combined, as they are often involved in applications. Thus, these six discretizations each have their place in representing fields, and sensible and complementary use should be made of them.

The precise selection of points, lines, areas, volumes, or hyper-volumes that are created in a discretization process impacts the accuracy of the representation of any geo-field with respect to the real world (Goodchild et al. 2007). For example, in representing an interval-scaled field on two dimensions, such as elevation, using a TIN, the sample points that form the triangle vertices should be located on peaks, pits, channels, and ridges. There are also algorithms developed for selecting very important points from raster digital elevation models (DEMs) in constructing TINs, aiming for tradeoff between accuracy and data compactness (Chen and Guevara 1987; Chou et al. 1999). Similarly, in representing a nominal-scaled field on two dimensions using polygons, it is desirable that the boundaries between areas be located along lines of rapid change in class, as discussed by Goodchild et al. (2007).

The discretizations described above create representations of fields as properties of discrete objects—polygons, triangular polygons, rectangular polygons, points, and polylines, respectively. However, there is an important distinction in that the discrete objects utilized to represent a field normally do not exist in reality, but are abstracted solely for the purpose of field discretization, as emphasized by Goodchild et al. (2007). We will discuss objects in the next section, which will help in appreciating the fundamental differences between field and object models.

2.3 GEO-OBJECTS

A geo-object is defined as an aggregation of points in space–time that have specified values for certain properties Z. Thus, geo-objects represent geo-atoms aggregated by the values of properties using specified rules, rather than by property alone as in the case of geo-fields (Goodchild et al. 2007). For example, the geo-object Zhangdu Lake in Wuhan is composed of all geo-atoms having the value Zhangdu Lake for the property lake. Many other geo-objects can be conceptualized as constituted from geo-atoms by aggregating them over Z values. Points are special geo-objects, although a point with a single property is formed from a single geo-atom.

We can represent an object with an identifier *ID* as (*ID*, {x}, *attri*), where {x} denotes the aggregated point set and implies a certain granularity, and *attri* denotes the value of the property. The identities of geo-objects (i.e., *ID*s) are established primarily by their locations and associated geometries, and only secondarily by their attributes, as explained by Goodchild et al. (2007). While homogeneity is the basis for the definition of many geo-objects, we can perform aggregation of geo-atoms using more complex rules, such as interaction in the definition of functional regions (Johnston et al. 2000; Goodchild et al. 2007). For geo-objects, we often use the synonym attribute in lieu of property to emphasize their differences from geo-fields. As in geo-fields, several terms in GIS convey similar meaning to that of geo-objects, including entity and feature.

The dimensionality of geo-objects is constrained by the space in which they are embedded (Goodchild et al. 2007). Objects in a two-dimensional space may be classified as points, lines, and areas, depending on the scale of the data and analysis. First, points may be entity points, label points, or nodes. As noted earlier, points may have finite footprints. Mount Everest's peak must have a nonzero dimension to facilitate its identification from a distance and hence surveys of its height. Many point objects are scale dependent. In other words, scale dependencies of data modeling by discrete objects should be considered. For instance, individual buildings that are represented as small rectangles or other shapes on large-scale maps may be shown as points or even selectively omitted on small-scale maps. To measure the distance between major cities, it would be appropriate to represent the cities as points.

However, to study urban sprawl, the cities should be represented as areas, and fine-grained urban fabrics can be seen from the dense street networks for a metropolitan city like London.

Second, lines represent linear features, such as rivers, roads, and pipelines. At fine scales, such features will be represented as areas bounded by lines, but at coarser scales they may be generalized to single lines of zero width. Each line is composed of a set of point vertices defining the shape of the line, together with mathematical functions defining how adjacent points are connected to form segments. Straight-line segments are the most commonly used, but circular and elliptical arcs, Bézier curves, and cubic splines are also encountered (Goodchild et al. 2007). A representation of a line as a collection of segments is termed a polyline. Unless a line can be represented fairly well by a mathematical function, the point set $\{x\}$ defining it must have been subjected to a process of approximation, selection, and generalization, which, in turn, implies scale dependence.

Third, areas may be represented as polygons in vector data structures or as blocks of contiguous gridded cells in a raster structure. Polygons are formed by boundaries, which keep track of their footprints. As in the case of polylines, the point set $\{x\}$ defining polygon boundaries is scale dependent, although the dependence is complicated by the boundary complexity and the internal homogeneity of the area. Raster structures represent features as a matrix of cells. The shape of an area object is implicit in its raster representation. Selection of cell size for rasters should preferably be based on the properties of the original data, although the cell size will have been predetermined at the outset in the case of remote sensing. So, for data captured from maps, cell size should be based on the original map scale and on the level of mapping detail, such as the size of the smallest area (the minimum mapping unit). Using too large a cell size will result in loss of information, whereas using a cell size smaller than necessary leads to data redundancy and inefficient processing, as discussed by Goodchild et al. (2007).

Consider the example of lakes in Finland to see more scale dependence and implications of inaccuracy in object representations. There are thousands of lakes in Finland; precisely, a total of 187,888 lakes, according to the Finnish Environment Institute (http://www.environment.fi). Lake is defined by the Finnish Environment Institute as a body of standing water larger than 5 acres (500 m²). By increasing the minimum mapping unit, the number of lakes larger than 1 ha (10,000 m²) is 56,000 and the number of lakes larger than 10 km² is 309. The annual variation of lake area is around 2350 km² against an estimated total lake area of 33,709 km² (http://www.environment.fi).

As described by Goodchild et al. (2007), the point sets formed by aggregations of geo-atoms are normally connected because of TFL. However, they may sometimes contain holes or enclaves. The Open Geospatial Consortium (OGC) Simple Feature Specification (documents available at http://www.opengeospatial.org), which deals with static geo-objects, allows in the two-dimensional planar case for multipart polygons (collections of disconnected islands that share the identity property, such as the US state of Hawaii), multipart polylines (collections of disconnected polylines with shared identity, such as contributing creeks upstream of a river), and multipoints (collections of points with shared identity, such as manhole cover locations for a sewer line).

As shown by Goodchild et al. (2007), the spatial extent of a geo-object may be established by fiat, as for example when counties are defined by administrative decision, or may be *bona fide* if the spatial extent reflects some form of internal homogeneity, as for example when geo-objects, such as dunes and wetlands, are represented. In the *bona fide* case, the rules defining membership in the constituent point set can be complex and vague, as for example in the rules defining membership in such features as moraines and suburbs.

Generally, any geo-object can be conceptualized as a membership field that is continuous scaled in the case of indeterminate boundaries (Burrough and Frank 1996) and binary in the case of determinate boundaries, with the latter case being a special type of the former. Thus, the isolines of the membership field are related to the boundary of the geo-object, and the gradient of the membership field at any location is related to the boundary's local degree of fuzziness or accuracy in positional description (Goodchild et al. 2007). This actually shows an example of complementarity of fields and objects in geographic representations. Because of this, scale dependence of objects may be explored usefully through their membership fields extrapolated from their atomic point sets $\{x\}$, although this will not be elaborated here.

Thus far, individual geo-objects have been described. Many geo-objects are often grouped into classes based on shared sets of properties (Goodchild et al. 2007). In the relational model, these populate tables that are linked with keys (Codd 1970); in the object-oriented model, they form classes that are linked by inheritance, composition, aggregation (not in the sense of geo-atom aggregation), and association relationships (Zeiler 1999). The six discretizations of a geo-field discussed previously all yield objects with common sets of properties that can be assembled into tables and classes. Thus, the relational and object-oriented models are well suited to the representation of geo-fields, as explained by Goodchild et al. (2007). It would be interesting to see how object representations and scale dependence may benefit from object-oriented modeling, by which scale metadata of geo-objects are aggregated and propagated into compound objects or classes of objects.

Before concluding this section, we discuss the links between geo-fields and objects briefly. Câmara et al. (2000) provide a comprehensive review of operations that create geo-fields from geo-objects and *vice versa*. For example, density estimation creates a continuous field of density from a collection of discrete objects; image processing and pattern recognition seek to extract discrete objects from a field of reflectance or radiation; and surface-specific points and lines (e.g., peaks and streams) need to be identified for terrain modeling, as discussed by Goodchild et al. (2007). It is of ecological interest to explore how discrete, homogeneous patches of a biogeographic landscape arise in a world that is defined by the field-like variation of physical properties such as temperature and precipitation (Goodchild et al. 2007). Phase space provides a conceptual framework for modeling and predicting area classes based on field-represented covariates (Goodchild et al. 2007, 2009). We will discuss this in greater detail in Chapter 10.

2.4   HIERARCHICAL DATA STRUCTURES

We often observe that terrain undulation or ruggedness is spatially varied. Vegetation composition and density in urban areas are also remarkably different from those in rural settings. Road networks are often denser in populous and thriving metropolitan areas than those in remote areas of wilderness. The information granularity will exhibit spatial variability, which should thus be accommodated in representations. It would be sensible to use differently sized rasters depending on the density of local information in the dataset: larger cells in relatively uniform or homogeneous areas, smaller cells in rapidly varying areas. Hierarchical data models provide multiresolution descriptions of spatial distributions and, thus, valuable constructs for geographic representations adaptive to local variability in the underlying forms and processes, as discussed by Goodchild and Yang (1992). These models can be classified depending on the shape of the underlying subdivision: quadtree-based models and hierarchical triangulated models, which may be usefully integrated for multiresolution representations (Birch et al. 2007). For example, Paredes et al. (2012) presented an efficient scheme for achieving interactive level-of-detail rendering of hybrid terrain models, without the need for a costly preprocessing or resampling of the original data that consist of regular grid data and local high-resolution TIN data. Their method achieved coherency in multiresolution terrain modeling that integrates heterogeneous data in different sub-areas of a study area through an adaptive tessellation of the sub-areas between their boundaries. We discuss quadtrees and hierarchical triangulated models below.

2.4.1   QUADTREES

Quadtrees provide hierarchical data structuring that is adaptive to local variability in spatial distributions (Samet 1981). They work by dividing the array into quadrants and numbering them 0, 1, 2, and 3 as in the Morton order. We subdivide a quadrant further into four quadrants if it is nonhomogeneous. We can recursively subdivide each quadrant using a rule of 4 until either a square is homogeneous or we reach the finest level of resolution (the pixel size). The resultant quadtree has the following characteristics: at each level there is a four-way branching, each branch terminates at a homogeneous block, and each terminal branch in the tree is known as a leaf (Samet 1981; Gargantini 1982). The octree is a three-dimensional version of the quadtree where the cube is divided recursively into eight pieces, having applications especially in mining and geology.

When successively partitioning a $2^n \times 2^n$ array ($n \geq 1$) into quadrants, we obtain a tree whose nodes are either leaves or have four progeny. Suppose that we code the pixels at the finest resolution as 1s or 0s, depending on whether or not they contain the specific objects or register the specific property values. A nonleaf node can be black (all quadrants 1s), white (all quadrants 0s), or gray (quadrants not entirely 0s or 1s), and is, in general, stored as a record with six fields, one pointing to its parent; four pointers to the NW, NE, SW, and SE quadrants; and the sixth field is the node's attribute (Samet 1981; Gargantini 1982). Linear quadtrees proposed by Gargantini (1982) have the following characteristics: (1) only black nodes are stored, (2) the encoding used for each node incorporates adjacency properties in the four principal directions, and (3) the node representation implicitly encodes the path from the root to the node. With linear quadtrees, space and time complexity depend only on the number of black nodes, without using pointers.

According to Gargantini (1982), we can encode black pixels in the following way:

1. Each black pixel is encoded in a weighted quaternary code, that is, with digits 0, 1, 2, and 3 in base 4, where each successive digit represents the quadrant subdivision (0—NW, 1—NE, 2—SW, and 3—SE) where it originates.
2. A pixel's integer coordinate pair (IX, IY) $(IX, IY = 0, 1 \ldots 2^n - 1)$ identifies the position of a pixel in the $2^n \times 2^n$ array and can be mapped into an integer K in base 4, which expresses the successive quadrants where the pixel belongs. To find K, IX and IY are written as $IX = a_{n-1}2^{n-1} + a_{n-2}2^{n-2} + \ldots + a_0$ and $IY = b_{n-1}2^{n-1} + b_{n-2}2^{n-2} + \ldots + b_0$, where a_{IX}, b_{IY} is either 0 or 1 and indicates which quadrant (IX, IY) belongs to at each level of subdivision. For example, suppose $n = 3$, for pixel (3, 2), we have $K = 031$ (as $3 = 0 \times 2^2 + 1 \times 2^1 + 1 \times 2^0$, $2 = 0 \times 2^2 + 1 \times 2^1 + 0 \times 2^0$).
3. After encoding all the black pixels into their corresponding quaternary codes, condensation can be applied. That is, if four pixels have the same representation except for the last digit, we eliminate them from the list and replace them with a code of $(n - 1)$ quaternary digits followed by some marker (Samet 1981). We refer to this sorted array as a linear quadtree (Gargantini 1982).

Global data present significant problems (Sahr et al. 2003; Ma et al. 2009). We might use a projection such as the Mercator and represent the data as a raster on this projection. The area and shape of each pixel would be significantly distorted, particularly near the poles, as shown by Goodchild and Kemp (1990) and Goodchild and Yang (1992). A more suitable approach is proposed by Dutton (1991). It projects the globe onto an octahedron, consisting of eight triangles: the vertices of the octahedron are at the poles and are 90° apart around the equator. Each triangle is recursively divided into four smaller triangles, by connecting the midpoints of the edges, with a numbering convention of 0 (central triangle), 1 (vertically above or below), 2 (diagonally to the left), and 3 (diagonally to the right). Resolution is automatically known from the length of the address: in the global scheme described above, a 21-digit address has a 1-m resolution; for a 1-km resolution, we need only a 13-digit address. In comparison, a latitude/longitude gridding system is fixed, and it is not always easy to tell the resolution from the latitude/longitude indexing scheme, as discussed by Goodchild and Kemp (1990).

2.4.2 Hierarchical TINs

Hierarchical triangulated models are resolution-adaptive geographic representations based on an irregular tessellation as opposed to the regular tessellations employed in quadtrees (De Floriani et al. 1984; Abdelguerfi et al. 1998). They can be further classified into strictly hierarchical triangulations, which are described by a domain partition tree, and multiresolution triangulations, which are basically sequences of TINs. These models are based on the recursive subdivision of an initial triangle (with vertices at data points and containing all the other points inside) into a set of nested sub-triangles with vertices at the data points (De Floriani et al. 1984).

In a ternary triangulation, a subdivision of a triangle t consists of joining an internal point p to the three vertices of t, whereas, in a quaternary triangulation, each triangle is subdivided into four triangles formed by joining three points, each lying on a different triangle side. The major problem with a ternary triangulation lies in the elongated shape of its triangles, which leads to inaccuracies in numerical interpolation. A quaternary triangulation, on the other hand, is only well suited when the data are regularly distributed and suffers from the same discontinuity problems at the boundaries of adjacent triangles as quadtree-based models, as discussed by De Floriani et al. (1984). For example, Goodchild and Yang (1992) encountered such discontinuities when implementing Dutton's scheme as a hierarchical representation of the globe in a GIS.

Techniques to obtain a TIN at a fixed level of resolution from a dense set of data have been described (Preparata and Shamos 1985). Below, we describe the method for building a hierarchical representation (e.g., terrain surface model) at various levels of resolution, based on De Floriani et al. (1984) and De Floriani and Magillo (1997). Assume that we have a comparatively exact representation of a surface, say, a dense regular grid of sampled data. The goal is to build our model by extracting points from it.

Let $\{\varepsilon_0, \ldots, \varepsilon_k\}$ be a given decreasing sequence of positive real values, called the tolerance values. First, the root of the hierarchical triangulation TGL^0 is built based on the initial dataset, which is a TIN at level of precision ε_0 (i.e., $Error\ (TGL^0) \leq \varepsilon_0$). Then each triangle tgl_i of TGL^0 is refined into a triangulation TGL_i^0 such that $Error$ $(TGL_i^0) \leq \varepsilon_1$. This process is repeated on the new triangulations with decreasing tolerance values, until each portion of the initial domain has been refined at precision ε_k, as described by De Floriani et al. (1984).

The basic step of the refinement procedure is as follows: for a triangulation (of precision ε_k) $TGL = \{tgl_1, \ldots, tgl_n\}$, where tgl_j $(\forall j = 1, \ldots, n)$ stands for a triangle in TGL such that $\varepsilon_{k+1} < Error\ (TGL) < \varepsilon_k$, compute a set of triangulations $\{TGL_1, \ldots, TGL_n\}$, where TGL_j $(\forall j = 1, \ldots, n)$ approximates the portion of the terrain corresponding to triangle $tgl_j \in TGL$ with precision ε_{k+1}. As shown by De Floriani et al. (1984), this task is performed in three steps:

1. Edge refinement: each edge e of TGL is refined by inserting new data points on it so that the terrain profile along e is approximated with precision ε_{k+1}.
2. First expansion: for all j, a triangulation TGL_j is computed that contains the vertices of tgl_j plus all vertices added to its edges.
3. Triangulation refinement: for all j, a Delaunay selector algorithm with precision ε_{k+1} is applied to TGL_j.

These three steps are further described below, following De Floriani et al. (1984). Edge refinement (Scarlatos and Pavlidis 1992) can be produced independently for all triangulations and for all edges in each triangulation. Given an edge e, its approximation error, *error(e)*, is the maximum difference between the terrain profile along *e* and its corresponding interpolated function. A refinement of *e* into a chain of edges to improve the approximation can be obtained through an iterative algorithm that splits *e* by inserting at each cycle the point that would otherwise cause the maximum error (De Floriani et al. 1984).

After edge splitting has been performed, the first expansion can by computed for each triangle tgl_j ($\forall j = 1, \ldots, n$) independently, by applying any Delaunay triangulation algorithm to the set of vertices corresponding to the chains that refine the edges of tgl_j. The triangulation refinement can be performed on each new triangulation independently through a Delaunay selector algorithm, that is, inserting the point, in turn, that would otherwise cause the maximum error in the approximation (for details, see De Floriani et al. 1984; De Floriani and Magillo 1997).

2.5 DISCUSSION

2.5.1 GEO-DIPOLES AND SPATIAL INTERACTIONS

The previous sections have described geographic representations by geo-fields and geo-objects. As manifestations of certain processes, geographic distributions and entities often need to be understood in terms of interactions. For example, urbanization, as an important dimension of land use and land cover change, is often accompanied by rural-to-urban flows of human populations, leading to modification of demographic distributions. Such variables as flows are properties of locations taken two at a time, or *spatially binary* variables, as distinct from the properties of single locations, or *spatially unary* variables, described in the previous sections.

Various methods have been devised for representing interactions to better understand dynamic processes and their effects (Fotheringham and O'Kelly 1989). As reviewed by Goodchild et al. (2007), these include object fields (Cova and Goodchild 2002), metamaps (Takeyama and Couclelis 1997), object pairs (Goodchild 1991), and association classes (Zeiler 1999). To unify various representational models for spatial interactions, Goodchild et al. (2007) introduced the concept of geo-dipoles, which, in relation to these methods, play a similar role to that of geo-atoms in relation to geo-objects and geo-fields. A geo-dipole is defined as a tuple connecting a property and value not to one location in space–time as in the case of the geo-atom but to two: $\langle x_1, x_2, Z, z(x_1, x_2) \rangle$, where Z represents properties for pairs (x_1, x_2), such as distance or direction in space, time interval, flow intensity, or direction, and $z(x_1, x_2)$ is their values. For example, by the *object field* model (Cova and Goodchild 2002), each point maps not to a value but to a geo-object, $O(x_1)$, which is defined as a set of points x_2, for which the value of a property Z with respect to x_1 (i.e., $z(x_1, x_2)$) meets certain criteria, such as accessibility or visibility.

Although distance and direction can be evaluated between points, the magnitude of a flow of migrants can only be observed between aggregate areas. Clearly, scale issues in geo-dipoles and their aggregates tend to be more complicated than those in geo-atoms and their aggregated constructs, geo-fields and geo-objects. However, we can address scale-related issues in represented spatial interactions through decomposition so that inference can be made from simpler geo-atoms.

2.5.2 Temporal Dimension

As described previously, geo-atoms link locations x in four-dimensional space–time to properties and values, although x will be two-dimensional in many applications. In geo-fields, time is often treated as a series of snapshots, that is, an ordered sequence of fields defined over the spatial dimensions only (Goodchild et al. 2007). There are four discretizations for the snapshots by which we can generate a time series over fixed networks of data supports: irregular points (e.g., weather station data), polygons (e.g., population statistics over irregular reporting zones), grids (e.g., satellite images over congruent rasters), and regular points (e.g., DEM data for surface monitoring), whereas the TIN and isoline representations often change between snapshots due to the change in the surface being monitored, disqualifying them as discretizations suitable for sequences of snapshot geo-fields. Nevertheless, TIN or isoline-based snapshots can be resampled to a shared discretization, such as regular points or polygons, by point or areal interpolation (Goodchild et al. 1993), as discussed by Goodchild et al. (2007).

Consider dynamic geo-objects whose changes through time are greater than those in the values of their attributes (to keep them distinct from dynamic geo-fields). Assume that *definitions* of geo-objects are persistent through time. We may characterize geo-objects through time based on three conditions: stationary vs. moving, rigid vs. elastic (i.e., shape changing or not), and uniform vs. evolving. These three sets of conditions produce eight combinations, as illustrated by Goodchild et al. (2007).

As shown above, the temporal dimensions of dynamic geo-fields and geo-objects, such as those involved in land use and land cover change research (Allen and Barnes 1985; Lambin and Geist 2006), are represented by snapshots and eight possible object metaphors, respectively. As established in the literature, although scales in space and time can be modeled in separate domains, valid and suitable models of spatio-temporal scale need to take care of space–time interactions (Porcu and Mateu 2007; Fonseca and Steel 2011). Thus, scale incongruities in space and time need to be properly accommodated in spatio-temporal scale representations of geographic distributions and entities. Also, the complex interactions of processes need to be understood and modeled at different spatio-temporal scales for research on such issues as the assessment of future socio-ecological consequences of land use policies for supporting decisions about what and where to invest for the best overall environmental and developmental outcomes (Le et al. 2010). These represent a major challenge for geographic information scientists and will provide great impetus for them to advance research on integrated scale metrics and scaling methods to fulfill the potential of GIScience for scale-explicit spatial problem solving, although the emphasis of this book is placed on space rather than on time.

2.5.3 Scale Dependence of Representations

In principle, both field and object conceptualizations are independent of scale. In practice, however, their digital representations always imply a certain scale dependence (Goodchild 2011). Both discrete objects and continuous fields can be represented in either raster or vector data structures. We discuss their scale dependence below.

As elaborated by Goodchild (2011), in the raster case, the size of the raster cells is an explicit indicator of spatial resolution. Discrete global grids are discussed in the literature, as a conventional raster format cannot be used for the Earth's curved surface (e.g., Sahr et al. 2003; Ma et al. 2009). In the three-dimensional case, the vertical dimension may be sampled differently from the two horizontal dimensions, leading to different resolution metrics in height vs. planimetry. Modifications to simple raster structures are sometimes made to allow for a degree of sub-cell representation, for example, in hierarchical quadtree structures (Samet 2006). Remotely sensed data often exist in raster format, enjoying widespread use in geospatial research and applications. In general, raster data are well suited for scale modeling and change of scale, as demonstrated in the literature and, indeed, in this book.

Resolution is more difficult to define in the vector case (Goodchild 2001, 2011). For data recorded at irregularly spaced points, although there may be some justification for using the distance between points as a measure of resolution, it is difficult to decide whether the minimum, mean, or maximum nearest-neighbor distance should be used, as discussed by Goodchild (2011). Areas are commonly represented as polygons, and volumes as polyhedra. Resolution for areas and volumes may be measured in several forms: the width of boundaries, the sampling density of boundaries, the within-area or within-volume variation, and the sizes of areas and volumes. For lines, such as polylines, similar issues arise with respect to scale measures, concerning whether line sampling density or line width should be used (Goodchild 2011).

For vector maps, in general, when spatial resolution becomes finer, there tend to be increasing numbers of areas and volumes, which tend to become more internally homogeneous with boundaries becoming more detailed. However, it would be incorrect to state that vector datasets have infinitely fine resolution because of the seemingly 0-size points or 0-width lines drawn on maps as opposed to a mosaic of pixels of finite size (Goodchild 2011).

In summary, resolution is not well defined in vector representations, and it is difficult, if not impossible, to infer resolution *a posteriori* from the contents of a vector dataset (Goodchild 2001, 2011). Although there is some guidance regarding the choice between raster and vector structures, such as the nature of existing data, the tools available for data handling, and the applications concerned, the use of raster data for rigorous scientific research is strongly recommended because of their resolution explicitness in representations and well-documented advantages in geoprocessing (Goodchild 2011).

3 Geospatial Measurements

Chapter 2 described geospatial conceptualization and scale dependence. This chapter will focus on geospatial measurement and related scale issues. Specifically, data acquisition and the associated measurement scale issues will be described in the context of spatial sampling and remote sensing so that scale is understood not only with respect to measurement *per se* but also from the perspective of the underlying processes that have generated the data.

With developments in remote sensing and other spatial information technology, an increasing amount of spatial data is derived from multiple sources, which are usually of different measurement spatial scales. Unsurprisingly, multiscale data acquisition and integration have long been recognized as major research topics in geography (Quattrochi and Goodchild 1997; Tate and Atkinson 2001). Studying the process and provenance of measurement (often at multiple scales) is constructive for enhancing the effectiveness of data integration, the subject of Chapter 8.

It is important to distinguish two kinds of spatial scale, as introduced in Chapter 1. One is concerned with scales of measurement, and the other with scales of spatial variation present in data (i.e., process scale). Since spatial data are obtained with particular scales of measurement, the process scale(s) observable in spatial data is (are) modulated by the scales of measurement. Thus, these two kinds of scales are often related, albeit in a complex way. We will focus on measurement scales in this chapter, whereas process scale will be discussed in the next two chapters.

Geographic reality is always observed through a particular spatial sampling framework to produce spatial data. Thus, the resultant data are a filtered version of reality, even when the data are obtained directly through the human senses. The filtering often implies degradation of the quality of the datasets obtained, since spatial sampling with limited coverage and finite support (see Skøien and Blöschl [2006] for details about sampling coverage and other concepts in spatial sampling) implies various processes of smoothing, generalization, selection, and approximation, as will be discussed. In other words, the measurements are subject to coarsening of resolution and generalization or smoothing due to convolution. Therefore, to understand how measurement scales are linked to the scale of the underlying phenomena and data-generating processes, we need to reevaluate the sampling framework and measurement processes with respect to the factors affecting sampling coverage and spatial resolution, respectively.

This chapter begins with a description of the spatial sampling frameworks that are used explicitly or implicitly in the process of measurement and provide the basis for discussing scales in different measurements. Spatial sampling is described briefly, followed by discussions about data support. Spatial data at irregular supports are described first, with remotely sensed data being covered as data of approximately regular supports. The definitions of scales in remotely sensed imagery are reviewed and compared, with their implications discussed. This provides some theoretical basis for later discussion about the resolutions attainable in the actual measurement processes, since the ideal performances indicated by these resolution measures are often degraded.

To discuss measurement scales, we need also to understand the processes through which measurements are undertaken and to examine the technical details concerning how data are derived. In a remote sensor, the upwelling radiance (reflected and/or emitted) is converted into an image of spatially distributed radiance, whereby the radiometric, spatial, and geometric properties of the radiance are transformed, as described by Schowengerdt (2007). Generally, the signal of interest (i.e., the portion of the total radiance containing information about the Earth's surface) is degraded in the measuring systems. Such degradation should be properly understood if we are to correctly interpret the measurements and their resolution (Schowengerdt 2007). Remote sensing in both optical and microwave modes will be described, with spatial scales being quantified with respect to both radiometric (signal and noise) and geometric characteristics. In this way, we will be able to understand the mechanisms by which scales in remotely sensed data are affected and determined by complicated interactions between the measurement systems and the underlying entities and distributions being measured, and the fact that a single measure such as instantaneous field of view (IFOV) is not adequate for describing the full spectra of spatial scales during the process of measurement, from radiation physics to the resultant data.

3.1 FRAMEWORK FOR SPATIAL SAMPLING

Measurement scales are determined by the sampling framework. The sampling framework can itself be divided into the spatial coverage of the *sample* and the spatial or geometric characteristics of each individual *observation*, as described by Skøien and Blöschl (2006). Below, we describe some important components of a sampling framework, such as sampling scheme, sampling density, and sample size, which define collectively the so-called coverage of samples (Skøien and Blöschl 2006). Another aspect of the sampling framework is data *support*. We will show how coverage of samples and supports of individual observations interact to determine measurement scales in spatial data acquisition. In discussing data support, we pay special attention to resolution in remotely sensed images, in particular, alternative definitions and measures.

3.1.1 SPATIAL SAMPLING DESIGN

Sampling aims to select subset(s) of individuals (e.g., locations or objects with characteristics of interest) from within a population (i.e., all individuals concerned in a study, such as all plots of forested land in a region or all households for a demographic survey of a municipality) to estimate the characteristics of the whole population (Cochran 1977). Because it is prohibitively expensive to conduct surveys of, and gather information from, every individual in a population, we often seek to find a representative sample of that population. A representative sample is one that is meant to provide accurate, in statistical terms, characterization of the underlying population. This is how a sampling design can contribute (Cochran 1977; Czaplewski 2003; Edwards et al. 2006; Stehman et al. 2011). For instance, spatial sampling in the context of forest surveys was described by Köhl et al. (2006).

As mentioned previously, sampling scheme, sampling density, and sample size define the spatial coverage of samples and boil down to the set of distance and direction vectors between the observations of a sample set, as described by Skøien and Blöschl (2006). Sampling scheme refers to the spatial pattern of the sample observations. Sampling density refers to the number of observations per unit area (Skøien and Blöschl 2006). Over vast spatial extents and, indeed, the globe, the Earth's curvature will have effects on areal, distance, and other measures. This will be discussed later in this subsection. Finally, sample size is the total number of observations, which should be determined on statistical grounds, but is often subject to compromise owing to practical constraints (e.g., time, money, and so on). Below, we will review major sampling schemes, followed by the sampling theorem as the theoretical basis for further discussion about spatial scales in measurement and the underlying data-generating processes.

For survey sampling, probability theory and statistical theory are employed to guide the practice of sampling design. Thus, probability sampling schemes are preferred, by which every element or unit in the population has a chance of being selected in the sample, enabling unbiased estimates of population totals by weighting sampled elements according to their probability of being selected. Probability sampling includes simple random sampling, systematic sampling, stratified sampling, and cluster sampling (Köhl et al. 2006), as described below.

In a simple random sample of a given size, all elements of the sample set are given an equal probability of inclusion. Sample statistics are relatively easy to compute. For instance, the sample variance of the mean calculated with n measurements is calculated as the variance in the underlying population divided by n, indicating a simple estimate for the accuracy of results. In a geospatial setting, the randomness of the sampling is enabled by locating randomly the positions of objects, whose coordinates can be derived using a selection of methods. For instance, random samples may be drawn by generating random degree lines (between 1° and 360°) that radiate out from the center of the community and selecting the households adjacent to these degree lines (i.e., within 80 m of these random degree lines) in the sample, as shown by Duranleau (1999). The sampling method of Duranleau (1999) would have undersampled the fringes without reckoning the fact that, for many human settlements, centers tend to be more densely populated than fringes. For more or less uniform distributions, we should perform random sampling by generating random X and Y coordinates within rectangular limits. Simple random sampling is usually conducted without replacement, meaning that each sample location can only be selected once, with every location having an equal chance of being selected. There are typically a huge number of pixels in remotely sensed data, suggesting the feasibility and relative simplicity of simple random sampling in studies assisted by remote sensing (Lo and Watson 1998). However, this kind of sampling is rarely easy to implement in practice, as randomly located sample sites may fall in locations that are inaccessible or more expensive to visit than elsewhere. Besides, the randomness of the selection may result in a sample that omits rare events (e.g., rare classes in vegetation mapping), which are nevertheless crucial to be sampled.

Systematic sampling works by arranging the population by some ordering scheme and then selecting elements at regular intervals. Systematic sampling may be implemented for entities or distributions already discretized into regular and uniform units, such as grids. For example, the proposed approach for the 2010 Forest Resources Assessment (FRA) of the UN Food and Agriculture Organization is a systematic sample of 10-km-wide squares at every 1° intersection of latitude and longitude (Steininger et al. 2009). As long as the starting point is randomized, systematic sampling is a type of probability sampling. However, systematic sampling may suffer from periodicity in the problem domain: although unlikely in natural environmental phenomena, if periodicity is present and the period is a multiple of the interval used, the sample may be nonrepresentative of the overall population, making the scheme potentially biased, unlike simple random sampling. Using digital data on deforestation patterns for the five Amazonian countries outside Brazil plus the Brazilian Amazon, Steininger et al. (2009) tested systematic sampling schemes by varying sample-unit size and spatial frequency: (1) 10-km-wide square sample units centered at each 1° intersection, (2) 10-km-wide units at each 0.25° intersection, and (3) 50-km-wide units at each 1° intersection, with sampling rates (the proportion of the area sampled) at 0.8%, 13.4%, and 20.6%, respectively. They calculated two estimates of sampling error: the standard errors, based on the size, variance, and covariance of the samples, and the actual errors, based on the differences between the sample-based estimates and the estimates from the full-coverage maps. Their results indicate that the systematic sampling scheme for FRA 2010 should have confidence intervals (CIs) of smaller than or close to 10% at the continental level. However, systematic sampling at the national level yields larger CIs unless the sample size is very large, especially if any subnational estimates are required.

Stratified sampling may be employed usefully for a population that consists of distinct categories, which form separate strata for the sampling frame. Each stratum is then sampled independently as a subpopulation, from which individual elements can be selected randomly. A stratified sampling method can lead to more efficient statistical estimates, and different sampling approaches can be applied to different strata to make the sampling cost-effective (Stehman et al. 2011). However, as reviewed by Zhang (2010), implementing such an approach can increase the cost and complexity of sample selection and statistical computing. As shown by Dymond et al. (2008), for cost-effective monitoring of changes in forest area over large areas, automated mapping by remote sensing is necessary, but is rarely of sufficient accuracy for estimation of changes in forest area when change is small. They showed how accurate area statistics of small change can be achieved by combining automated change mapping with stratified random sampling. The stratified random sampling can be made efficient by exhaustive sampling of a stratum with contiguous areas of apparent change greater than or equal to some lower limit (5 ha). A regional example from New Zealand confirmed acceptable accuracy (afforestation is estimated within ±7% of reported afforestation) at acceptable cost (10,702 sample points), representing an efficiency gain of 78% (in terms of sample size) over simple random sampling.

Dymond et al. claimed that the proposed method provided a more rapid alternative to multipurpose forest inventories, which depend on intensive and expensive field data.

Sometimes, it is more cost-effective to select sampling units in preselected clusters than from the whole population. As described by Köhl et al. (2006), a cluster sampling design is efficient owing to its balance between the cluster size (e.g., number of trees in a plot) and the number of sample locations to visit. Cluster sampling is commonly implemented as multistage sampling: clusters are selected first, and a sample of primary units is selected randomly from each cluster, with possibly secondary units selected from primary units (for example, see Köhl et al. 2006). Units selected at the last step are then surveyed. More often, combined schemes for sampling are cost-effective and practical. For instance, using as a test case the 21.9-million-ha cover map developed for Utah Gap Analysis, Edwards et al. (1998) implemented a mixed design by combining clustered and subsampled field data, stratified by ecological modeling unit and accessibility, to examine the efficiency of cluster sample designs relative to simple random designs. In terms of precision, their mixed design was more precise than a simple random design given fixed sample costs.

For spatial sampling, whether random, systematic, or stratified, we have implicitly assumed planar situations previously. However, there are issues of coordinates, locations of finite sizes, and gridding for sampling when very large areas, or indeed all, of the Earth are considered. These problems arise essentially from the classic map projection problem, as projecting from the Earth's surface to a plane results in area or shape distortion or both, especially over large areas, but the randomization implied in sampling requires that the sampled locations be regular and retain equal areas when projected on the surface of the Earth. Therefore, it is important to select an appropriately sized and shaped piece of the Earth on which to implement the sampling grid (White et al. 1992; Richards et al. 2000).

A number of approaches to regularly subdivide the surface of the Earth for large areas have been suggested, as reviewed by White et al. (1992) and Song et al. (2002). They are based on regular decompositions of map projections of the Earth to a plane (Tobler and Chen 1986; Goodchild and Yang 1992), decompositions of regular polyhedra fit to the Earth (Dutton 1989), or a hybrid of these. The objectives of a global gridding system for sampling include equal area, regular shape, compactness on the Earth's ellipsoidal surface, minimal scale distortion, and a hierarchical structure (White et al. 1992; Song et al. 2002). White et al. (1992) proposed the development of a sampling grid on the Lambert azimuthal equal-area map projection of the Earth's surface to the face of a truncated icosahedron fit to the globe. This geometric model has less deviation in area when subdivided as a spherical tessellation than any of the spherical platonic solids and less distortion in shape over the extent of a face when used for a projection surface by the Lambert azimuthal projection. Hexagon faces of the truncated icosahedron can be decomposed (by the vertices and center of a hexagon) into a triangular grid at an appropriate density for sampling. The geometry of the triangular grid provides for varying the density, and points on the grid can be easily addressed. Song et al. (2002) proposed an equal area global partitioning method based on small-circle edges on the Earth's surface, which appears to be the best developed approach to satisfying the essential criteria for a global grid.

Having briefly described sampling schemes, we attend to the issues of sampling density and sample size. As discussed by Skøien and Blöschl (2006), sampling density and sample size are closely related to each other, as the sampling density would need to be lowered if there is a limit for affordable sample size over a study area. Sampling density or sampling rate is the number of recorded samples (e.g., pixels) per unit distance, area, or volume when discretizing the underlying spatial distributions. Thus, the larger the sampling distance (e.g., the size of one pixel), the smaller the sampling density (Skøien and Blöschl 2006). For example, to determine the ideal sampling density for an imaging system, we may want to refer to the system's point spread function (PSF) and take samples accordingly. If sampling occurs at a density much lower than what the PSF indicates, we run the risk of not being able to reconstruct images of acceptable quality. Theoretically, there is the Nyquist–Shannon sampling theorem to guide the determination of optimal sampling rate (Shannon 1949).

The sampling theorem states that, if a function $z(t)$ contains no frequencies higher than B hertz (i.e., $z(t)$ is bandlimited to B), it is completely determined by sample values at a discrete sequence of points equidistantly spaced $1/(2B)$ second part (Shannon 1949). In other words, a bandlimited analog signal can be reconstructed perfectly from a sequence of equidistantly spaced samples if the sampling rate exceeds $2B$ samples per second in time (or meter in distance), where B is the highest frequency in the original signal.

Let $z(t)$ represent a continuous-time signal and $\mathbb{F}_Z(\omega)$ be the continuous Fourier transform of that signal:

$$\mathbb{F}_Z(\omega) = \int_{-\infty}^{\infty} z(t)e^{-I2\pi\omega t}\,dt \tag{3.1}$$

where ω stands for the frequency and I is the imaginary unit $I^2 = -1$. If the signal $z(t)$ is assumed to be bandlimited to a one-sided baseband bandwidth, B, we have

$$\mathbb{F}_Z(\omega) = 0 \text{ for } |\omega| > B \tag{3.2}$$

Then, according to the Nyquist–Shannon sampling theorem, the sufficient condition for exact signal reconstruction from samples at a uniform sampling rate f_s (in samples per unit time or space) is

$$f_s > 2B \tag{3.3}$$

$2B$ is called the *Nyquist rate* as a property of the bandlimited signal, and Nyquist (1928) showed that up to $2B$ independent pulse samples could be sent through a system of bandwidth B. The time interval between successive samples is referred to as the *sampling interval* $T = 1/f_s$. The samples of $z(t)$ are denoted by $z_s(t)$ (i.e., the discrete set of $z[n] = z(nT)$, with n being an integer), which can be written as the result of multiplication of the signal by the sequence of Dirac deltas (i.e., a comb function):

$$z_s(t) = \sum_{n=-\infty}^{\infty} z(nT)\delta(t-nT) = z(t)\mathrm{comb}(t;T) \tag{3.4}$$

where $\mathrm{comb}(t;T) = \sum_{n=-\infty}^{\infty} \delta(t-nT)$, as described in many textbooks, such as those by Jain (1989) and Marks (1991). Reconstruction of the original signal $z(t)$ (at times in-between the samples) is done through an interpolation process based on the discrete samples $z[n]$.

To understand the sampling theorem further, consider the Fourier transform of the samples, $z[n]$, $\mathbb{F}_{Z_s}(\omega)$:

$$\mathbb{F}_{Z_s}(\omega) = \mathbb{F}\left(z(t)\sum_{n=-\infty}^{\infty}\delta(t-nT) \right) = (f_s)\sum_{k=-\infty}^{\infty}\mathbb{F}_Z(\omega) * \delta(\omega - kf_s)$$

$$= (f_s)\mathbb{F}_Z(\omega) * \mathrm{comb}(\omega; f_s) \tag{3.5a}$$

$$= (f_s)\sum_{k=-\infty}^{\infty}\mathbb{F}_Z(\omega - kf_s)$$

which says that $\mathbb{F}_{Z_s}(\omega)$ consists of copies (scaled by $f_s = 1/T$) of $\mathbb{F}_Z(\omega)$ shifted by multiples of f_s and summed (Marks 1991; Unser 2000). Furthermore, as $\mathbb{F}_{Z_s}(\omega)$ is periodic with f_s, it can be expressed as a Fourier series:

$$\mathbb{F}_{Z_s}(\omega) = (f_s)\sum_{k=-\infty}^{\infty}\mathbb{F}_Z(\omega - kf_s)$$

$$= (f_s)\sum_{n=-\infty}^{\infty} coef_n e^{-12\pi nT\omega} \tag{3.5b}$$

$$= (f_s)\sum_{n=-\infty}^{\infty} Tz(nT)e^{-12\pi nT\omega}$$

where the Fourier coefficient $coef_n$ is found to be the product of $z(nT)$ and T (Hao 2011). For a bandlimited signal sampled with sufficient sampling rate as specified in Equation 3.3, there is no overlap of the scaled copies of $\mathbb{F}_Z(\omega)$ in $\mathbb{F}_{Z_s}(\omega)$. A "brick-wall" low-pass filter $FTR(\omega)$ can leave only the original spectrum, $\mathbb{F}_Z(\omega)$ from $\mathbb{F}_{Z_s}(\omega)$. This can be done by extracting only the $k = 0$ term of $\mathbb{F}_{Z_s}(\omega)$ in Equation 3.5a via the product

$$\mathbb{F}_Z(\omega) = FTR(\omega)(1/f_s)\mathbb{F}_{Z_s}(\omega) \tag{3.6a}$$

where the low-pass filter $FTR(\omega)$ can take the rectangular function:

$$FTR(\omega) = rect(\omega/f_s) = \begin{cases} 1 & |\omega| < f_s/2 \\ 0 & |\omega| > f_s/2 \end{cases} \tag{3.6b}$$

Therefore, Equation 3.6a can be rewritten as

$$\mathbb{F}_Z(\omega) = rect(\omega/f_s)(1/f_s)\mathbb{F}_{Z_s}(\omega)$$

$$= rect(T\omega)T \sum_{n=-\infty}^{\infty} z(nT)e^{-12\pi nT\omega} \tag{3.7}$$

$$= T \sum_{n=-\infty}^{\infty} z(nT)rect(T\omega)e^{-12\pi nT\omega}$$

where Equation 3.5b is referred to for the expression of $\mathbb{F}_{Z_s}(\omega)$. As the original signal's spectrum $\mathbb{F}_Z(\omega)$ contains all the information about the original signal $z(t)$, $z(t)$ can be recovered by an inverse Fourier transform of $\mathbb{F}_Z(\omega)$:

$$z(t) = \mathbb{F}^{-1}\left\{ T \sum_{n=-\infty}^{\infty} z(nT)rect(T\omega)e^{-12\pi nT\omega} \right\}$$

$$= T \sum_{n=-\infty}^{\infty} z(nT)\mathbb{F}^{-1}\left\{ rect(T\omega)e^{-12\pi nT\omega} \right\} \tag{3.8}$$

$$= \sum_{n=-\infty}^{\infty} z(nT)\, sinc\left(\frac{t-nT}{T} \right)$$

where we note the fact that the sinc function (defined as $sinc(t) = \sin(\pi t)/(\pi t)$) and the rectangular function form a Fourier pair along with the shifting and scaling properties of Fourier pairs (Marks 1991; Hao 2011):

$$sinc\left(\frac{t-nT}{T} \right) = T\mathbb{F}^{-1}\left\{ rect(T\omega)e^{-12\pi nT\omega} \right\} \tag{3.9}$$

Equation 3.8 is the Whittaker–Shannon interpolation formula, showing how the samples, $z(nT)$, can be combined to reconstruct the original signal $z(t)$.

For a two-dimensional spatial distribution $Z(X, Y)$, it is often convenient for data processing and spatial analysis to represent $Z(X, Y)$ by an array of sampled values taken on a discrete set of points in the (X, Y) plane. For bandlimited functions, whose Fourier transforms are nonzero over only a finite region of the frequency space, reconstruction can be accomplished exactly, provided the intervals between samples (along the X and Y directions) are not greater than the limits in the X and Y directions, respectively, as required by the Nyquist–Shannon sampling theorem (i.e., the reciprocals of *Nyquist rates* in the frequency space), as shown by Jain (1989) and Goodman (2005). The Nyquist–Shannon sampling theorem provides theoretical guidance for spatial sampling design, and extensions to nonuniform sampling, common in geography, would be usefully explored (Unser 2000).

3.1.2 DATA SUPPORTS

Remotely sensed imagery provides a complete coverage for a given study area. Thus, the sampling framework is determined largely by the spatial resolution (i.e., data support) that has been adopted. However, spatial sampling is often carried out with irregular and incomplete coverages in ground surveys and census data collection. Thus, we need to discuss irregular data supports as a general framework, for which the related spatial resolution definitions will be spatially varied and require special treatment in geo-computing.

The size, geometry, and orientation of the space on which an observation is defined are known in geostatistics as the data support (Atkinson and Tate 2000), as indicated in Chapter 1. This concept is entirely general. Pixels in images and grid cells in raster data are good examples of regular data support (care must be taken in using gridding systems over regional and global scales where distance or area distortions may result from map projections), whereas zones used in census data, such as enumeration districts (EDs) and wards in the UK census of population, represent irregular data supports for the values (e.g., population count) often associated with them. In the former case, the support is fixed, whereas for the latter the support is variable, as discussed by Atkinson and Tate (2000).

As an important concept and an indicator of measurement scale, data support is central to research on scale issues. The effect of the support on the scales of natural variation has been demonstrated most notably in physical geography and, in particular, remote sensing (e.g., Sèze and Rossow 1991) and in human geography through the modifiable areal unit problem (Openshaw 1983), which arises because census data are defined over variable supports and because analyses are often needed over supports different from the original ones whose data are available, as reviewed by Atkinson and Tate (2000).

To understand how the support, as a scale of measurement, affects the scales of spatial variation that are detectable from such measurement, it is necessary to examine how the support interacts with the underlying spatial variation to produce individual data values. We will see that the measurements often imply various convolutions, as described later in this chapter for understanding the process of measurement and, hence, measurement scale (Chapter 9 will review resolution issues regarding some of the freely available digital terrain datasets).

In practice, values are derived through measurement over a support v, a finite element of space with a specific size, geometry, and orientation (Atkinson and Tate 2000). Therefore, assuming an interval/ratio random field Z, an observation $Z_v(x_0)$

may be treated as a realization of the random variable, $Z_v(x_0)$, which is the spatial mean or integral of Z over v centered on x_0, that is,

$$Z_v(x_0) = \frac{1}{v} \int_{v(x_0)} Z(x) \, dx \qquad (3.10)$$

where $Z(x)$ is the property Z defined on a *punctual* (point) support x, as described by Journel and Huijbregts (1978). We will discuss these concepts in more detail in the next chapter. The important point is that the underlying spatial variation is averaged over a support, say, v (centered on x_0), to produce a *single mean value* $Z_v(x_0)$. A random field is often assumed, but other models could be treated in a similar way. For example, census data are averages over the space represented by EDs, wards, and so on, even though the individuals could be treated as representing points (Bailey and Gatrell 1995), as reviewed by Atkinson and Tate (2000).

The support may also vary in shape and orientation. For example, for remotely sensed imagery, the support may be approximated by a Gaussian or bell-shaped center weighting function (i.e., PSF) such that the underlying spatial variation at the center of the support receives more weight than that toward the edges. Where the support has an anisotropic shape (i.e., the support varies with the angle of orientation), the orientation in which the support is placed may be important, as discussed by Atkinson and Tate (2000).

3.1.3 SPATIAL RESOLUTION IN REMOTELY SENSED IMAGES

Spatial resolution is one of the most common measures of image quality. With the advent of airborne and satellite digital scanner data, pixel sizes are widely used as measures of spatial resolution despite the fact that a pixel size is only a measure of the spatial sampling rate and is usually not the same as the instrument's effective IFOV (Townshend 1981; Forshaw et al. 1983). Forshaw et al. (1983) found that, for a medium-contrast target under average lighting conditions, the equivalent photographic resolution of Landsat MSS imagery is of the order of 200 m, rather than the 79 m as usually quoted. We need to review and clarify some of the definitions of resolutions that are relevant in the context of remote-sensing images.

As indicated above, the IFOV of one detector in a sensor is a kind of definition of resolution based on geometric properties (Townshend 1981). IFOV is the solid angle subtended by a detector of an imaging sensor, assumed to be diffraction and aberration free. It determines the area of visibility on the Earth's surface which can be "seen" from the detector at a given altitude at a particular time and, thus, can be expressed as the "footprint" of the detector on the ground at a given instant.

In addition to IFOV being a geometric definition of spatial resolution, we can define spatial resolution of a remote sensor based on (1) the ability to distinguish between two point targets or to determine the size of the blurred image of a point source, (2) the ability to measure the periodicity of repetitive objects having nominally infinite extent, or (3) the ability to measure the extent of small but finite objects, as reviewed by Townshend (1981). Examples of classes 1, 2, and 3 are the Rayleigh criterion, optical transfer function (OTF) or modulation transfer function (MTF) measures (Lavin and Quick 1974), and the effective radiometric resolution element (ERRE) (Colvocoresses 1979; Strome 1979), respectively, as reviewed by Forshaw et al. (1983). By a suitable choice of units, all of these measures can be cast in the form of a linear distance on the Earth's surface, given the sensor altitude or range to the object on the ground (Townshend 1981). In the following subsections, we will review these three classes of definitions of spatial resolution.

3.1.3.1   Rayleigh Criterion

This criterion is based on diffraction effects, which assert that even if a lens is completely aberration free, the resultant image of a point source will not be a point but will show as an airy pattern (i.e., a central bright disk surrounded by faint dark and light rings). This criterion applies strictly only to the resolution of two mutually incoherent and equal-intensity point (or line) sources when imaged by an optical system having presumably an unobstructed and aberration-free circular aperture (or one having a nominally infinite extent in the direction parallel to the axes of the line sources). The image of each source is a diffraction pattern consisting of a central maximum surrounded by alternate light and dark rings or lines. According to the Rayleigh criterion, the sources are only resolved by the optical system if the peak of the image of one source lies on the first zero of intensity of the image of the other source. In other words, the two sources are resolvable if the separation of their images is at least the radius of the airy disk (Townshend 1981; Forshaw et al. 1983).

According to this criterion, two point sources will only be resolved if their angular separation $\sin \alpha = 1.22 \, \lambda/d$, where λ is the central wavelength of the spectral band being used and d is the diameter of the circular entrance pupil. For line sources and one-dimensional aperture, the separation $\sin \alpha = \lambda/d$ (Townshend 1981; Forshaw et al. 1983). The equivalent spatial resolution on the ground surface is derived from the product of α and the sensor altitude (or flight height in aerial photography). However, most remote-sensing targets are not point sources, lenses are not aberration free, and there are many other factors that will degrade the image and, hence, coarsen the resolution. Thus, this criterion gives an indication of the resolution achievable by a lens, as discussed by Townshend (1981) and Forshaw et al. (1983).

3.1.3.2   MTF and Its Equivalent Bandwidth

The use of MTF is based on the observation that if one takes an image of a set of parallel, linear objects, the contrast (or modulation) between them and their background will appear to be lower as their spacing decreases, until a stage is reached when the contrast is too small to make the linear objects distinguishable. The ratio of the image modulation to the object modulation is known as the modulation transfer factor, which, when plotted against spatial frequency, results in a curve called the modulation transfer function (Townshend 1981).

Mathematically, the MTF curve is the one-dimensional Fourier transform of the one-dimensional line spread function, which, for circularly symmetric imaging systems, is the Abel transform of the PSF (Forshaw et al. 1983). The MTF curve may be deduced from the PSF or measured directly, and provides a measure of the ability of the system to respond to periodic targets of any spatial frequency (Norton et al. 1977; Park et al. 1984; Primot and Chambon 1997; Lin and Liu 2011). The OTF is a more complete measure of system performance than the MTF in that the former includes the phase relationship as well as the amplitude degradation of the signal as the frequency changes. But phase shifts are often difficult to measure or predict and are therefore frequently ignored (Townshend 1981).

With MTF derived, we can calculate the effective instantaneous field of view, which is defined as the resolution corresponding to a value of spatial frequency (full or half cycle) for which the modulation of an object with a sinusoidal distribution of radiance has dropped to 50% of its original value as a result of the imaging system's MTF (Townshend 1981), as also described by Reulke et al. (2006). More useful is the equivalent bandwidth. This is obtained by replacing the MTF curve by a rectangle of equivalent area and giving the upper bound as the value of limiting resolution (Townshend 1981).

As mentioned previously, the IFOV is a measure of resolution for a sensor with discrete detectors. When this is not the case, the IFOV cannot be calculated directly. Thus, estimates of resolution based on MTF are especially appropriate (Townshend 1981). Also, for analysis of system performance, the MTF curves of individual system components (e.g., lens, detector, amplifier, transmission link, recording medium) can be multiplied together to produce an overall system MTF and, hence, a measure of the system resolution (Schowengerdt 2007), as will be shown in the next section (e.g., Equation 3.13). The MTF curve applies strictly only to sinusoidal targets, but can also be computed, with more difficulty, for square-wave or bar targets (Forshaw et al. 1983). The abscissa can be scaled to represent spatial frequency on the ground, with units of, for example, cycles per meter.

Although the MTF approach provides a more comprehensive description of system resolving power than measures based on geometric properties or the ability to distinguish between point targets, it has its limitations. It is based on high-contrast objects that are long compared with their width, which are rarely easy to find in reality. Hence measures derived from MTF curves tend to be optimistically biased, as remarked by Townshend (1981) and Forshaw et al. (1983).

3.1.3.3 Effective Radiometric Resolution Element

None of the previous resolution measures takes explicit account of the effects of noise on performance. Colvocoresses (1979) proposed a measure of spatial resolution, the effective ERRE, to take radiometric effects into account, and it was later made more precise by Strome (1979), as reviewed by Townshend (1981) and Forshaw et al. (1983).

The concept of ERRE is based on defining the minimum area of the Earth for which a single radiance or reflectivity value can be assigned with at least a certain specified confidence that this value differs by no more than a certain specified amount from the actual radiance or reflectivity of the same area under the given viewing conditions (Colvocoresses 1979; Forshaw et al. 1983). This concept is of interest for remote-sensing applications where relative radiometric accuracy of multispectral systems is important, and analysts need to determine the minimum size of a target whose radiometric properties can be sampled reliably (Colvocoresses

1979). Clearly, the concept of ERRE is radiometrically based and provides a kind of accuracy counting when employing this resolution measure for applications. For example, an ERRE measure will indicate the minimum size of crop fields that will yield at least one sample of sufficient accuracy for crop classification.

Consider a field of area *a* and of arbitrary shape with uniform radiance centered within a larger area of the same shape with an area of *A*, also with uniform radiance. The ERRE of the sensor system is defined by the dimensions of the smaller area *a* such that when the difference between the radiances of the inner and outer areas is at least *cntrt* of the full range of the measuring instrument, the radiance of the smaller area can be measured to a relative accuracy of *rac* (with respect to the full range of the instrument) with a confidence level exceeding *cnfl* (Strome 1979; Townshend 1981). For visible and near infrared sensors, a choice of *cntrt* = 30%, *rac* = 3%, and *cnfl* = 95% might be appropriate. For radar, it has been suggested that *cnfl* should be 80% and that *rac* should be expressed as 1 dB, rather than as a percentage, according to the report on effective resolution element (ERE) and related concepts by the Scientific and Technical Sub-Committee of the U.N. Committee on the Peaceful Uses of Outer Space (http://www.fas.org/irp/imint/docs/resunga.htm). The ERRE attempts to take account of the relative brightness of the area of interest and its surroundings, as, with very high contrast, a smaller area can be resolved compared with less contrast.

Wilson (1988) studied the system performance of the Landsat TM scanner through determining the ERE, a measure of its spatial resolution. Two methods have been used: simulation of the TM's spatial responses in an analytical model and measurements of water bodies on a selected Landsat-5 TM scene. By the former method, for TM bands 1–4, both Landsat-4 and Landsat-5 have an effective resolution element of 52 m. By the latter method, band 4 of the Landsat-5 TM scene has an ERE of 75 m approximately. This increase in the ERE (decrease in resolution) over the sensor-only value is attributed to additional factors within the images such as pixel sampling of the scene and atmospheric effects. The image-derived measure of the ERE will give a useful indication of the size of targets that can be recorded to a given radiometric accuracy and can, thus, determine the type of informational classes that may be utilized from present methods of automated classifications of remotely sensed data (Wilson 1988).

Having discussed three definitions for image spatial resolution, we will examine the processes of measurement in remote sensing, both optical and microwave modes, in the next section. We will see that the radiometric and spatial responses of the sensors, the geometric properties of sensing, and the topography interact to determine the effective resolution of the resultant data.

3.2 OPTICAL REMOTE SENSING AND RESOLUTION

3.2.1 Spatial and Spectral Responses

The spatial resolution of a sensor (or its image) is often quoted as the ground instantaneous field of view (GIFOV) or ground sampling interval (GSI) (Schowengerdt 2007). But there are also other factors affecting resolution in real imaging and the resultant imagery: sensor noise, the sample-scene phase (i.e., the relative location of the pixels and the target), nonuniform targets and backgrounds, and variable solar angle and topography, as discussed by Schowengerdt (2007). Thus, image resolution will not be as simple a matter as GSI indicates (Park and Schowengerdt 1982; Schowengerdt et al. 1984). This subsection reviews the processes of optical imaging and examines the various kinds of blurring involved therein, leading to an equation summarizing the combined effects of spectral and spatial responsivities on remote measurements, as elaborated by Schowengerdt (2007).

The general process of optical imaging is outlined here. In an electro-optical sensor, optical radiation carries the remotely sensed signal to the detectors, where it is transduced to an electronic signal. The continuous, temporal electronic signal is further processed up to the analog/digital (A/D) converter, where it is quantized into discrete digital number (*DN*) values representing the image pixels, as described by Schowengerdt (2007).

The radiation arriving at the sensor is transferred by the sensor optics to the detector focal plane where an image is formed. The spectral irradiance on a detector located on the optical axis, $ird_\lambda(X, Y)$, is related to the at-sensor radiance L_λ by the camera equation (Slater 1980)

$$ird_\lambda(X, Y) = \frac{\pi \tau_o(\lambda)}{4FN^2} L_\lambda(X, Y) \; (\text{Wm}^{-2}\mu\text{m}^{-1}) \qquad (3.11)$$

where $\tau_o(\lambda)$ is the optical system transmittance (excluding any filters); *FN* is the optics *f* number, being the optical focal length divided by the aperture stop diameter; and λ indicates wavelength, assuming the same (X, Y) coordinate system for both the scene and the image (Schowengerdt 2007). At this point, multispectral filters are introduced to separate the energy into different wavelength bands. Denote the product of the spectral filter transmittance and detector spectral sensitivity as the *spectral responsivity specres$_b$* (λ). The signal sgl_b measured by the sensor in band *b* is the spectral integration of the received radiance $ird_\lambda(X, Y)$:

$$sgl_b(X, Y) = \int_{\lambda_{\min}}^{\lambda_{\max}} specres_b(\lambda) ird_\lambda(X, Y) d\lambda \; (\text{Wm}^{-2}) \qquad (3.12)$$

where $\lambda_{\min}$ and $\lambda_{\max}$ define the range of sensitivity in the band and λ indicates wavelength (the integral is over wavelength), providing the total effective radiance in band *b* (Dereniak and Boreman 1996), as described by Schowengerdt (2007).

Clearly, the effective radiance measured by the sensor is the version of at-sensor radiance degraded by the spectral responsivity, as shown in Equation 3.12. As will be seen below, the effective radiance measurement is affected further by sensor blurring and geometric distortion in imaging. Blurring occurs because of the sensor's optics, detectors, and electronics, and generally occurs over a smaller spatial extent than the distortion (Schowengerdt 2007), as will be discussed in the next subsection.

The blurring can be characterized by the net (total) sensor PSF (PSF_{net}), which can be viewed as the *spatial responsivity* of the sensor (Schowengerdt 2007). According to Schowengerdt (2007), PSF_{net} consists of several components: the optical PSF, PSF_{opt}, blurring due to image motion during the integration time for each pixel; PSF_{IM}, the detector PSF caused by the nonzero spatial area of each detector in the sensor, PSF_{det}; and the electronics PSF occurring as the signal from the detectors is filtered electronically to reduce noise, PSF_{el}. By a simple theorem, the net PSF is the convolution of the component PSFs,

$$PSF_{net}(X,Y) = (PSF_{opt} * PSF_{IM} * PSF_{det} * PSF_{el})(X,Y) \qquad (3.13)$$

The width of the net PSF is the sum of the widths of each of the component PSFs (Schowengerdt 2007). Because of this blurring, the effective GIFOV of remote-sensing systems is larger than the oft-quoted geometric GIFOV. For example, the effective TM GIFOV was found to be 40 to 45 m, rather than 30 m (Anuta et al. 1984; Schowengerdt 2007). Similar analysis has been described for the Landsat MSS (Park et al. 1984) and TM (Markham 1985).

As described by Schowengerdt (2007), the net sensor PSF (PSF_{net}) is the weighting function for a spatial convolution, resulting in the electronic signal, $esgl_b(X, Y)$, out of the detectors,

$$esgl_b(X,Y) = \int\limits_{itv_X} \int\limits_{itv_Y} sgl_b(X', X')PSF_{net}(X - X', Y - Y')\,dX'\,dY' \qquad (3.14)$$

where itv_X and itv_Y refer to the coordinate ranges in the cross- and in-track directions, respectively. It is interesting to see in Equations 3.12 and 3.14 that the radiance measurements have to go through spectral and spatial averaging over nonzero spectral bandwidths and spatial areas. Note that the continuous spatial coordinates (X, Y) here will not be converted to discrete pixel coordinates until the signal is sampled (Schowengerdt 2007), as shown below.

The amplification process for the electronic signal, $esgl_b(X, Y)$, in band b is designed to provide sufficient signal level to the A/D converter for quantization, without incurring saturation. The electronics gain and offset values are then set to yield a full DN range out of the A/D converter, as described by Schowengerdt (2007). The amplified signal, $ampsgl_b$, is given by

$$ampsgl_b(X, Y) = gain_b esgl_b(X, Y) + offset_b \qquad (3.15)$$

The amplified and filtered signal is sampled and quantized into DNs, usually with a linear quantizer, so that the final DN at pixel (X, Y) in band b may be expressed as $DN_b(X, Y) = \text{int}\,[ampsgl_b(X, Y)]$ (i.e., taking the nearest integer value of $ampsgl_b(X, Y)$). The number of discrete DNs, determined by the number of bits/pixel Q, defines the *radiometric resolution* of the system. One can express the radiometric resolution as

2^{-Q} times the dynamic range in radiance (Schowengerdt 2007). Assuming constant spectral and spatial responses of the sensor (i.e., $specres_b(\lambda)$ and $PSF_{net}(X, Y)$), over an effective spectral band and an effective GIFOV, we can write

$$DN_b(X, Y) = int\left[K_b \iiint L_\lambda(X, Y)\,d\lambda\,dX\,dY + offset_b \right] \qquad (3.16)$$

where the sensor response functions are combined with the various other constants into the single constant K_b, and the integrals are over the effective spectral band and the effective GIFOV (for details, see Schowengerdt 2007).

However, for more rigorous treatment, we should revise the assumption of invariance in spectral and spatial responsivity. Spectral and spatial variation in responsivity can be accounted for by applying a convolution formula. Denote $KNL_b(X, Y)$ as a combined responsivity function at a particular pixel (X, Y) in band b. Equation 3.18 can be expressed as a convolution process

$$DN_b(X, Y) = KNL_b(X, Y)*L_b(X, Y) + \varepsilon_b(X, Y) \qquad (3.17)$$

where $\varepsilon_b(X, Y)$ is the noise term. This simplification relates the image DNs to the at-sensor radiance L_λ (the band-integrated version L_b, to be more accurate).

The simplification in Equation 3.17 implies integration over the effective spectral passband and GIFOV. The inversion of this equation to obtain band radiance values (e.g., L_b in Equation 3.17) from image DNs is known as sensor calibration (calibration-to-radiance). The process to convert radiance to reflectance is scene calibration (calibration-to-reflectance), which is more difficult as it requires knowledge or assumptions about atmospheric conditions and the terrain surface (Schowengerdt 2007). For example, in the solar-reflective spectral region (visible to shortwave infrared), the radiation source is the Sun. Part of the radiation received by a sensor is reflected at the Earth's surface and part is scattered by the atmosphere and does not even reach the surface. There are three significant components in the upwelling at-sensor radiation: the unscattered, surface-reflected radiation $L_\lambda^{unsctr,reflct}$, the downscattered, surface-reflected skylight $L_\lambda^{downsctr,reflct}$, and the up-scattered path radiance L_λ^{path}. Thus, for the solar reflective region, the energy collected by a remote-sensing system is, in general, proportional to the surface reflectance, which carries the signal of interest (i.e., the spatial–spectral reflectance distribution), contaminated with a spatially invariant, but spectrally dependent, constant bias term arising from atmospheric scattering (Schowengerdt 2007). Further understanding about optical radiation models is important for quantitative remote sensing and analytical approaches to scale-related research.

3.2.2 Geometric Characteristics

The previous subsection dealt with sensor characteristics that affect image radiometric quality. This subsection concerns the geometric characteristics of images by examining the controlling factors, such as orientation parameters and terrain undulation in aerial and space photography, and the orbit, platform attitude, and scanner properties in scanner-based imaging.

First, consider the case for aerial and space photography. Because of the tilted imaging plane and terrain undulation, the geometry for imaging is complex. For example, because of tilted aerial photography and a nonflat terrain, photo scale will not be simply a constant ratio of a camera's focal length to flight height over the photographed area. Similarly, in satellite scanner images, effective image resolution (spatial) will also be spatially varied. Below, we describe some of the elementary photogrammetric concepts and equations to provide a basis for understanding photogrammetric mapping and the spatially varied definition of resolution in raw images.

For proper orientation of an image, we need both inner and external orientation. There are three elements for inner orientation, which are used to locate the lens center with respect to an image. They are the principal point's coordinates (IX_{pp}, IY_{pp}) with respect to the image coordinate system's origin, which are defined through fiducial marks on aerial photographs taken by metric cameras, and focal length f. Inner orientation parameters are usually provided through camera calibration (Mikhail et al. 2001; Zhang and Goodchild 2002; Zhang 2010).

External elements are used to determine the position and attitudes of an image or a bundle of rays at the instant of exposure. There are six elements for external orientation, three linear and three angular. The three linear elements, $(X_{O'}, Y_{O'}, Z_{O'})$, are for locating the exposure center O' with respect to the object-space coordinate system, whereas the three angular elements are for framing the attitude of the bundle of rays at the instant of exposure. They are usually in the sequence of three rotation angles $\varphi - \varpi - \kappa$: primary rotation φ around the Y axis, secondary rotation ϖ around the once rotated X axis, and tertiary rotation κ around the twice-rotated Z axis (Zhang and Goodchild 2002). Elements for inner and external orientations are shown in Figure 3.1.

As shown in Figure 3.1, assume we have an image point js coordinates $(IX, IY, -f)$ in the image-space coordinate system (IX, IY, IZ), whose coordinates are (X, Y, Z) in the object-space coordinate system (but with shifted origin at $(X_{O'}, Y_{O'}, Z_{O'})$, i.e., the exposure station's location). These two sets of coordinates are related through rotational transformation:

$$
\begin{bmatrix} IX \\ IY \\ -f \end{bmatrix} = \mathbf{R}^{\mathrm{T}} \begin{bmatrix} X \\ Y \\ Z \end{bmatrix} = \begin{bmatrix} a_1 & b_1 & c_1 \\ a_2 & b_2 & c_2 \\ a_3 & b_3 & c_3 \end{bmatrix} \begin{bmatrix} X \\ Y \\ Z \end{bmatrix}
\tag{3.18}
$$

where $\mathbf{R}$ is a 3×3 orthogonal matrix, consisting of nine directional cosine quantities. Rotational transformation is well described in the photogrammetric literature (Mikhail et al. 2001).

After the transformation of image point j from the image-space coordinate system to a coordinate system parallel to the object-space coordinate system, we can easily derive the imaging equation for central projection, as in aerial or space photography.

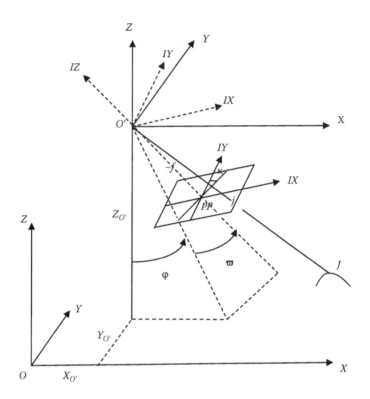

FIGURE 3.1 Image-plane coordinate system, image-space coordinate system, object-space coordinate system, and elements for inner and external orientations. (Adapted from Zhang, J., and M.F. Goodchild, *Uncertainty in Geographical Information*. New York: Taylor & Francis, 2002.)

As shown in Figure 3.1, projection center O', image point j, and ground point J lie in a line. Thus, there exists a simple relationship:

$$\begin{bmatrix} IX_j - IX_{pp} \\ IY_j - IY_{pp} \\ -f \end{bmatrix} = \frac{1}{M} \begin{bmatrix} a_1 & b_1 & c_1 \\ a_2 & b_2 & c_2 \\ a_3 & b_3 & c_3 \end{bmatrix} \begin{bmatrix} X_J - X_{O'} \\ Y_J - Y_{O'} \\ Z_J - Z_{O'} \end{bmatrix} \tag{3.19}$$

where (X_O, Y_O, Z_O) is projection center O's coordinates in the object-space coordinate system; $(IX_j - IX_{pp}, IY_j - IY_{pp}, -f)$ is image point js coordinates in the image-space coordinate system; (X_J, Y_J, Z_J) is ground point Js coordinates in the object-space coordinate system; $a_1 - a_3, b_1 - b_3, c_1 - c_3$ are nine elements of the rotation matrix $\mathbf{R}$; and M is a scale factor (Zhang and Goodchild 2002).

Clearly, the scale factor M in Equation 3.21 will be dependent on the particular point under consideration (J for this case), which is usually noninvariant unless we assume vertical photography and a flat terrain. Photogrammetric mapping is conventionally undertaken with properly reconstituted stereo-pairs of photographs of adequate overlap, from which geographic information can be extracted with positional (for spatial entities, e.g., landmarks) and hence thematic (for fields, e.g., terrain elevation, vegetation types, and so on) accuracy.

Now, consider scanner-based imaging. Although the orbital velocity of satellites can be considered constant in time, variation in ground speed is often too great to be neglected for airborne sensors, giving rise to much complexity in geo-referencing of airborne scanner imagers. Platform attitude (in angles of rotation) is also important for sensor geometric modeling, as hinted at by the rotation matrix shown in Equation 3.19. Because of the orbital satellite movement and the Earth's rotation, the projected Earth rotational velocity in the scan direction should be accounted for in geometric modeling of space-borne scanner images (Schowengerdt 2007).

Consider distortions in both whiskbroom and pushbroom scanners. Whiskbroom scanners with more moving parts are subject to more distortions than pushbroom scanners. For the former, pixel positioning cross-track is determined by the scan mirror motion, which couples with satellite motion to determine pixel positioning in-track. For the latter, of either the linear or array type, there is rigid cross-track geometry, which is essentially decoupled from the in-track geometry. Because these distortions are consistent over time, they can be corrected in a deterministic manner using orbital models and scanner calibration data (Schowengerdt 2007).

Consider and compare the nominal pixel sizes of both types of scanner images. According to Schowengerdt (2007), the cross-track pixel sampling in whiskbroom scanners is in fixed time increments, resulting in fixed angular increments, $\Delta\theta$ (θ being the scan angle from nadir), given a constant scan velocity. For whiskbroom scanners, the cross-track GSI with a scan angle of θ from nadir is

$$GSI(\theta) = [1/\cos(\theta)]^2 \, GSI(0) \qquad (3.20)$$

where $GSI(0)$ is the cross-track GSI at nadir, assuming a flat Earth (Richards 1993), as described by Schowengerdt (2007). In a pushbroom system (linear or area arrays), each cross-track line of the image is formed optically as if in a conventional frame camera. Assuming an equal spacing of the detector elements across the array (say *width*, also the detector element width), the cross-track IFOV changes across the array, (i.e., as a function of the cross-track view angle). The equal spacing of *width* in the focal plane corresponds to a uniform cross-track GSI on the ground,

$$GSI_f = width \times \frac{H}{f} \qquad (3.21)$$

but a changing cross-track IFOV,

$$IFOV(\theta)/IFOV(0) = [\cos(\theta)]^2 \qquad (3.22)$$

assuming a flat Earth and a distortion-free optical system (Schowengerdt 2007).

The collinearity equations (e.g., Equation 3.19) developed for camera photography may also be applied to scanner sensors, such as SPOT and Landsat. For example, a rigorous sensor model was proposed based on the classical collinearity equations and taking care of the specific characteristics of the acquisition of pushbroom linear scanners, showing its flexibility and subpixel accuracy in the checkpoints (Poli 2007). However, since the parameters of neighboring scanlines of scanners are highly correlated, it is possible to link the exposure centers and rotation angles of the different scanlines to integrate supplemental information, such as the ephemeris and attitude data using celestial mechanical laws for satellite images, or the GPS and inertial navigation system data for airborne images, as reviewed by Toutin (2004).

Although it is preferable to use rigorous three-dimensional models (i.e., collinearity equations) for geometric characterization of photography and scanner imaging processes (e.g., the platform attitude, altitude, and speed), the sensor distortions, and the Earth (rotation, Earth curvature, ellipsoid, and relief), we may resort to the so-called empirical models, such as polynomials, for geometric modeling, when the parameters of the acquisition systems or a rigorous physical model are not available (Bannari et al. 1995; Toutin 2004). Since empirical models do not reflect the source of distortions described previously, they do not require *a priori* information on any component of the total system (i.e., platform, sensor, and the Earth).

For example, polynomial functions can be written as

$$\left\{ \begin{array}{l} IX = \displaystyle\sum_{i_1=0}^{I_1} \sum_{i_2=0}^{I_2} a_{i_1 i_2} X^{i_1} Y^{i_2} \\[2em] IY = \displaystyle\sum_{i_1=0}^{I_1} \sum_{i_2=0}^{I_2} b_{i_1 i_2} X^{i_1} Y^{i_2} \end{array} \right. \tag{3.23}$$

for linking up image coordinates (IX, IY) with two-dimensional ground control points (GCPs) coordinates (X, Y), or

$$\left\{ \begin{array}{l} IX = \displaystyle\sum_{i_1=0}^{I_1} \sum_{i_2=0}^{I_2} \sum_{i_3=0}^{I_3} a_{i_1 i_2 i_3} X^{i_1} Y^{i_2} Z^{i_3} \\[2em] IY = \displaystyle\sum_{i_1=0}^{I_1} \sum_{i_2=0}^{I_2} \sum_{i_3=0}^{I_3} b_{i_1 i_2 i_3} X^{i_1} Y^{i_2} Z^{i_3} \end{array} \right. \tag{3.24}$$

for using elevation Z of GCPs in addition to those in Equation 3.23. There will be 6 and 12 unknowns for each pair of first- and second-order polynomial functions in the two-dimensional case, respectively, whereas 8 and 20 unknowns will be needed for the first- and second-order polynomial functions in the three-dimensional case (Toutin 2004).

It should be emphasized that such empirical models are applicable to small images, dependent on sufficiently numbered and well-distributed GCPs, very sensitive to errors, and lacking robustness and consistency in an operational environment. Their use is thus limited to systematically corrected images, where all distortions (except those due to terrain relief in the case of three-dimensional polynomial functions) have been precorrected (Toutin 2004).

We have seen through both the previous subsection and this subsection how the sensor modifies the signal of interest in remote sensing. The sensor affects the spatial and radiometric quality of the signal: (1) scene spectral radiances received at the sensor are weighted by the sensor spectral response in each band; (2) scene spatial features are weighted by the sensor spatial response, which includes blurring by the optics, image motion, detector, and sensor electronics; (3) the overall imaging process can be modeled as a convolution process; and (4) geometric distortion arising from the geometry of imaging and topography leads to a spatially varied definition of local IFOV and, thus, effective spatial resolution (Schowengerdt 2007).

3.3   MICROWAVE REMOTE SENSING AND RESOLUTION

3.3.1   RESOLUTION

In addition to optical systems, microwave remote-sensing systems are common. Both active and passive modes are useful for microwave remote sensing. The emphasis is placed on the range and azimuth components of the definitions of spatial resolution for the former, whereas the implications for relatively coarser spatial resolution for the latter are discussed in the third subsection below on passive microwave remote sensing. Because of the relative technical complexity of microwave remote sensing for most data users, we provide a general introduction to microwave remote sensing and its applications.

Radar is an important technology in microwave remote sensing. Radar is an acronym for radio detection and ranging, and is an active sensor system. As reviewed by Chan and Koo (2008), a typical radar system consists of a transmitter, switch, antenna, receiver, and data recorder. The transmitter generates electromagnetic (EM) waves at radio wavelengths. The switch directs the pulses to the antenna and the returned echoes to the receiver. The antenna transmits the EM pulses toward the area to be imaged and collects returned echoes, which are often only a fraction of the energy reflected off the target. The returned signal is converted, by the receiver, to digital numbers, which are stored at the data recorder for later processing and display. Information about entities and properties at, and just below, the surface or from within the atmosphere, is derived by analyzing radar signal modifications, as surface structural and electrical properties are known to control radar backscattering from terrain (Dobson et al. 1995; Short 2005). Radar operates in part of the microwave region of the electromagnetic spectrum that is in the frequency interval from 40,000 to 300 MHz (or 0.8 cm to 1 m in wavelength). The latter frequency extends into the higher frequencies of the broadcast-radio region. Commonly used frequencies and their corresponding wavelengths are Ka band (40,000–26,000 MHz, 0.8–1.1 cm), K band (26,500–18,500 MHz, 1.1–1.7 cm), X band (12,500–8000 MHz, 2.4–3.8 cm), C band (8000–4000 MHz, 3.8–7.5 cm), L band (2000–1000 MHz, 15.0–30.0 cm), and P band (1000–300 MHz, 30.0–100.0 cm) (Short 2005).

By supplying its own illumination, a radar can function day and night and, for some wavelengths, without significant interference from atmospheric conditions (e.g., clouds), being especially useful for areas with thick clouds or dense vegetation (Short 2005). Radar has widespread applications, as shown in the following examples. Elachi (1980) illustrated the applications of Seasat's imaging radar for geologic mapping and ocean surface pattern monitoring, and showed its potential for extracting structural and topographic features such as lineaments, anticlines, folds and domes, drainage patterns, stratification, and roughness units, and for observing ocean surface waves, internal waves, current boundaries, and large-scale eddies. Radar systems are effective in soil moisture detection (Engman 1991; Kerr 2007), flooding monitoring (Hess et al. 1990), land cover classification, and aboveground biomass estimation (Dobson et al. 1995). Some radar systems are multiband and multipolarization radars, allowing for color-composite construction and quantitative analysis (Pierce et al. 1994). Radar backscatter information was shown to be complementary to optical sensor data and can increase terrain classification accuracy (Rebillard and Evans 1983; Paris and Kwong 1988). An interferometric radar technique was developed for topographic mapping, leading to the generation of digital elevation models (DEMs) (Zebker and Goldstein 1986). In radar interferometry, two images are made of a scene by simultaneously flying two physically separated antennas, with the phases of corresponding pixels differenced, and altitude formation deduced from some simple computation and image rectification (Gabriel et al. 1989). Use of a single satellite in a nearly repeating orbit reduces operational cost and spaceborne hardware complexity, and also permits inference of changes in the surface from the correlation properties of the radar echoes (Zebker et al. 1994).

Consider a typical radar system, as shown in Figure 3.2. For an airborne system, as the aircraft moves forward at some altitude above the terrain in an *azimuthal direction*, the pulses spread outward in the *range (or look) direction*. Any line of sight from the radar to some ground point within the terrain strip defines the *slant range* to that point, as shown in Figure 3.2. The distance between the aircraft nadir (directly below) line and any ground target point is its *ground range*. The ground point closest to the aircraft flight trace, where sensing begins, is the *near range* limit. The pulsed ground point at the greatest distance normal to the flight path is the *far range*, as also illustrated by Short (2005).

At the radar antenna, the angle between a horizontal plane (essentially, parallel to a level surface) and a given slant range direction is called the *depression angle* for any point along that directional line, as shown in Figure 3.2. We refer to the complementary angle (measured from a vertical plane) as the *look angle* β (i.e., $\beta = 90° - depression\ angle$). The *incidence angle* at any point within the range is the angle between the radar beam direction (of look) and a line perpendicular (normal) to the surface, as also described by Short (2005). The duration of a single pulse, radar antenna length, and radar system wavelength determine the two types of resolution at a given slant range, range resolution, and azimuth resolution, denoted as R_r and R_ω, respectively, as shown in Figure 3.2.

The range resolution is the ability of the radar to differentiate or resolve two targets that are close together in range. Two distinct targets on the surface will be

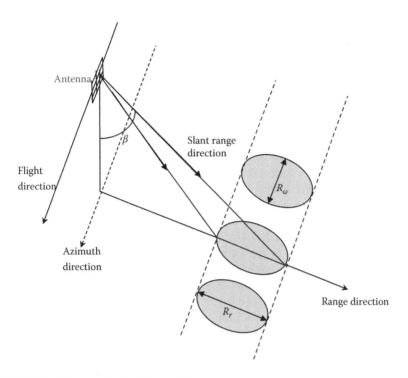

FIGURE 3.2   Range and azimuth resolution.

resolved in the range dimension if their separation is greater than half the pulse length. The formula for range resolution is

$$R_r = T_{\text{pulse}} c / 2 \sin\beta \tag{3.25}$$

where T_{pulse} is the pulse length (duration), c is the speed of light (3×10^8 m/s) and β is the look angle (Short 2005; Chan and Koo 2008). In Equation 3.25, $T_{\text{pulse}} c / 2$ represents the slant range resolution, which is projected to its ground dimension by denominator $\sin\beta$. The ground range resolution R_r will decrease (i.e., get coarser) when looking outward from near to far range (because pulse travel times T_{pulse} increase outward from near to far range), as shown in Figure 3.2.

The second measure, the *azimuth resolution*, is the minimal size of an object along the direction of the flight path that can be resolved, as shown in Figure 3.2. The azimuth or along-track resolution is determined by the angular width of the radiated microwave beam and by the slant range distance. This beamwidth is a measure of the width of the illumination pattern (Canada Center for Remote Sensing [CCRS] 2007). As the radar illumination propagates to increasing distance from the sensor, the azimuth resolution becomes coarser. The radar beamwidth is inversely proportional to the antenna length (also referred to as the aperture), meaning that a longer antenna will produce a narrower beam and finer resolution. The *azimuth resolution* is given by

$$R_\omega = range_{\text{slant}} \lambda / length \tag{3.26}$$

where $range_{slant}$ is the slant range, λ is the system wavelength, and *length* is the effective length of the antenna (Chan and Koo 2008).

As can be seen from Equations 3.25 and 3.26, respectively, finer range resolution can be achieved by using a shorter pulse length T_{pulse}, which can be done within certain engineering design restrictions, whereas finer azimuth resolution can be achieved by increasing the antenna length *length*. However, the actual length of the antenna carried on an airborne or spaceborne platform is limited: for airborne radars, antennas are usually limited to 1 to 2 m; for satellites, they can be 10 to 15 m in length. To overcome this size limitation, the forward motion of the platform and special recording and processing of the backscattered echoes are used to simulate a very long antenna and, thus, increase azimuth resolution (Short 2005; CCRS 2007). This is the approach used by synthetic aperture radar (SAR) (Currie 1991; Cumming and Wong 2005), which is used on moving platforms, such as SIR-C (Spaceborne Imaging Radar-C) (Jordan et al. 1995). In fact, the SIR-C, X-Band SAR (SIR-C/X-SAR) was the first multifrequency and multipolarization SAR system launched into space (Schmullius and Evans 1997).

The range of brightness levels that a remote-sensing system can differentiate is related to radiometric resolution and is referred to as the dynamic range. Although older optical sensors, such as those carried by satellites such as Landsat and SPOT, typically are 8-bit (radiometric resolution) and produce 256 intensity levels, radar systems can differentiate intensity levels up to around 100,000 levels. To deal with this excessive range of intensity, most radars record and process the original data as 16 bits (65,536 levels of intensity), which are then sometimes scaled down to 8 bits (256 levels) for visual interpretation and digital computer analysis (CCRS 2007; Gad-el-Hak 2008).

3.3.2 RADAR SIGNALS, IMAGING, AND DISTORTIONS

As mentioned previously, radar backscattering from terrain is controlled by surface structural and electrical properties. Two material properties provide clues about composition and surface state by the manner in which they interact with the incoming pulses, as described by Short (2005) and Gupta (2011). One property is the *dielectric constant*, which is the ratio of the capacitance of a material to that of a vacuum (the value is arbitrarily set to 1 for a vacuum), being a measure of both the conductivity and reflectivity in terms of the electrical response of materials (Hallikainen et al. 1985; Short 2005). Radar waves penetrate deeper into materials with low dielectric constants and reflect more efficiently from those with high constants. Thus, variation in reflected-pulse intensities may indicate differences in soil moisture, if other factors are constant (Short 2005; Gruhier et al. 2010).

The second material property is *roughness*. Roughness, in this sense, refers to minute irregularities that relate to textures either of the surfaces or of objects on them (such as, closely spaced vegetation that may have a variety of shapes) (Short 2005). Surface irregularity, together with radar wavelength and grazing angle at the point of contact, determines the behavior of a surface as smooth (specular reflector), intermediate, or rough (diffuse reflector), as shown below. As discussed by Short (2005), a smooth surface at low angles to the look direction scatters most of the pulses away from the receiver, so that its image appears dark. A rough surface, in contrast, scatters the pulse beam over many directions, with only a fraction of it returned to the radar. The amount of returned signal determines the relative lightness (related to signal amplitude) of the image tone. The quantity of returned energy (as backscatter echoes) also depends on the size of the target relative to the signal's wavelength. Objects with dimensions similar to the wavelength appear bright (as though rough), whereas smaller ones are dark (smooth). By the same token, radar wavelengths influence the penetrability of the surface. Depth of penetration increases with wavelength, λ, and thus L and P bands penetrate deeper than K or X bands (Gupta 2011). In forests, for example, shorter wavelengths, such as C band, reflect mainly from the canopy tops, whereas longer wavelengths will penetrate canopies to varying degrees, providing valuable information for vegetation discrimination (Dobson et al. 1995).

In addition to the general description of radar backscattering, it is important to be able to perform quantitative analysis. This requires some basic understanding about how radar works and how information about target position and reflectivity is derived from radar range and Doppler measurements.

Consider a radar system working over a scene littered with targets assumed to be point sources. We begin with a brief mention of the waveform used for the radar transmitter. For example, the linear frequency modulation waveform (so-called chirp signal), whose instantaneous frequency increases (or decreases) linearly over the duration of the signal, is the most widely used in practice (Chan and Koo 2008). For the chirp waveform, the transmitted signal $trns(t)$ can be written as

$$trns(t) = \text{Re}\{\exp[I2\pi(f_0 t + kt^2/2)]\} \tag{3.27}$$

for $-T/2 \le t \le T/2$ (where T is chirp signal duration; $trns(t)$ is 0 outside of this interval), Re{} is the real operator for a complex function, f_0 is the carrier frequency of transmitted waveform, and k is the chirp rate of the waveform (so the bandwidth of the signal is kT) (Chan and Koo 2008).

Suppose a target located at a range of $range_x$ from the transmitter is traveling with *constant velocity v* along track (relative to radar) and has *reflection coefficient reflect* (τ_r, ψ_v), where $\tau_r = 2\ range_x/c$ is the round-trip propagation time between the transmitter and the target, $\psi_v \approx 2\psi_0 v/c$ is the Doppler shift, and $\psi_0 = 2\pi f_0$ is the angular carrier frequency. The complex scene reflectivity in terms of delay and Doppler shift is denoted by $\{reflect(\tau_r, \psi_v)\}$ (Herman and Strohmer 2009; Potter et al. 2010).

After transmitting signal $trns(t)$, the radar receiver observes the reflected signal. The total received signal backscattered from the scene is given by

$$msmt(t) = \text{Re}\left\{\iint reflect(\tau_r, \psi_v)\exp[I(\psi_0 - \psi_v)(t - \tau_r)]\,d range\,dv\right\} + noise(t)$$

$$\tag{3.28}$$

where $noise(t)$ may be assumed to be white Gaussian noise arising from thermal noise in the transmitter and receiver hardware. Upon receipt, quadrature demodulation yields the complex baseband signal

$$msmt_B(t) = \iint reflect(\tau, \psi)trns(t - \tau)\exp(-I\psi t)\,d\tau\,d\psi + noise_B(t) \tag{3.29}$$

where the constant phase terms have been absorbed into the reflectivity, and $noise_B$ represents the circular white complex Gaussian baseband noise. The basic radar problem is to estimate the reflectivity $reflect(\tau, \psi)$ of the scatterers in the scene (Potter et al. 2010).

With a matched filter at the receiver, the reflected signal is correlated with a time-frequency shifted version of the transmitted signal via the cross-ambiguity function (Chan and Koo 2008; Herman and Strohmer 2009). The output of the matched filter is

$$
\begin{aligned}
test(\tau', \psi') &= \int msmt_B(t)[trns(t - \tau')] * \exp(I\psi't)\,dt \\
&= \iint reflect(\tau, \psi)ambg(\tau - \tau', \psi - \psi')\,d\tau\,d\psi \\
&+ \int noise_B(t)[trns(t - \tau')] * \exp(I\psi't)\,dt
\end{aligned}
\tag{3.30}
$$

which is actually a test statistic for the existence of a single target with delay τ' and Doppler frequency ψ', where the radar ambiguity function $ambg(\tau, \psi) = \int trns(t)[trns(t - \tau)] * \exp(I\psi t)\,dt$, and $[trns]*$ is the complex conjugation of $trns$. Clearly, the output of the matched filter is the convolution of the reflectivity $reflect(\tau, \psi)$ with the radar ambiguity function plus a filtered copy of the baseband noise (Potter et al. 2010).

The scene reflectivity field $reflect(\tau, \psi)$ may be discretized over range and Doppler shift on a grid of points $\{\tau_{tn}, \psi_{tm}\}$. We can represent the discretized field (matrix) as a vector, say **refl**. Accordingly, we can sample the received baseband signal $msmt_B$ at times $\{tn\}$ to obtain measurements, which are, again, represented as a vector **mst**; thus, we obtain the linear system of equations

$$
\textbf{mst} = \textbf{KNL}_A\textbf{refl} + \textbf{ns}
\tag{3.31}
$$

where $\textbf{KNL}_A$ is a matrix taking care of the convolution of $reflect(\tau, \psi)$ with the radar ambiguity function in Equation 3.30 and **ns** stands for the vector representing the filtered copy of the baseband noise in Equation 3.30 (Potter et al. 2010). Note the similarity and difference between Equations 3.31 and 3.17.

Now, consider speckling. Speckle appears as a grainy "salt and pepper" texture in an image and is a universal property of coherent imaging, as in radar. This is caused by interference from the multiple scattering returns that occur within each resolution cell, as shown below.

Specifically, when incident waves are scattered by a random distribution of scatterers, the resulting scattered field contains random interference terms. Provided that the scattering volume has a depth greater than the wavelength, the phase of the individual contributions will vary randomly over 2π (Oliver 1991). The reflectivity at a resolution cell pix can be expressed as the sum of a large number of randomly phased contributions:

$$
reflect(pix) = \sum_{j=1}^{N} amptd(x_j)\exp(I\psi(x_j))
\tag{3.32}
$$

where *amptd*(x_j) and $\psi(x_j)$ represent the scattering amplitude and phase at position (x_j), respectively, and N represents the total number of scatterers. As *amptd*(x_j) is assumed to be a random variable that is uncorrelated for different scatterers, we can, with large N, expect that both the real and the imaginary components of *reflect* have a Gaussian probability density function (PDF) from central limit theorem. Thus, the intensity *itnst* = |*reflect*|² has a negative exponential PDF. Furthermore, because the standard deviation of the detected intensity is equal to the mean value, speckle is often described as 100% multiplicative noise (Oliver 1991).

We have tried to establish some technical details regarding radar signal processing, radar imaging, and image characteristics. This concerns how an imaged scene, say, **refl**, is reconstructed from radar measurements and why speckle is characteristic of radar images. It is hoped that, by doing so, we have some concrete and tractable mathematical tools for understanding radar and its concepts of resolution and information content, which will be part of the subject of Chapter 11.

In the remainder of this subsection, we discuss some geometric aspects of radar images. Slant-range distortion exists in radar images, in which features in the near range are compressed relative to features in the far range owing to the slant-range scale variability, as can be seen from Figure 3.2. For most applications, it is necessary to correct for this distortion, to enable true distance measurements between features based on distortion-corrected images (implying that such corrected images can be used as maps of uniform scale), similar to the case of optical remote sensing as described in Section 3.2.2. This can be done by converting the slant-range image to one displaying correct ground range through proper transformation.

In radar images over mountainous regions, there are also significant distortions due to height differences in the across-track direction, known as layover, foreshortening, and shadow (the first and the second occur for hills or ridge slopes facing the radar), as shown by Loew and Mauser (2007). Layover results when the radar wavefront reaches the top of a slope before the bottom, whereas foreshortening is the other way around. Slopes on the opposite side of mountains with ridges will generally not be illuminated and are thus rendered as shadows, as no signals are returned from them (for illustration, see Short 2005).

Geometric distortions can be corrected through proper geocoding. The objective of the geocoding process for radar images is to reconstruct the correct imaging geometry and, thus, to find, for each image pixel, its corresponding position on the Earth (Curlander 1982; Naraghi et al. 1983; Curlander and McDonough 1991; Loew and Mauser 2007). The standard outputs from geocoding are image products corrected for geometric distortions and usually resampled to specified map projections (Curlander et al. 1987; Loew and Mauser 2007). For radar images, the Doppler and range equations are the physical models as collinearity equations are for aerial and space photographs. Interestingly, the collinearity equations were historically adapted as radargrammetric equations to process radar images and later as a unified mathematical framework to process both optical and radar images (Toutin 1995, 2004).

Suppose that the image coordinates of a radar system are given by the slant range distance and the zero-Doppler position. The targets and sensor positions should be referred to a common, typically Cartesian, reference frame (e.g., WGS84). This enables the calculation of the look vector from the sensor to the target and the appropriate estimation of the Doppler frequency shift as shown below.

For a given sensor position x_{sns} and target position x_{tgt} in an Earth-centered fixed Cartesian coordinate system, the vector from target to sensor is

$$range_{t2s} = x_{sns} - x_{tgt} \tag{3.33}$$

and the slant range distance to the target position is calculated as $|range_{t2s}| = |x_{sns} - x_{tgt}|$.

The Doppler frequency shift $f_{Doppler}$ is calculated using the sensor and target position vectors (x_{sns} and x_{tgt}) and velocity vectors (v_{tgt} and v_{sns}):

$$f_{Doppler} = \frac{2f_0(v_{tgt} - v_{sns})}{c} \frac{range_{t2s}}{|range_{t2s}|} \tag{3.34}$$

where c is the speed of light and f_0 is the carrier frequency of the sensor (Curlander 1982; Loew and Mauser 2007). The target velocity vector v_{tgt} is related to its position vector through $v_{tgt} = \omega_E x_{tgt}$, where ω_E is the Earth's rotational velocity (Curlander et al. 1987).

Generally, there are two approaches to geocoding image pixels, namely, forward and backward geocoding (Curlander and McDonough 1991). As reviewed by Loew and Mauser (2007), in a forward geocoding approach, the position of each image pixel on the Earth's surface is calculated separately, using the range-Doppler equation above. The backward scheme is *vice versa*: the image pixel with the nearest range-Doppler coordinate is calculated for each ground position.

Taking the backward scheme, Loew and Mauser (2007) described precise terrain geocoding using a fine-resolution DEM and additional information about the orbit of the sensor platform. By their geocoding strategy, both precise georectification of the image pixels and rigorous radiometric terrain correction were performed on SAR images. The radiometric terrain correction is based on the integration of the radar brightness of multiple image pixels, corresponding to an element of a reference DEM. Thus, this strategy preserves the energy contained in the image data throughout the geocoding process, leading to added value in quantitative image analysis.

3.3.3 Passive Microwave

Although active microwave systems (i.e., radar) are the more commonly used sensors for this region of the spectrum, *passive microwave* (PM) sensors, both airborne and spaceborne, have also provided information about the Earth's land surface, its oceans, and its atmosphere. PM sensors measure directly the radiation incited by thermal states in these media: such radiation is at longer wavelengths that extend into the microwave region (hence, passive). For instance, the advanced microwave scanning radiometer (AMSR-E) onboard the MODIS-Aqua satellite provides measurements of land, oceanic, and atmospheric properties for the investigation of global water and energy cycles, including precipitation rate, sea surface temperature, sea ice concentration, snow water equivalent, soil moisture, wind speed, and water vapor, as documented by the US National Snow and Ice Data Center (NSIDC) (http://nsidc.org/daac/projects/passivemicro/).

PM sensors detect the naturally emitted microwave energy within their field of view, unlike active microwave sensors that provide their own illumination. This kind of natural emissive radiation is generally much weaker in intensity than that at shorter wavelengths, but is still detectable by sensitive instruments and is not much attenuated by the atmosphere (Short 2005), making PM remote sensing attractive, as shown by the data products archived by NSIDC. The wavelength employed in passive microwave detectors is generally between 0.15 and 30 cm. The signal beamwidth is usually wide, in order to gather sufficient amounts of the weak radiation. Thus, the spatial resolution of PM sensors tends to be coarse (commonly, in kilometers from space and in meters from the air), allowing for large sampling areas to provide enough radiation for detection (Short 2005).

PM sensors are typically radiometers or scanners that use an antenna to detect and record the microwave energy. PM sensing works on the basis of the spectral curves showing relative intensities of radiances as a function of wavelength for materials with different intrinsic temperatures. With scanning radiometers on moving platforms, the variation in intensity detected can be converted to photographic gray levels, resulting in images that resemble those formed in visible, near-infrared, and thermal-infrared regions (Short 2005).

The temperatures measured by PM instruments are *brightness temperatures*. PM remote measurements can be used for mapping spatial distributions and for retrieval of process fields. Below, we review some of the PM sensing applications to provide an idea about how relatively coarser spatial resolution, fine temporal resolution, and the unique sensing capabilities of PM sensors are valuable for environmental monitoring.

On land, passive microwave surveys are particularly effective for detecting soil moisture and temperature, because of sensitivity to water. Because of the PM sensor's coarse-resolution data, spatial heterogeneity is often present within their footprints, especially over land. With the use of aircraft-based 1.4-GHz (L band) radiometer data collected at several altitudes over test sites, soil moisture was measured concurrently (Jackson 2001). The use of multiple flightlines at lower altitudes allowed the direct comparison of different spatial resolutions using independent samples over the same ground location. Results showed that the brightness temperature data from a 1.4-GHz sensor in the study region provide the same mean values for an area regardless of the spatial resolution of the original data. The relationship between brightness temperature and soil moisture was similar at different resolutions. These results suggest that soil moisture retrieval methods developed using fine-resolution data can be extrapolated to satellite scales (Jackson 2001).

Using passive microwave brightness temperatures from the Defense Meteorological Space Program (DMSP) Special Sensor Microwave/Imager (SSM/I), McFarland et al. (1990) sought to determine the surface temperature over land areas in the central plains of the United States. A regression analysis comparing all of the SSM/I channels and minimum screen air temperatures (representing the surface temperature) produced large correlations, with root mean squared errors (RMSEs) of 2°C–3°C.

Plot-scale brightness temperature measurements at 6.9, 19, 37, and 89 GHz were acquired in forest, open, and lake environments near Churchill, Manitoba, Canada, with mobile sled-based microwave radiometers during the 2009–2010 winter season (Derksen et al. 2012). Detailed physical snow measurements within the radiometer footprints were made to relate the microwave signatures to the seasonal evolution of the snowpack and to provide inputs for model simulations with the Helsinki University of Technology snow emission model. The snow emission model accuracy increased when simulations were run regionally at the satellite scale (using three proportional land cover tiles: open, forest, and lake) and compared to AMSR-E measurements. Snow water equivalent retrievals produced using forward brightness temperature simulations with the snow emission model in combination with AMSR-E measurements produced RMSE values below 25 mm for the intensive study area.

3.4   DISCUSSION

Geospatial research and applications, such as urban housing policy (Qian 2003), cadastral mapping of forestlands (Vogiatzis 2008), deforestation (Skole and Tucker 1993), soil mapping (Summerell et al. 2009), ecosystem monitoring (Pfeifer et al. 2012), land use and global change (Foley et al. 2005), and many more, are usually dependent on the availability and quality of data to extract useful information or perform spatial analysis to address pressing issues in resources management, environmental modeling, and decision support. Geospatial data are usually derived from measurements (*in situ* or remote) or obtained from secondary sources (e.g., data archives and digital library), with the latter being derived originally from the former. It is thus important to understand the statistics of spatial sampling, the physics of remote sensing, and the overall data acquisition processes that have modulated the effective resolutions of resultant data.

As has been shown previously, sampling coverages are often irregular and incomplete for field surveys and other data collection tasks under constraints or with consideration of cost-effectiveness. Data supports may be difficult to describe mathematically, but need to be dealt with properly in data processing and spatial analysis. Spatial variation in sampling density and local definition of effective data supports are commonly encountered in practice, as we will see, for example, in the context of DEM data production in Chapter 9. Below, we extend the discussion about sampling density and effective data supports and their effects on adequacy of sampling for spatial data acquisition in the light of the preceding reviews regarding sampling design and remote measurements.

We start with the seemingly simple case of remote sensing. Remotely sensed images are often presented over uniform grids. The Shannon sampling theorem provides the guidance for uniform sampling of band-limited signals (i.e., spatial distributions and entities). Consequently, there will be aliasing (i.e., overlapping of a signal's frequency spectra) and loss of information due to inadequate sampling rate, as dictated by the sampling theorem in Equation 3.3. The Shannon sampling theorem can also be extended to higher-dimensional space (Euclidean) so that a function (of multidimensional arguments) whose spectrum is restricted to a finite region in the frequency space may be reconstructed from its samples taken over a uniform grid (or lattice) at a necessary sampling rate, although the most efficient grid (i.e., requiring minimum sampling points per unit hypervolume) is not, in general, rectangular, nor is a unique reconstruction function associated with a given sampling grid (Petersen and Middleton 1962). The optimum sampling density for remote sensing should be determined based on the spectra of the underlying spatial distributions. However, this is seldom performed in practice. Given the arrays of multiresolution images currently available and increasingly cumulating, such a situation may lead to inefficiency in remote measurements.

Although at approximately nadir, remote sensing generates data over approximately regular supports, the effects of sampling imply that spatial resolutions for remote-sensed images are far from simple, as almost all remote-sensing systems are subject to blurring, not to mention various sources of noise occurring in images, as shown in Sections 3.2 and 3.3 where both optical and microwave imaging are described. Therefore, despite the availability of three single-figure measures of spatial resolution for images, the effects of sampling coupled with errors in imaging and image preprocessing mean that the assumption of spatially invariant resolution across the whole image field will not hold true. In other words, effective resolution will rarely be spatially invariant. Effective resolution in data products derived from image data will also be spatially varied and will likely become more complicated than that of the input image data. To reiterate, the theoretic requirements for sampling density (i.e., image GSI) and the effective data support (i.e., pixel size) in reality may not be aligned, resulting in suboptimal or sometimes unsuitable choice of spatial resolution for image data acquisition.

It should be noted that SAR is able to achieve large increases in effective spatial resolution by post-processing of the data. In such cases, it will be necessary to modify some definitions of spatial resolution. Given the complementarity of optical and radar remote-sensing data, there is merit in integrating multisource images for information extraction and spatial analysis. For this, both signal processing and geometric modeling (for image co-registration or rectification) are essential (Palubinskas et al. 2010). It is hoped that an insightful understanding of optical and radar remote measurements and their relationships with the underlying signals of interest will help develop integral approaches to multisource image fusion, which will be discussed in more detail in Chapter 8.

In addition to imaging-based sampling that often implies uniform sampling grids, spatial sampling is also undertaken at irregular nodes or on nonuniform grids, such as weather station networks and ground surveys to support satellite mapping of spatial entities and distributions. Unsurprisingly, the conventional probability sampling schemes appear largely oriented for nonuniform grids with the exception of systematic sampling schemes. In parallel to the Shannon sampling theorem for uniform sampling, there is increasing literature on nonuniform sampling (Landau 1967; Marvasti 1996). As reviewed by Unser (2000), similar to the case of uniform sampling, for the reconstruction of band-limited signals in t, there is also the requirement for a (lower) sampling density (i.e., the minimum average number of samples per unit length in t). Although we will not elaborate this here, further discussion about nonuniform sampling is continued in Chapters 4 and 11.

Such sampling strategies as simple random sampling are based on classical sampling theory, whereby it is assumed that data are identically and independently distributed (i.i.d.). They are thus called design-based strategies, as they describe the underlying population itself. On the other hand, there are the so-called model-based approaches, which are based on geostatistics and seek to describe the data-generating processes (Brus and de Gruijter 1997). Model-based strategies are becoming increasingly popular, and their advantages relative to design-based strategies are dependent on various factors including the application objectives, the problem domains, sample sizes, and so on. For example, Brus and de Gruijter (1997) compared design-based and model-based estimates of spatial means by the design-based quality criteria using simulated data. Performance was compared for two sample sizes (140 and 1520), two block sizes (8×6.4 and 1.6×1.6 km^2), and two strategies (stratified simple random sampling combined with the Horvitz–Thompson estimator and systematic sampling combined with the block kriging predictor). Point estimates of spatial means by the model-based strategy were more accurate in all cases except the global mean (8×6.4-km^2 block) estimated from the small sample, whereas interval estimates were, in general, more accurate with the design-based strategy, except when the number of sample points in the block was small.

Both uniform and nonuniform spatial sampling imply finite data support. The classic sampling theorem for continuous signals assumes infinite sample support in space and time. We need to account for the interactions between sampling rate (density) and finite data support in determining the optimum sampling protocols for spatial applications. For instance, it is of theoretical and practical interest to investigate the influences of sampling density and data aggregation (data support) on the accuracy of spatial interpolation, as demonstrated by Vucetic et al. (2000). Further research on spatial sampling and geostatistics would be valuable, given what geostatistics has to offer in this regard and its methodological support of discussions on scale issues. In Chapter 4, geostatistics will be described for scale modeling based on variograms and their regularizations, which can handle the issue of finite data support in the determination of measurement scale. Certainly, geostatistics will have much to offer to scale issues in spatial sampling by exploring the relationships between spatial dependence, information redundancy, and surface reconstruction accuracy (Modis and Papaodysseus 2006).

By now, it should be evident that there are two main components to describe scale of measurement. The first is spatial coverage of the samples and the second is data support. The importance of these scales of measurement will become clear when spatial variation is examined in the context of spatial dependence in the next chapter.

4 Geostatistical Models of Scale

Chapter 3 described geospatial measurements and the related scale issues by reviewing spatial sampling (e.g., sampling density and data support) and some of the technical details underpinning remote sensing, providing a basis for understanding measurement scale and the underlying data-generating processes. This chapter moves to the modeling of process scale by elucidating geostatistical models of scale, furnishing the characterization of scale of the underlying phenomena under study and, thus, determination of a commensurate scale for measurement.

Geostatistics is well known for its capabilities of characterizing spatial dependence and of spatial prediction. Quantification of spatial dependence is required to characterize the spatial scales in the underlying spatial phenomena of interest or at least in measurements of spatially varying properties. This chapter describes geostatistics as it pertains to variogram modeling, which provides the basis for modeling spatial scales.

The scale model (e.g., a variogram model) of a geographic distribution is estimated from sample data, which are often of finite support themselves. The resultant experimental or sample variograms are thus regularized versions of the underlying point-support variograms, which are prerequisite for cross-scale data analyses and applications. To obtain point-support variogram models, it is necessary to perform deconvolution or de-regularization, which aims to estimate point-support variograms from finite-support variograms through an inversion process. Once point-support variograms are inverted from finite-support variograms, it is relatively easy to compute regularized variograms for different supports, thus facilitating spatial problem solving with multiscale datasets.

Characterization of scales should also consider spatial heterogeneity (e.g., through a nonstationary model), as the interactions between spatial dependence and spatial heterogeneity have been shown previously to alter local definitions of scales, thus leading to spatial variation of scale ranges over space. Nonstationary variogram models have been developed to deal with local variation in the spatial structure of the underlying phenomena, which can be explored usefully with or without reference to covariates. Local models of scale, as extensions to global and stationary models, can thus be fitted from experimental data to capture spatial heterogeneity in spatial dependence.

This chapter will describe variograms in terms of their definition and interpretation, in particular as they relate to the characterization of spatial scale, after introducing random fields, a fundamental concept in geostatistics, and related concepts, such as random variables, regionalized variables, and the hypothesis of stationarity, which is necessary for characterizing random fields. Variograms are used as general terms for both semivariograms and covariance functions here, although variograms often refer to semivariograms. Sometimes, it is necessary to write out semivariograms for clarity of formulas. Regularization of variograms is then discussed to make the link between variograms and data support. A more challenging issue is de-regularization (i.e., inversion of point-support variograms from finite-support ones). Methods for de-regularization of variograms are reviewed based on both regular and irregular grids. Section 4.3 will review some methods for determining measurement scale based on variograms, local variance charts, and dispersion variance decomposition. Lastly, local methods for variogram modeling are discussed, specifically nonstationary variograms, to handle spatial heterogeneity in the local pattern of spatial dependence. In addition, issues related to local modeling of spatial dependence, such as variogram regularization and de-regularization, indicator and transition probability-based models for categorical fields, and multivariate geostatistics, will also be addressed.

4.1   GEOSTATISTICAL FUNDAMENTALS AND VARIOGRAMS

4.1.1   RANDOM FIELDS

Geoscientists are interested in spatial distributions of biophysical and other environmental properties. These properties exhibit spatial continuity, either in continuous variables, such as elevation, or in categorical variables, such as soil type. An exhaustive spatial description of a property is often required for problem solving. This is accomplished by predicting the spatial behavior of the property at unsampled locations (Machuca-Mory 2010). For example, methods for spatial interpolation were reviewed by Lam (1983), models of topography–landslide correlation were used for landslide susceptibility mapping (Atkinson and Massari 1998; van Westen et al. 2008), and models of cover type transition were used for predicting future land-cover changes (Lambin and Geist 2006; Ouedraogo et al. 2011).

We can build such models of spatial distributions from sparse data and incomplete knowledge of the settings, which can then be used to map domains where the properties are deemed reasonably homogeneous, as discussed by Machuca-Mory (2010). Deterministic approaches may be taken for interpolation of unknown values between the sampled locations by distance weighting or other surface-fitting techniques. The rationale behind spatial interpolation (or prediction to be more accurate) is Tobler's first law of geography, which states that points closer together in space are more likely to be similar in values of a property, say, Z, than those further apart.

As an alternative, geostatistics considers that the true, but unknown, value of property Z at each unsampled location x in a problem domain D is one of a range of possible values (Journel and Huijbregts 1978). Thus, location-specific variable $Z(x)$ (with x denoting the unsampled location) is conceptualized as a random variable (RV). The probability of the true value (of Z at x) lying within certain intervals is modeled by the probability distribution of the RV $Z(x)$. The values of Z over space most often exhibit some spatial continuity; therefore, the RVs at different locations may be dependent (Cressie and Chan 1989). The ensemble of all RVs at all possible locations within a homogeneous domain, denoted $\{Z(x), x \in D\}$ or simply Z without causing ambiguity in the context, is known as a random function (RF). The multivariate probability distribution of the RF is inferred by pooling the available data within the domain and selecting an appropriate mathematical model (e.g., those to be described in Section 4.1.2). This pooling of data values at different locations is made possible by the property of stationarity (Journel and Huijbregts 1978; Wackernagel 2003), a concept that will be discussed later in this subsection.

If a variable is spatially autocorrelated, it is said to be *regionalized*. Many variables encountered in the geosciences can be regarded as regionalized, such as terrain variables, rainfall, snow depth, population density, and land cover. In mathematical terms, we interpret this regionalized variable as a realization of the underlying RF. Consider the regionalized values at all points in a given region D. The associated variable $Z(x)$ for $x \in D$ is a regionalized variable. The set of values $\{Z(x), x \in D\}$ can be viewed as one draw from a random vector (RVe), a subset of the infinite set of RVs (one RV at each point of the domain). A family of RVs $\{Z(x), x \in D\}$ is called a RF, as discussed previously when the concept of RF was introduced (Wackernagel 2003).

The significance of RF models is twofold. One is that the data values are in some way dependent on their location in the region (so they are regionalized). The other is that the regionalized sample values $z(x_\alpha)$ cannot be modeled deterministically but only stochastically via a RF $\{Z(x), x \in D\}$, as discussed by Wackernagel (2003).

As in many situations where the data-generating mechanism cannot be determined, a probabilistic approach is chosen and the data values are viewed as outcomes of the random mechanism (Wackernagel 2003). An uppercase Z (in $Z(x)$) is used to denote a RV, whereas a lowercase z is used for a realization of Z. The concept of a RF combines both regionalization and randomness, because of the spatial dependence and stochasticity in realized Z values, respectively.

In this model, the RV $Z(x_0)$ at a given location x_0 in D generates realizations following a probability distribution F

$$P(Z(x_0) \leq z_0) = F_{x_0}(z_0) \tag{4.1}$$

where P is the probability that an outcome of Z at the point x_0 is smaller than a fixed value z_0. A multiple distribution function for n random variables at n locations can be defined

$$F_{x_1,\ldots,x_n}(z_1, \ldots, z_n) = P(Z(x_1) \leq z_1, \ldots, Z(x_n) \leq z_n) \tag{4.2}$$

where P is the probability that outcomes of $Z(x_1)$ through $Z(x_n)$ are simultaneously smaller than values z_1 through z_n (Journel and Huijbregts 1978; Chilès and Delfiner 1999; Wackernagel 2003).

In this manner, we have a general model able to describe any spatial process. In practice, however, we generally possess only a few data as realizations of the RF and it is impossible to infer all the univariate and multivariate distributions for any set of points. Simplification is needed and this is facilitated by the concept of stationarity, as discussed by Journel and Huijbregts (1978) and Wackernagel (2003). Stationarity means that the characteristics of a RF stay the same when a set of n locations is shifted by any vector h. Although lag h is a vector having length and direction, we take the same exception as for location x to use lowercase italic h to denote it. To be more specific, a RF $Z(x)$ is strictly stationary if a translation of a point configuration does not change the multivariate distribution (because of multiple locations), that is, $F_{x_1,...,x_n}(z_1, ..., z_n) = F_{x_1+h,...,x_n+h}(z_1, ..., z_n)$ (Matheron 1971; Deutsch and Journel 1998; Wackernagel 2003).

Strict stationarity requires the specification of the multipoint distribution for any set of points $\{x_1, ..., x_n\}$. This decision may be too rigid to accommodate local changes in lower-order distributions and their statistics. Relaxed forms of stationarity can be valuable for both theory and practice (Machuca-Mory 2010). A flexible strategy is to consider only pairs of points in the domain and to try to characterize only the first two moments. This is second-order stationarity. Such a strategy is ideal for Gaussian distributions described entirely by the first two moments, as discussed by Wackernagel (2003). A second possibility is to assume the stationarity of the first two moments of the difference of a pair of values at two points, the so-called intrinsic stationarity, a short form for hypothesis of intrinsic stationarity of order 2 (Wackernagel 2003), leading to the notion of a variogram (Journel and Huijbregts 1978; Chilès and Delfiner 1999). Issues of stationarity will be discussed further in the following subsection.

4.1.2 VARIOGRAMS AND COVARIANCE FUNCTIONS

As models of spatial dependence and thus spatial scale, variograms are important geostatistical tools. They are defined based on hypotheses of stationarity concerning the underlying RFs' distributions. These are the intrinsic hypothesis of stationarity and the hypothesis of second-order stationarity, which underlie the definitions of semivariograms and covariance functions, respectively, with the former being more relaxed than the latter. Stationarity is also necessary for estimating experimental variograms because of often sparsely sampled data, as shown below.

Suppose there is a set of samples $z(x_\alpha)$ ($\alpha = 1, ..., n$) for the problem domain D. We can compute the experimental variogram from them, which is the sequence of averages of dissimilarity measures for different distance classes (Wackernagel 2003), as will be shown below. Experimental variograms are convenient tools for quantification of spatial variability and the analysis of spatial data. Many properties can be adequately described by them, including properties exhibiting unbounded spatial variation. When the spatial variation is bounded, the variogram is equivalent to a covariance function (Wackernagel 2003), as described below.

The spatial variation of a RF $Z(x)$ can be described by taking the differences between values at pairs of locations x and $x + h$: $Z(x + h) - Z(x)$, referred to as increments. The theoretical variogram $\gamma(h)$ is defined by the intrinsic hypothesis, based on two assumptions about the increments. One is that the mean $m(h)$ of the increments is invariant for any translation of a given lag h within D and is zero. The other is that the variance of the increments has a finite value of $2\gamma(h)$ depending on the length and orientation, but not on the position, of a given lag h, as explained by Journel and Huijbregts (1978) and Wackernagel (2003). This is further discussed below.

The intrinsic hypothesis of stationarity, for any pair of points $x, x + h \in D$, is further described through the following equations:

$$E[Z(x + h) - Z(x)] = m(h) = 0 \qquad (4.3a)$$

$$\mathrm{var}[Z(x + h) - Z(x)] = 2\gamma(h) \qquad (4.3b)$$

These two properties of a RF lead to the definition for the theoretical variogram (Journel and Huijbregts 1978)

$$\gamma(h) = \frac{1}{2} E[(Z(x + h) - Z(x))^2] \qquad (4.3c)$$

which is defined as half of the variance (hence semivariance) of the difference between RF Z values at two locations across realizations of the field Z and, for this reason, is also known as the semivariogram (Cressie 1993). Without causing any ambiguity, variogram and semivariogram are used interchangeably in this book. The existence of the first two moments (i.e., mean and variance) of the increments does not imply the existence of the first two moments of the RF itself, as an intrinsic RF can have an infinite variance, although the variance of its increments is finite for any lag h (Wackernagel 2003), as will be shown later when linking up semivariance and covariance models. Thus, an intrinsically stationary RF does not need to have a constant mean or a constant variance, because the intrinsic hypothesis of stationarity is more general than the hypothesis of second-order stationarity (for more details, see Journel and Huijbregts 1978; Wackernagel 2003).

We can define a covariance function as a second-order moment between RFs $Z(x)$ and $Z(x + h)$ at locations x and $x + h$:

$$\mathrm{cov}(Z(x), Z(x + h)) = E[Z(x) \, Z(x + h)] - E[Z(x)]E[Z(x + h)] \qquad (4.4)$$

which is dependent on x and h. When $h = 0$, $\mathrm{cov}(Z(x), Z(x + 0) = E[Z(x)^2] - [E[Z(x)]]^2$, which is equal to the *a priori* variance $\mathrm{var}(Z(x))$.

The stationarity hypothesis about the invariance of a RF $Z(x)$ distribution to shifting applies to all moments of $Z(x)$ (i.e., multipoint distribution for the set of points $\{x_1, \ldots, x_n\}$ discretizing RF Z), as mentioned in the previous subsection. But for linear geostatistics, first and second moments are usually considered:

$$E[Z(x)] = m, \, \forall x \qquad (4.5a)$$

$$C(h) = \mathrm{cov}(Z(x), Z(x + h)) = E\{[Z(x) - m][Z(x + h) - m]\}$$
$$= E\{Z(x) \cdot Z(x + h)\} - m^2 \qquad (4.5b)$$

where m is the stationary mean of Z. So, by assuming the stationarity of the first two moments (mean and covariance) of the RF $Z(x)$, we can define the covariance function $C(h)$ as above (Journel and Huijbregts 1978; Wackernagel 2003).

The relationship between variogram $\gamma(h)$ and covariance $C(h)$ functions can now be determined. Given a covariance function $C(h)$, a variogram function $\gamma(h)$ can be obtained readily by the following formula:

$$\gamma(h) = C(0) - C(h) \qquad (4.6)$$

where $C(0)$ is the *a priori* variance of RF $Z(x)$, $\text{var}(Z(x))$ (Journel and Huijbregts 1978). But the reverse is generally not true, as the variogram is not necessarily bounded. In other words, an unbounded variogram model does not have a covariance function counterpart, as discussed by Wackernagel (2003).

Consider the parameters of variogram models. As shown in Figure 4.1, the semi-variance increases with lag and tends to flatten out at an upper limit known as the sill at a fixed lag known as the range. Nugget, sill, and range are the parameters defining the RF. Their interpretations with respect to semivariograms are as follows.

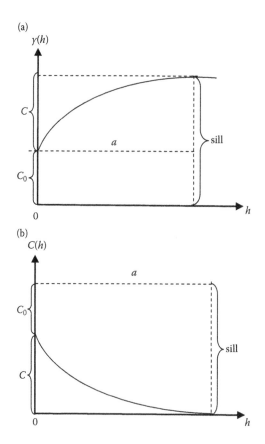

FIGURE 4.1   (a) Semivariogram, (b) covariance function.

Nugget (c_0) represents a discontinuity at the origin of the semivariogram, resulting from measurement error and microscale variation (i.e., variation at a lag less than the measurement scale), both of which lead to different measurement values at locations very close to each other (i.e., when distance $|h|$ is very small). Sill refers to the height of the semivariogram when the lag increases to a certain limit, denoted as $c_0 + c$, where c is the so-called structural variance resulting from spatial correlation. Range is the lag distance at which a semivariogram reaches its sill, reflecting the scale distance of spatial autocorrelation of a RF, denoted as a, beyond which locations are no longer correlated, unless semivariograms appear periodic. At lags shorter than a, spatial correlation increases as the lag distance is reduced. Thus, the range also relays some information on which to base a choice of search radius for spatial interpolation, beyond which sampled locations will be of limited value for spatial interpolation. Numerous textbooks on geostatistics, such as Journel and Huijbregts (1978) and Wackernagel (2003), may be consulted for details about the interpretation of variogram parameters.

To estimate a semivariogram, we require multiple realizations for the pair of RFs $Z(x)$ and $Z(x + h)$ at a lag h apart. However, we often have only one realization for the pair of variables (i.e., a pair of measurement values at locations x and $x + h$). Thus, the intrinsic hypothesis is necessarily introduced (Journel and Huijbregts 1978), as shown previously, indicating that the increment of $Z(x)$ (i.e., $Z(x) - Z(x + h)$) depends only on vector h not on x so that all measurement pairs $\{z(x), z(x + h)\}$ at lag h apart can be considered as realizations of the variable pair $\{Z(x), Z(x + h)\}$. Therefore, $\gamma^*(h)$ can be estimated as

$$\gamma^*(h) = \frac{1}{2N(h)} \sum_{i=1}^{N} [z(x_i) - z(x_i + h)]^2 \qquad (4.7)$$

where $N(h)$ is the number of sample data (i.e., experimental pairs$\{z(x), z(x + h)\}$) approximately h apart.

Now, we describe some theoretical models of variograms, to which experimental variograms may be fitted through various statistical approaches, such as weighted least squares (Cressie 1993). Journel and Huijbregts (1978) and other classic books may be consulted for further details about variogram modeling. There are two types of models: those with finite sills and those without sills.

Spherical model:

$$\gamma(h) = \begin{cases} c_0 & h = 0 \\ c_0 + c\left(\dfrac{3}{2}\dfrac{h}{a} - \dfrac{1}{2}\dfrac{h^3}{a^3}\right) & 0 < h \le a \\ c_0 + c & h > a \end{cases} \qquad (4.8)$$

where c_0 is the nugget variance, $c_0 + c$ is the sill, and a is the range.

Exponential model:

$$\gamma(h) = \begin{cases} c_0 & h = 0 \\ c_0 + c\left(1 - e^{-\frac{h}{a}}\right) & h > 0 \end{cases} \tag{4.9}$$

where c_0 is the nugget variance, $c_0 + c$ is the sill, and a is the range. When $h = 3a$, $1 - e^{-\frac{3a}{a}} = 1 - e^{-3} = 0.95 \approx 1$. Thus, the effective range of an exponential model is $3a$.

Gaussian model:

$$\gamma(h) = \begin{cases} c_0 & h = 0 \\ c_0 + c\left(1 - e^{-\frac{h^2}{a^2}}\right) & h > 0 \end{cases} \tag{4.10}$$

where c_0 is the nugget variance, $c_0 + c$ is the sill, and a is the range. When $h = \sqrt{3}a$, $1 - e^{-\frac{r^2}{a^2}} = 1 - e^{-3} = 0.95 \approx 1$. Thus, the effective range of a Gaussian model is $\sqrt{3}a$.

Power model:

$$\gamma(h) = ch^a \tag{4.11}$$

where a is a dimensionless quantity with typical values of between 0 and 2, and the parameter c has dimensions of the variance, as in the other models. There is no sill for the power model.

In practice, a linear combination of the variogram models mentioned above is often used to model the variogram for a domain. This kind of extended variogram model is called a linear model of regionalization (Wackernagel 2003; Garrigues et al. 2006) and is a weighted sum of nst elementary variogram models, $\gamma_\alpha(rng_\alpha, h)$ $(\alpha = 1, \ldots, nst)$: $\gamma(h) = \sigma^2 \sum_{\alpha=1}^{nst} c_\alpha \gamma_\alpha(rng_\alpha, h)$, where σ^2 is the total amount of spatial variability present in the domain, $\gamma_\alpha(rng_\alpha, h)$ is the elementary variogram model with range rng_α, and c_α is the fraction of the total variance of elementary model $\gamma_\alpha(rng_\alpha, h)$.

Multiple RFs that are cross-correlated are often involved in spatial structural analysis and spatial prediction, which are likely to benefit from the optimal use of the covariation inherent in multiple RFs as well as in univariate spatial correlation (Lark 2003; McBratney et al. 2003; Li and Heap 2011). Below, we mention cross-variogram and cross-covariance briefly, whereas multivariate geostatistics will be discussed in more detail in Chapter 8.

Consider a bivariate case. Suppose two RFs $Z_1(x)$ and $Z_2(x)$ are relevant for a domain D. We can define the cross-variogram $\gamma_{Z_1 Z_2}(h)$ as

$$\gamma_{Z_1 Z_2}(h) = \frac{1}{2} E\{[Z_1(x+h) - Z_1(x)][Z_2(x+h) - Z_2(x)]\} \tag{4.12a}$$

and cross-covariance as

$$C_{Z_1 Z_2}(h) = \text{cov}(Z_1(x), Z_2(x+h)) = E\{[Z_1(x) - m_1][Z_2(x+h) - m_2]\} \quad (4.12\text{b})$$

where m_1 and m_2 represent the stationary means for Z_1 and Z_2, respectively (Wackernagel 2003). The definitions above for cross-variable spatial dependence can be extended to inter-class situations, as shown in the next subsection.

4.1.3 INDICATOR VARIOGRAMS AND TRANSITION PROBABILITY-BASED MODELS

Indicator geostatistical methods are becoming increasingly popular in the Earth sciences because many types of spatial data are categorical, for example, facies and soil classifications. Furthermore, geospatial data of a continuous nature may not conform to Gaussian models, and nonparametric approaches to probabilistic mapping and analysis of non-Gaussian RFs may become necessary (Journel 1983; Zhang and Yao 2008b).

Below, we describe two components for indicator geostatistics: indicator variable definitions and indicator variogram modeling. For categorical variables, the indicator variable $I_k(x)$ can be defined, in general, over a region D by

$$I_k(x) = \begin{cases} 1, \text{ if category } k \text{ occurs at location } x \\ 0, \text{ otherwise} \end{cases} \quad (4.13\text{a})$$

where $x \in D$ and $k = 1, \ldots, K$. The categories are mutually exclusive area classes such as those of land cover or soil types. For originally continuous variables, an indicator variable can be defined for cutoff value z_k

$$I_k(x) = \begin{cases} 1, \text{ if } Z(x) \le z_k \\ 0, \text{ otherwise} \end{cases} \quad (4.13\text{b})$$

where $Z(x)$ is a continuous variable (Journel 1983; Zhang and Yao 2008b).

Indicator auto/cross-variograms and indicator auto/cross-covariances are used traditionally in geostatistics to measure the spatial continuity of indicator variables (Journel 1983; Deutsch and Journel 1998). The variogram for $I_k(x)$ is defined as in Equation 4.3c:

$$\gamma_k(h) = \frac{1}{2} E\{[I_k(x+h) - I_k(x)]^2\} \quad (4.14\text{a})$$

whereas its covariance function is a modified version of Equation 4.5b:

$$C_k(h) = E\{[I_k(x) - p_k(x)][I_k(x+h) - p_k(x+h)]\} \quad (4.14\text{b})$$

where $p_k(x)$ and $p_k(x + h)$ are the probabilities of class occurrences for class k (i.e., the means of indicator $I_k(x)$) at locations x and $x + h$.

For a combination of two classes k and k' (i.e., two indicator variables), we can define cross-variogram $\gamma_{kk'}(h)$ as

$$2\gamma_{kk'}(h) = E\{[I_k(x) - I_k(x + h)][I_{k'}(x) - I_{k'}(x + h)]\} \tag{4.15a}$$

and the cross-covariance $C_{kk'}(h)$ as

$$C_{kk'}(h) = E\{I_k(x)I_{k'}(x + h)\} - E\{I_k(x)\}E\{I_{k'}(x + h)\} \tag{4.15b}$$

where h denotes a lag vector and $I_{k'}(x)$ and $I_k(x)$ are indicator variables defined for classes k and k', respectively. Note the similarity between Equations 4.12a and 4.14a, and between Equations 4.12b and 4.14b, respectively.

Traditionally, sample indicator auto/cross-variograms can be fitted empirically with permissible variogram models. In practice, sampling data alone may not suffice for spatial structural modeling, and some combination of empirical data and subjective judgment can be used to assist model fitting (Carle and Fogg 1996). In the remainder of this subsection, we show how the transition probability provides an alternative strategy for spatial structural modeling, complementing indicator geostatistics based on auto/cross-variogram/covariance (Carle and Fogg 1996).

For measuring spatial continuity, we first define the transition probability $tp_{kk'}(h)$ (Carle and Fogg 1996)

$$tp_{kk'}(h) = p \text{ \{category } k' \text{ occurs at } x + h \mid \text{ category } k \text{ occurs at } x\} \tag{4.16a}$$

which, by applying indicators, becomes

$$tp_{kk'}(h) = p\{I_{k'}(x + h) = 1 \mid I_k(x) = 1\}$$
$$= p\{I_{k'}(x + h) = 1 \text{ and } I_k(x) = 1\} / p\{I_k(x) = 1\} \tag{4.16b}$$

Assume stationarity in one-location marginal probabilities $p_k(x) = p\{I_k(x) = 1\} = E\{I_k(x)\}$ and in two-location joint probabilities $p_{kk'}(x, h) = p\{I_k(x) = 1 \text{ and } I_{k'}(x + h) = 1\} = E\{I_k(x)I_{k'}(x + h)\}$. We have

$$E\{p_k(x)\} = p_k \ \forall x \in D \tag{4.17a}$$

where p_k denotes a stationary probability of occurrence for class k, and

$$E\{p_{kk'}(x, h)\} = p_{kk'}(h) \ \forall x \in D \tag{4.17b}$$

where $p_{kk'}(h)$ denotes a joint probability for classes k and k' depending only on lag h.

By applying Equations 4.17a and 4.17b, the indicator cross-variogram definition (Equation 4.15a) becomes

$$\gamma_{kk'}(h) = p_{kk'}(0) - [p_{kk'}(h) + p_{kk'}(-h)]/2 \tag{4.18}$$

whereas the indicator cross-covariance definition (Equation 4.15b) becomes

$$C_{kk'}(h) = p_{kk'}(h) - p_k p_{k'}$$ (4.19)

and the transition probability definition (Equation 4.16b) becomes

$$tp_{kk'}(h) = p_{kk'}(h)/p_k$$ (4.20)

Equations 4.18 and 4.19 can be rewritten as

$$\gamma_{kk'}(h) = p_k\{tp_{kk'}(0) - [tp_{kk'}(h) + tp_{kk'}(-h)]/2\}$$ (4.21)

and

$$C_{kk'}(h) = p_k[tp_{kk'}(h) - p_{k'}]$$ (4.22)

respectively, using the transition probability in Equation 4.20 (Carle and Fogg 1996).

The spatial continuity measures $C_{kk'}(h)$, $\gamma_{kk'}(h)$, and $t_{kk'}(h)$ for indicator variables can be used for the same dataset. When working with sparse measurements, proportions, mean length, and spatial juxtapositioning patterns are attributes that may be utilized for model fitting in indicator geostatistics (Carle and Fogg 1996). For this, spatial continuity models based on $t_{kk'}(h)$ are advantageous, as shown below.

Class proportions, which typically can be inferred directly from the indicator data, can guide the fitting of the sill of a spatial continuity model, whether expressed as $C_{kk'}(h)$, $\gamma_{kk'}(h)$, or $tp_{kk'}(h)$, to ensure that model sills are consistent with proportions known *a priori*. The sill of $tp_{kk'}(h)$ approaches p_k, the proportion of class k, whether $k = k'$ or $k \neq k'$. In contrast, the sills of $\gamma_{kk'}(h)$ and $C_{kk'}(0)$ approach $p_k(1 - p_k)$. The sill of $\gamma_{kk'}(h)$ and $C_{kk'}(0)$ for $k \neq k'$ approaches $-p_k p_{k'}$, an even more ambiguous situation (Carle and Fogg 1996).

4.2 VARIOGRAM REGULARIZATION AND DEREGULARIZATION

4.2.1 REGULARIZATION

Very often, measurement data $z_v(x)$ are defined on a certain support $v(x)$ centered on a point x, with v being a point sample or, more generally, the area or volume of a sample, as mentioned previously. $z_v(x)$ is then considered as the mean value of the point-support values $z(x')$ in $v(x)$, that is,

$$z_v(x) = \frac{1}{v} \int_{v(x)} z(x') \, dx'$$ (4.23)

as also discussed in the context of finite data supports in Chapter 3. The mean value $z_v(x)$ is said to be the regularization of the point variable $z(x')$ over the block $v(x)$.

Suppose $Z(x)$ is a second-order stationary RF, with mean m and covariance $C(h)$ or semivariogram $\gamma(h)$. The regularization of the point RF $Z(x)$ over the block $v(x)$

is also a RF denoted by $Z_v(x)$, as expressed in Equation 4.23. It can be shown that the regularized RF $Z_v(x)$ of a second-order stationary point RF is also second-order stationary, with a mean identical to that of the point RF $Z(x)$,

$$E[Z_v(x)] = m \qquad (4.24a)$$

and a variogram $\gamma_v(h)$,

$$\gamma_v(h) = \frac{1}{2} E\{[Z_v(x+h) - Z_v(x)]^2\} \qquad (4.24b)$$

as described by Journel and Huijbregts (1978).

To derive the regularized variogram $\gamma_v(h)$ from the point-support variogram $\gamma(h)$, we can consider the regularized variogram as the variance of the estimation of $Z_v(x)$ by $Z_v(x + h)$, with the latter being separated from the former by a lag h. This prediction variance is then given by the general rule described by Journel and Huijbregts (1978):

$$\gamma_v(h) = \bar{\gamma}(v(x), v(x+h)) - \frac{1}{2}\bar{\gamma}(v(x), v(x)) - \frac{1}{2}\bar{\gamma}(v(x+h), v(x+h)) \qquad (4.25a)$$

where $\bar{\gamma}(v(x), v(x+h))$ stands for the mean value of the point variogram $\gamma(h)$ when one of the extremities of the vector h describes the support v centered at x and the other independently describes the translated support v centered at $x + h$, whereas $\bar{\gamma}(v(x), v(x))$ and $\bar{\gamma}(v(x+h), v(x+h))$ stand for the mean value of the point variogram $\gamma(h)$ within the supports $v(x)$ and $v(x + h)$, respectively. Since the point-support variogram $\gamma(h)$ is stationary, the last two terms in Equation 4.34 are equal and, thus,

$$\gamma_v(h) = \bar{\gamma}(v, v_h) - \bar{\gamma}(v, v) \qquad (4.25b)$$

where $\bar{\gamma}(v, v_h)$ equals $\bar{\gamma}(v(x), v(x+h))$ in Equation 4.25a, v_h denotes an identical support v translated from support v by the vector h, and $\bar{\gamma}(v, v)$ is the within-block variance (Journel and Huijbregts 1978).

For distances $|h|$ that are very large in comparison with the dimension of support v, the mean value $\bar{\gamma}(v, v_h)$ is approximately equal to the value $\gamma(h)$ of the point variogram. We then have the useful approximation below:

$$\gamma_v(h) = \gamma(h) - \bar{\gamma}(v, v) \quad (h \gg v) \qquad (4.26)$$

which states that, at large distances, the regularized variogram is simply derived from the point variogram by subtracting a constant term $\bar{\gamma}(v, v)$ related to the dimension and geometry of the support v of the regularization (Journel and Huijbregts 1978).

The within-block variance $\bar{\gamma}(v,v)$ in Equations 4.25b and 4.26 is commonly estimated using the following arithmetical average

$$\bar{\gamma}(v,v) = \frac{1}{N \times N} \sum_{i=1}^{N} \sum_{i'=1}^{N} \gamma(x_i, x_{i'}) \qquad (4.27)$$

where N is the number of points used to discretize the block v. As a rough guide, a minimum of 8 by 8 points is commonly used for block discretization.

In many applications, blocks are often of irregular shapes. For regularization of the point support semivariogram over irregular blocks, some elaborations are necessary, as shown below. In addition to the assumption of stationarity, Equation 4.26 is derived under the assumption that all the blocks in the study area have the same size and shape. Therefore, the within-block variance $\bar{\gamma}(v,v)$ is constant. To handle irregularities in blocks, a more general expression for the regularization is proposed:

$$\gamma_v(h) = \bar{\gamma}(v,v_h) - \bar{\gamma}_h(v,v) \qquad (4.28)$$

where the within-block variance $\bar{\gamma}_h(v,v)$ is not a constant but varies as a function of lag h, since the size of the irregular blocks is varied relative to the distance between them, unlike $\bar{\gamma}(v,v)$ in Equation 4.25b (Goovaerts 2008). Below, we describe how $\bar{\gamma}_h(v,v)$ and $\bar{\gamma}(v,v_h)$ in Equation 4.28 are computed with irregular blocks.

$\bar{\gamma}_h(v,v)$ is estimated as the arithmetic average of within-block variances for blocks separated by a given vector h

$$\bar{\gamma}_h(v,v) = \frac{1}{2N(h)} \sum_{\alpha=1}^{N(h)} [\bar{\gamma}(v_\alpha, v_\alpha) + \bar{\gamma}(v_{\alpha+h}, v_{\alpha+h})] \qquad (4.29)$$

where $\bar{\gamma}(v_\alpha, v_\alpha)$ and $\bar{\gamma}(v_{\alpha+h}, v_{\alpha+h})$ are estimated according to Equation 4.27, $N(h)$ is the number of block pairs separated by a given lag h, and between-block distance $|h|$ is computed according to

$$|h| = \text{distance}\ (v_\alpha, v_\beta) = \frac{1}{N_\alpha \times N_\beta} \sum_{i=1}^{N_\alpha} \sum_{i'=1}^{N_\beta} \|x_i - x_{i'}\| \qquad (4.30)$$

where N_α and N_β are the number of points used to discretize the two blocks v_α and $v_{\alpha+h}$, and $\|x_i - x_{i'}\|$ stands for the distance between points x_i (in block v_α) and $x_{i'}$ (in block $v_\beta = v_{\alpha+h}$) (see Goovaerts [2008] for details). If all the blocks are the same, then $\bar{\gamma}(v_\alpha, v_\alpha) = \bar{\gamma}(v_{\alpha+h}, v_{\alpha+h})$ and Equation 4.29 is equal to $\bar{\gamma}(v,v)$ in Equation 4.27. Similarly, the first term $\bar{\gamma}(v,v_h)$ in Equation 4.28 is estimated as

$$\bar{\gamma}(v,v_h) = \frac{1}{N(h)} \sum_{\alpha=1}^{N(h)} \bar{\gamma}(v_\alpha, v_{\alpha+h}) \qquad (4.31)$$

where $N(h)$ is, again, the number of block pairs separated by a given lag h, and $\bar{\gamma}(v_\alpha, v_{\alpha+h})$ is computed as

$$\bar{\gamma}(v_\alpha, v_{\alpha+h}) = \frac{1}{N_\alpha \times N_{\alpha+h}} \sum_{i=1}^{N_\alpha} \sum_{i'=1}^{N_{\alpha+h}} \gamma(x_i, x_{i'}) \tag{4.32}$$

where N_α and $N_{\alpha+h}$ are the number of points used to discretize the two blocks v_α and $v_{\alpha+h}$, and $\gamma(x_i, x_{i'})$ represents the semivariance between points x_i (in block v_α) and $x_{i'}$ (in block $v_\beta = v_{\alpha+h}$), as described by Goovaerts (2008).

Regularization of point variogram models over irregular blocks can be performed on the basis of Equation 4.28, where the terms $\bar{\gamma}_h(v, v)$ and $\bar{\gamma}(v, v_h)$ are both lag specific and can be computed using Equations 4.29 and 4.31, respectively. As spatial data are often on irregular supports, the methods reviewed above for variogram regularization are general but can be simplified to what is described in Equation 4.25b for cases of regular supports.

4.2.2 Deconvolution of Regularized Variograms

Block-support variograms can be derived through regularization of point or quasi-point support variograms, which is a kind of convolution of point variograms, as shown in the previous subsection. The reverse process is de-regularization or deconvolution of variograms estimated from block support data (areal data in the two-dimensional case) and is extremely important for quantifying the scale intrinsic to the processes underlying geographic distributions being observed or having generated the data at hand. However, the inference of a point-support variogram from the variogram of areal data is not as straightforward as the regularization procedures described earlier.

For many applications, the data support is often very small with respect to the size of the blocks to be estimated (prediction support). In these cases, the support v can be approximated to a quasi-point support ($|v| \approx 0$), and the experimental semivariogram of areal data $\hat{\gamma}_v(h)$ can be treated as an estimator of the point-support model $\gamma(h)$ (Journel and Huijbregts 1978; Goovaerts 2008). These assumptions are clearly unrealistic for applications in the social sciences, where the blocks can be very large and highly irregular (e.g., census units).

Journel and Huijbregts (1978, p. 90) proposed an iterative approach to deconvolution, as reviewed by Goovaerts (2008):

1. Define a point-support model from inspection of the semivariogram of areal data $\hat{\gamma}_v(h)$ and estimate the semivariogram model parameters using basic deconvolution rules.
2. Compute the theoretically regularized model $\gamma_v(h)$ using approximation (Equation 4.26) and compare it with the experimental curve $\hat{\gamma}_v(h)$.
3. Adjust the parameters of the point-support model so as to bring the regularized model in line with $\hat{\gamma}_v(h)$.

With conventional deconvolution methods, it was not clear how the closeness between the experimental and theoretically regularized semivariogram models should be assessed, as questioned by Goovaerts (2008). Below, we review the deconvolution procedure proposed by Goovaerts (2008), which can accommodate areal units of different sizes and shapes, not just for deconvolution of experimental semivariograms estimated over regular pixels (Collins and Woodcock 1999).

According to Goovaerts (2008), the iterative approach proceeds as follows:

1. Compute the experimental semivariogram of areal data $\hat{\gamma}_v(h)$ and fit a model $\gamma_v^{\text{exp}}(h)$ using weighted least-square regression (Pardo-Igúzquiza 1999).
2. Use the model fitted to areal data, $\gamma_v^{\text{exp}}(h)$, as an initial point-support model $\gamma^{(0)}(h)$.
3. Regularize the model $\gamma^{(0)}(h)$ according to Equation 4.28.
4. Quantify the deviation between the "data-based" ($\gamma_v^{\text{exp}}(h)$) and the theoretically regularized ($\gamma_v^{(0)}(h)$) semivariogram models using the average relative difference between these two curves over L lags h_l

$$dif^{(0)} = \frac{1}{L} \sum_{l=1}^{L} \frac{\left| \gamma_v^{(0)}(h_l) - \gamma_v^{\text{exp}}(h_l) \right|}{\gamma_v^{\text{exp}}(h_l)} \tag{4.33}$$

5. Consider the initial point-support model $\gamma^{(0)}(h)$, the regularized model $\gamma_v^{(0)}(h)$, and the associated difference statistic $dif^{(0)}$ as "optimal" at this stage:

$$\gamma^{\text{opt}}(h) = \gamma^{(0)}(h), \gamma_v^{\text{opt}}(h) = \gamma_v^{(0)}(h), dif^{\text{opt}} = dif^{(0)} \tag{4.34}$$

6. For each lag h_l, compute the experimental values for the new point-support semivariogram through a lag-specific rescaling of the optimal point-support model $\gamma^{\text{opt}}(h_l)$

$$\overline{\gamma}^{(l)}(h_l) = \gamma^{\text{opt}}(h_l) \times w^{(l)}(h_l) \tag{4.35a}$$

$$w^{(l)}(h_l) = 1 + \frac{\left(\gamma_v^{\text{exp}}(h_l) - \gamma_v^{\text{opt}}(h_l) \right)}{s_{\text{exp}}^2 \sqrt{\text{iter}}} \tag{4.35b}$$

where s_{exp}^2 is the sill of the model $\gamma_v^{\text{exp}}(h)$ fitted to areal data, and the parameter "iter" indicates the iteration number (iter = 1 at this stage). The denominator $\sqrt{\text{iter}}$ causes a gradual attenuation of the magnitude of changes as the deconvolution proceeds.

7. Fit a model $\gamma^{(1)}(h)$ to the rescaled values using weighted least-squares regression (same procedure as in Step 1).

8. Regularize the model $\gamma^{(1)}(h)$ according to Equation 4.28, that is,

$$\gamma_v^{(1)}(h) = \overline{\gamma}^{(1)}(v, v_h) - \overline{\gamma}_h^{(1)}(v, v) \tag{4.36}$$

9. Compute the difference statistic in Equation 4.33 for the new regularized model $\gamma_v^{(1)}(h)$:
   - If $dif^{(1)} < dif^{\mathrm{opt}}$, use the point-support model $\gamma^{(1)}(h)$ and the associated statistic $dif^{(1)}$ as the new optimum. Repeat steps 6 through 8.
   - If $dif^{(1)} \geq dif^{\mathrm{opt}}$, repeat Steps 6 through 8 using the same optimal model but the new rescaling coefficients should be computed as

$$w^{(l')}(h_l) = 1 + \frac{(w^{(l)}(h_l) - 1)}{2} \tag{4.37}$$

which makes the new candidate point-support model closer to the optimal model found so far.

10. Stop the iterative procedure after the ith iteration whenever one of the following three criteria is met: (1) the difference statistic reaches a sufficiently small value, for example, $dif^{(i)} < dif^{(0)} \leq 0.05$; or (2) the maximum number of allowed iterations has been tried, for example, iter > 25; or (3) a small decrease in the difference statistic dif was recorded a given number of times, for example, $|dif^{(i)} - dif^{\mathrm{opt}}| / dif^{\mathrm{opt}} \leq 0.01$ was recorded three times.

There are alternative approaches to deconvolution of variograms estimated over nonpoint data supports. Vargas-Guzmán et al. (2000) proposed a method based on spatial variance for recovering point support variograms in cases where point data cannot be measured, but their spatial variances can be known. Spatial variance is conceptualized as an estimation variance between two physical regions or a region and itself. If such a spatial variance could be measured within several sizes of windows, such variances allow the computation of the sample variogram. This approach can be extended to cross-variable covariances and can also take care of the case of nonpoint sample support of the blocks or elements composing each window.

4.3 STATISTICS FOR DETERMINING MEASUREMENT SCALES

We have shown previously in this chapter how the variograms may be modeled from empirical data and interpreted in real applications. However, it is not clear how the variograms might be useful for selecting a size of support or pixel size (i.e., measurement scale) for a particular application. In this section, attention is turned to the problem of choosing appropriate measurement scales, in particular pixel size in remote-sensing applications, through computing certain statistics related to pixel size.

Variograms are useful for selection of an appropriate combination of pixel size and estimation technique for a given application such as image-based land cover classification (Woodcock and Strahler 1987). For example, to investigate spatial scaling in ground-based and airborne hyperspectral data for canopy-to-watershed level ecosystem studies of southern California chaparral and grassland vegetation, Rahman et al. (2003) used three optical reflectance indices, namely, normalized difference vegetation index, water band index, and photochemical reflectance index, as indicators of biomass, plant water content, and photosynthetic activity, respectively. Variograms and local variance were used to determine the spatial scale of these indices. The results indicate that a pixel size of 6 m or less would be optimal for studying the functional properties of southern California grassland and chaparral ecosystems using hyperspectral remote sensing. A problem with using variograms in this way is that it may be difficult to interpret a variogram model that does not have a defined range (such as the power model) (Curran and Atkinson 1999). Thus, some modifications or alternatives are necessary. Below, we review some useful statistics linking observed spatial variation and spatial resolution, which can be employed to choose an optimal measurement scale (pixel size) for a domain given sampled data of finer spatial resolution.

Consider the methods for spatial scale determination based on image coarsening and regularization of variograms. The experimental semivariogram $\gamma_v(pix)$ at a lag of one-pixel interval pix may be computed from $i = 1, 2, \dots, N(pix)$ pairs of pixel values $\{z_v(x_i), z_v(x_i + pix)\}$ defined on a pixel of area $|v| = pix \times pix$ at locations $\{x_i, x_i + pix\}$:

$$\hat{\gamma}_v(pix) = \frac{1}{2N(pix)} \sum_{i=1}^{N(pix)} [z_v(x_i) - z_v(x_i + pix)]^2 \qquad (4.38)$$

The pixel size of the imagery is coarsened in successive integer multiples, and $\gamma_v(pix)$ is estimated each time. Then, the plot of $\hat{\gamma}_v(pix)$ against pixel size (i.e., pix) usually indicates a peak, which helps to identify the predominant scale of spatial variation in the image. Thus, the peak in this plot is related to the range of the variogram model and conveys similar information (Curran and Atkinson 1999). As $\gamma_v(pix)$ can be estimated readily by regularizing the punctual variogram as an alternative to degrading the imagery, it can be estimated for any pixel size, not just integer multiples of the original pixel size. Also, it is possible to deal with measurement error and the point spread function (as mentioned in Chapter 3) more appropriately (Atkinson and Curran 1997; Curran and Atkinson 1999).

The average local variance has been used previously to help select a suitable pixel size (Woodcock and Strahler 1987; Woodcock et al. 1988). The average local variance $\bar{\sigma}_v^2$ may be estimated from a moving (3×3) window w applied to an image of NR rows by NC columns of pixels with support v using

$$\bar{\sigma}_v^2 = \frac{1}{NR \cdot NC} \sum_{nr=1}^{NR} \sum_{nc=1}^{NC} \frac{1}{9} \sum_{j=-1}^{+1} \sum_{k=-1}^{+1} [\bar{z}_v(nr, nc) - z_v(nr + j, nc + k)]^2 \qquad (4.39)$$

where $\bar{z}_v(nr, nc)$ is the mean for a 3×3 window centered on (nr, nc) and it is assumed that there is a buffer of one pixel surrounding the image (Curran and Atkinson 1999).

The local variance $\bar{\sigma}_{vw}^2$, expressed as a function of pixel size, is estimated for a range of integer multiples of the original pixel size. As for $\gamma_v(v_1)$ mentioned earlier, the plot of $\bar{\sigma}_v^2$ against pixel size often rises to a peak and thereafter decreases with increasing pixel size. The peak in the plot may help to identify the predominant scale of spatial variation in the image (Woodcock and Strahler 1987; Curran and Atkinson 1999), as also reviewed in Chapter 1.

Woodcock and Strahler (1987) examined local variance vs. resolution plots for forested, agricultural, and urban/suburban environments (scenes), confirming that the spatial structure of each type of scene is a function of the sizes and spatial relationships of the objects contained in the corresponding scene. For instance, a peak of local variance was identified at a resolution of 6 m for a 1:15,000 aerial photograph of a forested area in South Dakota and 240 m for a Landsat TM image (band 3) of an agricultural area near Dyersburg, Tennessee (Woodcock and Strahler 1987). They concluded that, at the spatial resolutions of SPOT and Landsat TM images, local image variance is relatively high for forested and urban/suburban environments, suggesting that information extraction utilizing texture, context, and mixture modeling is well suited for these sensor systems, whereas, for agricultural environments, local variance is low, indicating the appropriateness of more traditional classifiers. Hyppänen (1996) investigated the effect of tree species and forest age on the optimal spatial resolution of optical remote-sensing images in a forested landscape. Variograms were used to quantify spatial autocorrelation, whereas local variance curves were used to define the spatial resolution that maximizes the variance between neighboring pixels. The range of the variograms was 5 m for Scots pine and 7 m for Norway spruce, whereas the sill was found to be strongly dependent on tree species and forest age class. The local variance maximum in the infrared and green bands was obtained with a spatial resolution of 3 m and in the red channel with a resolution of 2 m. Like the semivariance, the local variance was at a higher level in spruce and old forests.

Theoretically, there should be an obvious peak in the plot of average local variance against pixel size. However, in practice there may be no recognizable peak in the plot. To handle this problem, Ming et al. (2011) proposed a modified local variance method based on variable window size and variable resolution. To determine the optimal spatial resolution for urban form detection from remote-sensing images, Tran et al. (2011) applied the local variance method and identified optimal spatial resolution at two levels of observation: individual urban elements and urban districts in two agglomerations in Western Europe (Strasbourg, France) and in Southeast Asia (Da Nang, Vietnam). For this, they experimented with three categories of interval values of spatial resolutions for identifying optimal resolution: from 0.8 to 3 m for isolated objects, from 6 to 8 m for vegetated areas, and equal to or coarser than 20 m for urban districts. Their approaches were claimed to be robust and transferable to different urban contexts.

In addition to variogram regularization and local variance, the dispersion variance can be used for determining the suitable measurement scale for images. The dispersion variance $\sigma^2(v/V)$ of a variable defined on a support v within a block V, as defined by Journel and Huijbregts (1978), may also be used to relate information pertinent to the choice of pixel size, which is the expectation of empirical variance dispersed in space $S^2(x)$:

$$\sigma^2(v/V) = E\left\{ S^2(x) = \frac{1}{nblk} \sum_{i=1}^{nblk} [Z_v(x_i) - Z_V(x)]^2 \right\} \tag{4.40a}$$

where $Z_v(x_i)$ and $Z_V(x)$ are the means of RF $Z(x)$ over the set of support vs (i.e., $nblk$ equally sized units $v(x_i)$ centered at points x_i, such as pixels in an image) and block V centered at x (i.e., the image under study). $S^2(x)$ depends on x, but $\sigma^2(v/V)$ does not, assuming second-order stationarity (Journel and Huijbregts 1978; Curran and Atkinson 1999; Wackernagel 2003). The dispersion variance $\sigma^2(v/V)$ may be obtained from the regularized variogram using

$$\sigma^2(v/V) = \frac{1}{|V|^2} \int_V \int_V \gamma_v(v(x), v'(x')) \, dv \, dv' \tag{4.40b}$$

that is, the integral of the regularized (over support v) variogram over the block V with an area of $|V|$ (Atkinson 2001).

Suppose that the fine-spatial-resolution image with N pixels is partitioned into $nblk$ congruent blocks $\{v_j, j = 1, \ldots, nblk\}$, each with n pixels. According to Myers (1997) and Garrigues et al. (2006), the image spatial variability can be decomposed at three spatial scales:

1. The experimental dispersion variance of the N fine-spatial-resolution pixel values within the image (i.e., image variance)

$$S^2(x/V) = \frac{1}{N} \sum_{i=1}^{N} [Z(x_i) - Z_V(x)]^2 \tag{4.41}$$

   where $Z_V(x)$ is the spatial average of $Z(x)$ over the image.
2. The experimental dispersion variance of the $nblk$ block values within the image (i.e., inter-block variability)

$$S^2(v/V) = \frac{1}{nblk} \sum_{j=1}^{nblk} [\bar{Z}_v(x_j) - Z_V(x)]^2 \tag{4.42}$$

   where $\bar{Z}_v(x_j)$ is the average of $Z(x)$ over the domain of $v(x_j)$.
3. The average experimental dispersion variance of the n fine-resolution pixel values within each of the blocks (i.e., intra-block variability)

$$S^2(x/v) = \frac{1}{nblk} \sum_{j=1}^{nblk} \frac{1}{n} \sum_{iv=1}^{n} [Z(x_{iv}) - Z_v(x_j)]^2 \tag{4.43}$$

where $S^2(x/v)$ quantifies empirically the loss of image variability when aggregating the pixels to a coarser resolution.

The decomposition of the image-wise spatial variability is thus written as

$$S^2(x/V) = S^2(v/V) + S^2(x/v) \qquad (4.44)$$

where each quantity can be computed from the corresponding theoretical dispersion variance, which is, in turn, related to a proper variogram model. For instance, $S^2(x/v)$ can be computed directly from the variogram of the fine-resolution image. As shown in Equation 4.40b, assuming second-order stationarity, the dispersion variance of $Z(x)$ within the block v, $\sigma^2(x/v)$, is defined as the expectation of $S^2(x/v)$: $\sigma^2(x/v) = E\{S^2(x/v)\}$, which is equal to the double integration of the variogram over the domain v (Chilès and Delfiner 1999; Garrigues et al. 2006)

$$\sigma^2(x/v) = \gamma(v/v) = \frac{1}{|v|^2} \int_v \int_v \gamma(x, x') \, dx \, dx' \qquad (4.45)$$

where x and x' are points exhausting the domain of v, and $|v|$ is the area of v.

As shown by Garrigues et al. (2006), dispersion variance can be computed from variogram models to quantify the spatial heterogeneity within a moderate-resolution pixel, such as that of MODIS images. Based on the analysis of 18 landscape sites, it was estimated that a pixel size of less than 100 m was sufficient to capture the major part of the spatial variability of vegetation cover at the landscape scale (Garrigues et al. 2006).

Thus far, we have discussed optimum spatial resolution for regular sampling as in remote-sensing image-based information extraction and mapping. As implied in Chapter 3, irregular or nonuniform sampling is also important for spatial sampling design, since it is often the only feasible and practical way of obtaining accurate, but sparse, sample data, as in field surveys.

Based on the classic sampling theorem, Modis and Papaodysseus (2006) computed the power spectrum of various representative models of stationary RFs with finite *a priori* variance and showed that the underlying RF models are approximately band limited. Then, they asserted that the range of influence, interpreted from a variogram, is a measure of the maximum variability frequency observed in a problem domain. Thus, a simple application of the well-known sampling theorem shows that, under certain assumptions, it is possible to define a critical sampling density. For this, an approximate rule of thumb can then be stated: that critical sampling interval is half the range of influence observed in the variogram. Thus, a RF $Z(x)$ can be reconstructed by its samples if the sampling rate is greater than or equal to the above-specified value.

Nonuniform sampling theory has been established and can shed light on how signals of interest may be sampled irregularly and reconstructed from nonuniform sample data (Unser 2000; Marvasti 2001; Baraniuk 2007). Nonuniform and sparse sampling and reconstructed data will be further discussed in the light of information theory in Chapter 11.

4.4 DISCUSSION

So far in this chapter, we have reviewed variogram models that are based on the assumption of stationarity. Such models assume global stationarity across the problem domain and, thus, do not allow local changes in the variograms (Myers 1989). To overcome the limitations of the traditional forms of stationarity, local stationarity is proposed for modeling RFs with local variogram models, thus reflecting better the local spatial structure of the RFs under study. For local modeling of spatial dependence, we need to rebuild the parameters of the RF at many locations and to have abundant data available if local statistics are to be reliable (Machuca-Mory 2010). This section reviews some methods for local variogram modeling and provides links to some related issues.

Local variogram models imply certain forms of nonstationarity (Cuba et al. 2012). Nonstationarity of the mean may be handled by the intrinsic assumption or by decomposing the RF into a local mean and a residual. McLennan (2007) discussed two main components in handling nonstationarity in means: delineation of homogeneous domains and modeling of a locally varying mean within these domains. However, local variation in other statistics can also be incorporated in geostatistical modeling: locally changing definitions of variance, locally changing measures of spatial continuity, and locally changing bivariate correlation between properties owing to certain processes, as reviewed by Machuca-Mory (2010).

According to Machuca-Mory (2010), there are currently local and global approaches to nonstationary spatial correlation: the stationarity of the RF is restricted locally in the former, whereas nonstationarity is dealt with for the entire problem domain in the latter (Schabenberger and Gotway 2005). Sampson and Guttorp (1992) proposed a global approach to nonstationarity, which is based on spatial deformation: the coordinate space is transformed so that the covariance structure becomes stationary and isotropic in the transformed space, as also discussed by Machuca-Mory (2010). Below, we review some local approaches to nonstationary scale modeling.

Local variogram models can be developed based on segmentation of the space of interest (i.e., division of the problem domain into smaller and homogeneous regions, or segments, within which the variogram can be modeled as stationary) (Atkinson and Lloyd 2007). As a moving window approach, the experimental local variograms or covariances are calculated using only the samples that fall within a window centered on the location to be estimated. At each location, the window size is adjusted automatically to contain a minimum number of observations necessary for calculating experimental semivariograms and automatic model fitting. Kriging is then applied using the fitted local variogram model and considering only those observations within the moving window. However, this approach is computationally more expensive owing to the need for local variogram model fitting at each location being analyzed. Other drawbacks include the difficulty of supervising the modeling process, unreliability of local variogram modeling with scarce data, and artifacts resulting from including or excluding individual observations from one estimation location to another, as discussed by Machuca-Mory (2010).

As reviewed by Machuca-Mory (2010), the process convolution approach proposed by Higdon (1998) decomposes the RF $Z(x)$ as a Gaussian process $Z_G(x)$ plus an independent error process $e(x)$ where $Z_G(x)$ is defined as a convolution of a zero-mean white noise with a Gaussian convolution kernel locally parameterized. The covariance function can be made nonstationary by changing the kernel parameters at each location. There is yet another type of approach based on weighted stationary processes. It is assumed that the nonstationary RF $Z(x)$ can be represented as a linear combination of k stationary RFs $Z_1(x), ..., Z_k(x)$, which are defined at k partitions of the domain and are uncorrelated outside each partition (Fuentes 2002). Lastly, we have adaptive approaches that use an image analysis gradient algorithm to iteratively and progressively tune the local anisotropy ranges and directions (Stroet and Snepvangers 2005). With a global search radius, the gradient algorithm is able to identify locally the directions of maximum continuity. These directions are used to modify the anisotropy of a local linear variogram model. Then, at successive iterations, finer-scale details in anisotropy are tuned by decreasing the search radii. No local variogram fitting is required, since the single parameter needed for the linear variogram model is derived from the gradient algorithm. This approach is limited to the use of the linear variogram model, and the gradient algorithm may become unreliable if data are scarce (Machuca-Mory 2010).

In the light of local models of spatial dependence discussed above, regularization and de-regularization of variograms may be usefully reflected upon. Regularization in a domain partitioned into k subregions each with a stationary RF model should be performed separately for each subregion, accounting for different point-support variograms in different subregions. For de-regularization, on the other hand, the pooled experimental variogram should be deconvoluted separately for individual subregions. De-regularization is inherently more complicated than regularization, especially when the point-support variogram models for subregions are not known *a priori*. Methods for deconvolution of block-support variograms should be developed in a nonstationary domain believed to consist of k subregions of local stationarity, drawing upon the existing methods of deconvolution described in Section 4.2.2 but with adaptation for different variograms deemed representative over individual subregions.

Determination of measurement scale over a nonstationary domain consisting of local stationarity should be undertaken in line with variogram models and other statistics computed over subregions. Obviously, there will be no single optimal measurement scale for a domain that is better described by multiple local variogram models than by a single model. As local models of semivariance and covariance can be ascribed to a nonstationary domain, spatial statistics such as local variance and dispersion variance can also be made to vary locally to reflect the spatial nonstationarity in spatial dependence over a domain. Methods for optimization of measurement scale should be developed, taking advantage of developments in local modeling of spatial dependence and geostatistical sampling design.

In principle, many of the methods for variogram modeling and measurement scale optimization can be applied, with proper modification, to categorical variables. Indicator auto- and cross-variograms can be employed for quantifying spatial dependence in area-class occurrences and distributions, and so can transition probability-based models. Methods for modeling spatial structure in area classes can be usefully addressed on the basis of phase space (Goodchild et al. 2007), which is characterized by the process covariates that have shaped the area-class distribution observed. We will discuss this and related issues in Chapter 10.

As shown in Section 4.1.3, indicator and transition probability-based models necessitate the use of multivariate geostatistics, which is valuable for quantifying auto- and cross-covariance in multiple RFs involved in a problem domain. We need to extend models of univariate regionalization to those of co-regionalization (Wackernagel 2003). For example, to monitor surface processes at regional and global scales using fine temporal but moderate spatial resolution sensors (from 250 m to 7 km) and to correct for biases in nonlinear estimation processes of land surface variables derived from remote-sensing data, it is important to quantify surface spatial heterogeneity (Garrigues et al. 2008). For this, we need to properly characterize the scale of spatial variation of the processes shaping the landscape. Since various spectral bands are often used for the description of land-surface processes, a multivariate approach to characterize the surface spatial heterogeneity from multispectral remote-sensing observations has to be established. Garrigues et al. (2008) used auto- and cross-variograms modeled together with the geostatistical linear model of co-regionalization to quantify the landscape spatial heterogeneity captured by red and near-infrared fine spatial resolution images. Mutivariate geostatistics will be discussed in the context of data fusion in Chapter 8.

Geostatistical models of scale have been described in this chapter. As summarized above, variograms are versatile in being able to characterize RFs of both continuous and categorical nature, extensible to multivariate cases, and flexible for global and local quantification of scale. In the discourse, geostatistical modeling has implicitly assumed continuous indexing of locations over space. In addition to this kind of "geostatistical data," there is another type of data known as lattice data, that is, data that are aggregated over finite areal units. For this kind of data, alternative modeling approaches are necessary and complement the geostatistical ones, as explained in the next chapter.

5 Lattice Data and Scale Models

Chapter 4 described geostatistical modeling of scale based on variograms, which assume continuous indexing of locations over space. As mentioned toward the end of Chapter 4, unlike geostatistical data, lattice data result from aggregation of field variables over finite areal units, between which there is no space where additional sample points could be obtained. This chapter will discuss lattice data and their scale models, based on measures of spatial autocorrelation, for which neighborhood relationships play a central role in data analysis and scale.

As shown in Chapter 2, fields and objects, as aggregations of geo-atoms, are the two complementary data models for representing geographic distributions and entities, respectively. Both continuous (interval/ratio) and categorical (nominal/ordinal) variables are relevant for fields, with elevation and land cover being respective examples, whereas discrete objects (in two-dimensional space) can be points, lines, and areas in geometric terms, as exemplified by survey monuments, roads, and land parcels, respectively. In terms of data support, spatial data sampled from fields can be called geostatistical data (point-referenced) and lattice data (areal or zonal), as discussed below.

Geostatistical data result from sampling a field over a continuous domain. With geostatistical data, the spatial domain D is fixed and continuous so the number of locations for making observations is uncountable. Between any two sample locations x_i and x_j are an infinite number of other potential sample points. We can lay down a grid and sample at the grid nodes, or we can select points randomly. $Z(x)$, the outcome of the process, can be categorical or interval/ratio, as mentioned above. Geostatistical models of spatial dependence have been described in the previous chapter. A variogram quantifies the degree and range of spatial autocorrelation in a particular dataset (and the underlying RF, say, $\{Z(x), x \in D\}$, if the variogram is point supported) and can be used as a model of scale.

For lattice data, on the other hand, although the domain D is fixed as for geostatistical data, the number of lattice units (cells or faces) that is imposed on a continuous field is countable (usually finite). For instance, nationwide health statistics by counties, housing surveys on a neighborhood level, and species richness measured on contiguous quadrants (e.g., Goodall et al. 2006) are examples of lattice data. With lattice data, there is no space between lattice units where additional sample points could be obtained; in short, lattice cells or faces exhaust the space and do not overlap. Lattice data can be regular or irregular in terms of tessellation (e.g., Landsat TM image pixels resampled to 30-m × 30-m regular squares vs. ZIP code or census tract-based irregular areal units for mapping homeownership rates), or discrete or continuous in terms of scale of measurement as in the case of geostatistical data.

The characteristics of lattice data were also described by Schabenberger and Pierce (2002) and Saveliev et al. (2007).

The primary tool in geostatistical data analysis is the variogram as described previously. Geostatistical data can be aggregated by areas to generate lattice data. Regularization of variograms may be usefully applied for quantifying spatial dependence in the resultant lattice data. The transformation from lattice to geostatistical data is, unfortunately, not so straightforward. We may assume that the values of $Z(x)$ on a lattice occur at the centers of the lattice units and then treat D as if it were continuous to permit spatial prediction at intermediate locations (Bracken and Martin 1989). This arbitrariness of assigning the data to the centroid, however, raises concerns because variogram modeling depends heavily on the distances between observations. There are implications if we apply geostatistics and variograms to lattice data; the effects of lattice data values placed randomly within their associated regions have been documented by Zhao and Wall (2004). However, geostatistical models may be adapted for handling lattice data properly.

Spatial autocorrelation has long been discussed in the geographic literature (see e.g., Moran 1950; Geary 1954; Cliff and Ord 1969; Ord and Getis 1995; Karlström and Ceccato 2002; Haining 2009). As discussed in the context of geostatistics in Chapter 4, the essence of spatial autocorrelation lies in spatial dependence, which refers to the tendency for observations at locations (i.e., area units or cells) nearer to each other to be more (positive spatial autocorrelation) or less (negative spatial autocorrelation) similar than those farther apart. For the former case (positive spatial autocorrelation), units with high values (above average) tend to be located near other units with high value, and units with low values (below average) are likely to be adjacent to other units with low values. In the latter case of negative spatial correlation, there is alternation of neighboring units with high and low values, as described by Saveliev et al. (2007). Although there has already been much research on, and application of, positive spatial autocorrelation, work on negative cases is growing (Griffith and Arbia 2010).

A key element in measuring and understanding spatial autocorrelation is a description of the relationship between the similarity of observations and the distance separating them (Goodchild 1986). This relationship is the same key concept underlying geostatistics, an area parallel to the work on spatial autocorrelation (Goovaerts 1997), as has been described in the preceding chapter where the concepts of RFs and variograms (including both semivariograms and covariograms, based on semivariance and covariance, respectively) were introduced. Correlograms, which are used to quantify spatial autocorrelation and will be described later in this chapter, are the equivalent of variograms for lattice data, although the former can be used for geostatistical data. Thus, whereas geostatistics uses variograms to predict quantities across distances, the lattice approach uses neighborhood relationships to predict quantities across distances, as discussed by Goodchild (1986). The definitions of neighborhood play a central role in lattice data analysis, in particular with respect to scale determination.

However, spatial autocorrelation statistics are typically global descriptors of data, meaning that they measure the average spatial dependence across a region. Interesting spatial variation may well exist in the degree of spatial dependence in the data, which are obscured by globally averaged measures. Through the computation of local measures of spatial autocorrelation, such as those proposed by Anselin (1995), we can measure spatial variation at any level of spatial dependence of a field. Thus, parallel to developments in local geostatistical models, research exists on local models of spatial autocorrelation. Such methods share a common goal of identifying subsets whose characteristics are statistically significant in some way (Boots and Okabe 2007). Putting this in the context of scale modeling, locally defined scale models are more valuable than global ones in being able to capture spatial variability of scales over the extent of a problem domain D, as for the geostatistical setting discussed in Chapter 4.

This chapter will describe the key characteristics of lattice data, especially how they are different from geostatistical data so that they necessitate special treatment. The description is based on, but with extensions to, the field and object models described in Chapter 2. We then discuss quantification of spatial autocorrelation in lattice data, with Moran's I and Geary's c introduced. Local indicators of spatial autocorrelation, which can accommodate spatial heterogeneity in the strength of spatial autocorrelation in a region, are then described. Finally, we discuss how geostatistics and statistics of lattice data may be explored usefully for modeling scales in spatial data that are increasingly heterogeneous in terms of sources and semantics (Gelfand et al. 2001; Evans and Kelley 2004; White et al. 2009).

5.1 LATTICE DATA

As mentioned previously, unlike geostatistical data that are indexed over continuous domains, lattice data are indexed as finite sets of areal units, which are called cells or faces. We elaborate on the characteristics of lattice data in this section.

As a partition of the plane, a lattice samples a field with the resultant field values attached to individual faces. Commonly, there are three cases for the evaluation of the underlying field properties and hence the interpretation of lattice data semantics: (1) the values in lattice cells are integrals over the field, such as population count being an integral of population density (Moon and Farmer 2001); (2) the value in each cell of the lattice is the mean value of the field (e.g., population density, housing density, per-capita income, etc.); (3) the value in a lattice cell is the class label assumed to be constant over the cell (e.g., soil type, land cover, etc.). Cases 1 and 2 are termed spatially extensive and spatially intensive, respectively.

Lattice data can be sampled from continuous field domains with regularly or irregularly shaped cells (Cressie 1993; Schabenberger and Pierce 2002), as mentioned in the previous section. A *regular* lattice is formed if all of the cells have the same form and size, which is usually obtained if a region is divided into a regular grid of cells (although regularity of the cells' areal extents should account for the Earth's curvature when working with lattice data over the globe or large regions, as discussed in Chapter 3). If a region is divided into cells based on natural regionalization, such as plant community (Kent et al. 1997) and river basins (Alexander et al. 2002), or human geographies, such as administrative boundaries, census tracts, reporting zones, and postal codes (Grubesic 2008; Werneck 2008), an *irregular* lattice results, as discussed by Saveliev et al. (2007). For irregular lattices, *bona fide* and *fiat* boundaries are employed for framing the lattice faces for the former and the latter, respectively (Smith and Varzi 2000; Smith and Mark 2001).

A regular lattice consists of a grid of cells, usually square, although other arrangements are possible, such as triangles and hexagons. Regular lattices are common in the analysis of image data and mathematical modeling of spatially autocorrelated processes, as implied previously. There is a large literature on regular lattice data analysis (e.g., Besag 1974; Dubes and Jain 1989; to name a few). A typical example of spatial data arranged in a regular lattice pattern is remotely sensed satellite sensor images (Townshend and Justice 1988; Marceau and Hay 1999; McRoberts 2006), where the region of interest is discretized as a finite collection of locations (pixels) with observations of reflected or emitted electromagnetic radiation recorded, as shown in Chapter 3 where remote measurements were discussed. Remote sensing can be used for inventory of natural resources (such as crop yields and timber volumes), monitoring the effects of deforestation or urbanization, observing the dynamics of snow and ice, and other applications through interpretation and analysis of the images captured in a discrete, digitized format, as described in Chapter 3. These image data can be collected at a wide range of spatial resolutions (e.g., fine-resolution [1 m and submeter] IKONOS and QuickBird images and coarse-resolution [56 km] AMSR-E 6.9 GHz measurements; Lobl 2001). A wide variety of statistical techniques and models are used to process satellite sensor images to extract information about topography, land cover, hydrology, and other properties, as also remarked by Saveliev et al. (2007).

On the other hand, irregular lattices of administrative units and reporting zones are used commonly in human-geography research. For example, most data used in epidemiologic and health research are irregular lattice data, such as rates of disease, cancer incidence, and socio-economic indicators measured at the county, neighborhood, or census-tract level (Johnson 2004; Werneck 2008). A simple way to display lattice data is to use a *choropleth map*, which shows regions or areas by single values or integrals. For instance, in an epidemiological study of a particular cancer incidence in a city or a state, we could record the number of cancer cases in ZIP codes (areal units) and then look at the rates in each unit (number of cases over population in each unit) (e.g., Johnson 2004). Counties with similar rates will then be shaded the same color on a choropleth map. Thus, in epidemiologic studies, health statistics are often reported in the form of an irregular lattice.

In many biophysical studies, irregular lattice data are also common, as field data with explicit or implicit areal supports are collected at sites that occur in some irregular configurations. Also, some ecosystem variables, like species richness (Jetz and Rahbek 2002) and plant diversity (Borer et al. 2012), are often estimated over areal units through data aggregation (Aarts et al. 2012). Aggregation can be done in two ways: using natural spatial units like landscape patches or using arbitrary spatial units like counties or forest inventory stands (Saveliev et al. 2007). The effects of these two aggregation methods need to be examined (Goodchild 1980b).

As mentioned in Chapter 3 where the concept of data support was described, for data analysis, the *modifiable areal unit problem* (MAUP) (Openshaw and Taylor 1981) can result if arbitrary units are used for aggregation, since the size and shape of the areal units used for aggregation influence the correlation between the variables and may bias the inferences drawn from the correlation analysis, as discussed by Saveliev et al. (2007). For this reason, area units used in studies such as population health should be delineated so as to represent the *a priori* scale of the expected scale interactions in processes and forms, such as those between neighborhood processes and health. Moreover, a thorough understanding of the role of the MAUP in lattice data analyses is necessary for statistical modeling of disease processes and for decision making concerning the optimal scale of interventions in disease control, as discussed by Parenteau and Sawada (2011).

It is important that the size of the cells for aggregation match the process scale of the underlying field to guide a meaningful (dis)aggregation of lattice data. For this, scale modeling is essential. In Chapter 4, we described some geostatistical models that may be used to examine the effects of scale in lattice data of both regular (e.g., remotely sensed data) and irregular structures (e.g., biomass data derived from forest surveys and image analysis) (this issue will be revisited in the discussion section). Such models built on measures of spatial autocorrelation are necessary for describing spatial scale in lattice data and for analyzing their scale effects. In the next section, we will describe how spatial autocorrelation in lattice data is quantified and what links may be made between spatial autocorrelation and scale.

5.2 SPATIAL AUTOCORRELATION AND ITS MEASURES

Spatial autocorrelation means that the spatial variation of a field is not random but correlated as there is some structure in the spatial variability of the underlying distribution (Cliff and Ord 1970; Goodchild 1986; Choo and Walker 2008) and is synonymous to spatial dependence. Spatially autocorrelated data violate the assumptions usually implied in traditional statistics, which is oriented to nonspatial data without consideration of spatial dependence, which is, however, central to spatial data analysis (Haining 2003). As remarked by Haining (2009), the presence of spatial autocorrelation, when properly modeled, makes spatial interpolation possible, may increase the precision of small area estimates based on small samples, and helps to design efficient spatial sampling schemes. In addition, spatial autocorrelation is associated with the scale at which processes operate in space, providing the basis for scale modeling.

Spatial statistics includes descriptive statistics used to identify spatial structures or patterns (e.g., clustering, dispersion, or anomaly) in the data (and hence the underlying spatial distributions), and analytical methods are concerned with identifying and measuring the relationships among spatially distributed properties, as in spatial regression analysis (Fotheringham et al. 2002; Fortin et al. 2012), and were designed specifically for use with spatial data.

Geostatistics is a type of spatial statistics. Kriging, which we will cover in later chapters, is a powerful geostatistical technique that seeks to predict at unsampled locations by utilizing spatial autocorrelation and multivariate covariation (Cressie 1993; Goovaerts 1997). Thus, geostatistics has been used to analyze environmental data with variograms and to create surfaces from sampled datasets through kriging (Chapter 6 will describe kriging-based spatial prediction in more detail). Geostatistics has become sophisticated for analysis and prediction associated with nearly any kind of spatially distributed property, such as rainfall, soils, terrain, or vegetation.

For lattice data, extensions or alternatives to geostatistical modeling are usually necessary, as definitions of neighborhood and measures of proximity need to be adapted for them. Spatial autocorrelation can be interpreted as a descriptive index, measuring patterns of spatial distributions, which can also be seen as a causal process, measuring the degree of influence of locations over their neighbors (Goodchild 1986). To quantify spatial autocorrelation in lattice data, we need to introduce some suitable descriptors measuring the degree and scale of spatial dependence existing in the dataset under study, as follows.

We use the following notation: n denotes the number of sampled locations or lattice units, i, j any two of the units, z_i the value of the property of interest Z for i, c_{ij} the similarity of is and js properties, and w_{ij} the similarity of is and js locations ($w_{ii} = 0$ for all i), which is also known as a measure of spatial proximity, contiguity, or weight (Goodchild 1986).

A variety of approaches have been devised for measuring spatial proximity through an appropriate matrix of w_{ij}s (Griffith 1996; Getis 2009). For areas, a common boundary between areas is a simple, binary indicator of proximity. For example, one simple way of defining the w_{ij} for rasters is to assign a value of 1 to pairs of raster cells that share a common boundary and 0 otherwise. Sometimes, pairs of cells that join at a corner are also defined as adjacent and given a value of 1. Thus, with a square raster, we can define the four-neighbor and eight-neighbor cases, or rook's case and Queen's case, respectively, as described by Goodchild (1986).

The length of the common boundary (Goodchild 1980b) is another possible definition of w_{ij} for areas, which is often a better representation of the proximity of two areas than simple adjacency. Another option is to represent each area by a control point and to measure the distances between points. One might use the geographic centroid as a control point (Bracken and Martin 1989), which is easy to calculate for an area represented as a polygon with known vertices, although with the disadvantage that the centroid is sometimes located outside the area (Goodchild 1986). With distances between lattice cells computed, the weights can be calculated using some suitable functions, such as a negative power, $w_{ij} = d_{ij}^{-b}$, or a negative exponential, $w_{ij} = \exp(-bd_{ij})$, where b is a parameter that affects the rate at which weight declines with distance (Goodchild 1986). It may, in some applications, be appropriate to have a measure of proximity based on an index of social or economic interaction, rather than on distance (Gatrell 1979; Goodchild 1986), although this is not elaborated in this chapter.

A variety of approaches have also been devised to measure the similarity of z values, c_{ij}, depending on the types of properties involved. For nominal data, the usual approach is to set c_{ij} to 1 if i and j have the same value and to 0 otherwise. For ordinal data, similarity is usually based on comparing the ranks of i and j, whereas for interval data both the squared difference $(z_i - z_j)^2$ and the product $(z_i - \overline{z})(z_j - \overline{z})$ are commonly used, as described by Goodchild (1986). In the next paragraphs, we review Geary's c and Moran's I indices for quantifying spatial autocorrelation possibly present in a dataset and hence the underlying spatial distribution (Cliff and Ord 1970).

Geary's (1954) index is a measure of spatial autocorrelation for areas and interval/ratio data and, as such, has found its most suitable applications in human geography in the analysis of data assembled for statistical reporting zones (Goodchild 1986). Property Zs similarity c_{ij} is calculated from the squared difference between z values, that is,

$$c_{ij} = (z_i - z_j)^2 \tag{5.1}$$

Locational similarity may be measured in a binary fashion: w_{ij} being given the value of 1 if i and j share a common boundary and 0 otherwise. Property similarity and locational similarity as defined above are used to construct Geary's index c:

$$c = \sum_i \sum_j w_{ij} c_{ij} / 2s^2 \sum_i \sum_j w_{ij} \tag{5.2}$$

where $s^2 = \sum_i (z_i - \bar{z})^2/(n-1)$ denotes the sample variance, which is incorporated for normalizing the range of c, and $\bar{z}$ denotes the mean of z values (Goodchild 1986). Geary's c index compares the differences between neighboring cells to provide a measure of dissimilarity. Greater values of c correspond to larger values of w_{ij} (indicating pairs of areas in contact) and to larger values of c_{ij} (i.e., larger differences in values). In terms of interpretation, Geary's index c has a value of 1 when properties are distributed independently of location, dropping below 1 when similar Z values coincide with similar locations (i.e., when there is positive spatial autocorrelation), and above 1 otherwise (Goodchild 1986).

Moran's (1950) index, which compares the differences between neighboring cells and the mean to provide a measure of local homogeneity, provides an alternative to Geary's index for measuring spatial autocorrelation. The property-similarity measure used by Moran's index is analogous to a covariance between the values of a pair of locations:

$$c_{ij} = (z_i - \bar{z})(z_j - \bar{z}) \tag{5.3}$$

where $\bar{z}$ denotes the mean of the variable, as before, and c_{ij} measures the covariance between the value of a variable at one place and its value at another. Moran's index I is defined on the basis of c_{ij} as

$$I = \sum_i \sum_j w_{ij} c_{ij} / s^2 \sum_i \sum_j w_{ij} \tag{5.4}$$

where $s^2 = \sum_i (z_i - \bar{z})^2/(n-1)$ denotes the sample variance, and the w_{ij} terms represent the spatial proximity of locations i and j, which can be calculated in any suitable way as discussed previously. Moran's index is positive when nearby areas tend to be similar in Z values, negative when they tend to be more dissimilar than one might expect, and approximately zero when Z values are arranged randomly and independently in space (Cliff and Ord 1973; Goodchild 1986).

Consider the issue of hypothesis testing about the spatial autocorrelation in a sample of n cases with a null hypothesis that the sample was drawn from a population without autocorrelation. As reviewed by Goodchild (1986), there are two different interpretations of the null hypothesis: the randomization null hypothesis that the sample is one possibility chosen randomly from among the $n!$ possible arrangements of the observed Z values among the n locations, and the resampling null hypothesis that each of the n values is drawn independently from a population of normally distributed values. Both tests can follow the format of statistical hypothesis testing. We can compute the probability p that the null hypothesis could yield results as extreme as those observed, which can also be interpreted as the probability that the null hypothesis is true. If this probability is small, less than, say, 5%, the null hypothesis is rejected and we conclude that, indeed, spatial autocorrelation is present in the population (Goodchild 1986).

The probability p is relatively easy to calculate because the sampling distributions of both I and c are asymptotically normal under both null hypotheses, as proved by Cliff and Ord (1973, 1981). Below, we use the subscripts RSP and RND to denote resampling and randomization null hypotheses, respectively. According to Cliff and Ord (1969, 1970) and Goodchild (1986), the equations for calculating the expected value and variance of I for samples of size n under each null hypothesis are

$$E_{\text{RSP}}(I) = E_{\text{RND}}(I) = -1/(n-1) \tag{5.5}$$

$$\text{var}_{\text{RSP}} = \left(n^2 s_1 - n s_2 + 3 s_0^2\right)/s_0^2 (n^2 - 1) \tag{5.6a}$$

$$\text{var}_{\text{RND}} = \left\{ n\left[(n^2 - 3n + 3)s_1 - n s_2 + 3 s_0^2\right] \right.$$
$$\left. - b_2 \left[(n^2 - n)s_1 - 2 n s_2 + 6 s_0^2\right] \right\}/s_0^2 (n-1)^{(3)} \tag{5.6b}$$

where $s_0 = \sum_i \sum_j w_{ij}$, $s_1 = \sum_i \sum_j (w_{ij} + w_{jt})^2$, $s_2 = \sum_i \left[\sum_k w_{ik} + \sum_k w_{ki}\right]^2$, b_2 is the sample kurtosis coefficient (having an expected value of 0 under the resampling null hypothesis implying normality in distribution), $m_4 = \sum_i [z_i - z]^4/n$, $m_2 = \sum_i [z_i - \bar{z}]^2/n$, and $(n-1)^{(3)} = (n-1)(n-2)(n-3)$. On the basis of the appropriately calculated values for expectation and variance, the test statistic is

$$Z\_\text{test}(I) = [I - E[I]]/\sqrt{\text{var}[I]} \tag{5.7}$$

Although autocorrelation statistics described above provide an indication of the local (in)homogeneity of a dataset, it is sometimes interesting to understand how that autocorrelation decreases as distance increases. One can examine this using the correlogram (Cliff and Ord 1981), which plots an autocorrelation statistic calculated at various lag distances displayed against the lag. In other words, we can plot Moran's I or Geary's c across increasing discrete lag distance classes, as discussed below.

The concept of scale is implicit both in the definition of lattice cells, particularly if these are arbitrarily defined, and in measures of spatial autocorrelation, such as Moran's I and Geary's c statistics. This suggests that it would be useful to make scale explicit in the measurement of spatial autocorrelation (Goodchild 1986). This can be done through adjacencies as in the original Moran and Geary indices. We could define a second-order adjacency as one that exists between two nonadjacent areas separated by exactly one intervening area. In other words, two areas i and j are second-order adjacent if there exists some third area k such that pairs i and k and pairs k and j (but not pairs i and j) are adjacent. The same idea can be extended up to any order of adjacency. By measuring autocorrelation at each of these levels, one can construct a correlogram showing the performance of the index across scale (Goodchild 1986).

A plot of the Moran index with scale is referred to as a correlogram, which plots an autocorrelation statistic calculated at various lag distances displayed against the lag, as mentioned above. Similar to the use of variograms for spatial analysis, correlograms are useful for quantifying and analyzing spatially autocorrelated distributions. They are clearly straightforward with rasters and irregularly placed point samples as in geostatistics. On the other hand, use of correlograms with irregular lattice data must consider the special treatments that might be needed to handle arbitrariness in lattice cell partitions.

We now turn to ordinal variables where the zs are set equal to an integer indicating the ordinal class. The similarity measures c_{ij} can be based on the absolute difference in ranks between each pair (Goodchild 1986). The ordinal data can also be treated as if they were nominal, as described below. For nominal data, as a pair of lattice cells can either be the same or different, the c_{ij}, which measure the similarity between Zs values at locations i and j, can take one of only two values. The spatial proximity measures w_{ij} are binary: two raster cells are joined or not (one or zero) if they do or do not share a common boundary. Most of the measures devised for nominal data are based on join count statistics: the join count between color k and color l is defined as the number of times a cell of color k is joined to a cell of color l, as described by Goodchild (1986). Measures of spatial autocorrelation can then be devised based on join count statistics, which are reduced forms of Moran's I for indicators. In terms of interpretation, a higher than expected (in a random distribution of colors) incidence of joins of the same colors indicates positive autocorrelation, whereas a higher than expected incidence of joins between different colors indicates negative autocorrelation (Goodchild 1986).

5.3 LOCAL MODELS

We have described two types of spatial autocorrelation statistics known as Moran's I and Geary's c, which are global in the sense that they indicate the overall patterns between the proximity of lattice cells and the similarity of Z values under study. In other words, such global statistics describe the spatial autocorrelation in the field under study as a whole.

Consider the mean rainfall measured across China in spring 2011. This is a global statistic as it summarizes what is happening across the country in a single value. The localized variants of this statistic could be a local mean rainfall value of subregions (e.g., individual lattice cells), down to the individual rain gauge measurements themselves across the country. Such local statistics are often represented by surfaces, whereas global statistics are single-valued; the latter are spatial aggregations of the former, as discussed by Fotheringham (2009). This argument about global vs. local statistics is also applicable to air temperature in China in summer 2013 when much of southeastern China, in particular Zhejiang Province, endured an unusually hot summer, whereas for residents elsewhere the summer was just typical of the season.

Local models and methods have become increasingly popular in spatial analysis and geography (Getis and Ord 1996; Fotheringham 1997; Unwin and Unwin 1998; Karlström and Ceccato 2002; Boots 2003; Jetz et al. 2005). Their development has come, in part, in response to the recognition that, for many study regions, it is inadequate to summarize the data with a single global value (of the underlying property Z or a statistic computed for Z) or to use a single set of global values for parameterizing process models over the study area. Moreover, for a spatially heterogeneous domain, global values will not be applicable throughout the domain or to any part of it, as reviewed by Boots and Okabe (2007). With local values computed for subregions, they can be used for examining spatial variation in the local statistic computed via a moving window in the study area, assessing the appropriateness of stationarity, identifying the existence and extent of subregions of distinctive data values, and illustrating how spatial autocorrelation (strengths and ranges) varies spatially (Getis and Ord 1996; Boots and Okabe 2007).

As reviewed by Boots and Okabe (2007), some of the local measures of spatial autocorrelation or spatial association take the form of scaled-down versions of their global measures. These include local Moran's I and local Geary's c for continuous data (Anselin 1995) and local composition and join counts for categorical data (Boots 2002, 2003). There are also other measures, such as the G statistics of Getis and Ord (1992), which measures local clustering of high and low values. All these measures are typically calculated for local neighborhoods associated with individual locations (e.g., lattice cells). Often, these local statistics can be computed and output as additional map layers that can be used for assessing the appropriateness of stationarity and for identifying the scale (spatial extent) at which there is no discernible association of data values (Boots 2003; Boots and Okabe 2007).

The Gamma or Γ index provides a general framework for discussing models of spatial autocorrelation, as originally outlined by Mantel (1967). Such an index consists of the sum of the cross products of the matching elements w_{ij} and c'_{ij} in two matrices of similarity, say, $\mathbf{W}$ and $\mathbf{C}'$, such that

$$\Gamma = \sum_i \sum_j w_{ij} c'_{ij} \tag{5.8}$$

as reviewed by Anselin (1995).

Measures of spatial association are obtained by expressing spatial similarity in one matrix (e.g., a contiguity or spatial weights matrix) and Z value similarity in the other, as shown in the preceding section where Moran's I and Geary's c indices were described. Different indices for spatial association are defined using different measures of value similarity: $c'_{ij} = (z_i - \bar{z})(z_j - \bar{z})$ for Moran-like measures and $c'_{ij} = (z_i - z_j)^2$ for a Geary-like index (Anselin 1995; Getis and Ord 1996).

A local Gamma index for a location i may be defined as

$$\Gamma_i = \sum_j w_{ij} c'_{ij} \tag{5.9}$$

As for the global measure of spatial association Γ in Equation 5.8, different measures of value similarity will yield different indices of local spatial association. It is easy to see that the Γ_i statistics sum to the global measure Γ (Anselin 1995).

As a special case of the local Γ, a local Moran statistic for location i may be defined as

$$I_i = z_i \sum_j w_{ij} z_j \tag{5.10}$$

where z_i and z_j are deviations of Z values at locations i and j, respectively, from the mean, and j refers to only those cells neighboring cell i. For ease of interpretation, the weights w_{ij} may be in row-standardized form, although this is not necessary, and by convention, $w_{ii} = 0$ (Anselin 1995).

It can be seen readily that, for local Moran's I_i, the corresponding global statistic is equivalent to Moran's I by a factor of proportionality, as follows. The sum of local Moran's I_i is

$$\sum_i I_i = \sum_i z_i \sum_j w_{ij} z_j \tag{5.11a}$$

whereas Moran's I is

$$I = (n/s_0) \sum_i \sum_j w_{ij} z_i z_j \Big/ \sum_i z_i^2 \tag{5.11b}$$

where $s_0 = \sum_i \sum_j w_{ij}$. The equivalence of Equations 5.11a and 5.11b is established through the factor of proportionality between the sum of the local and the global Moran indices:

$$\lambda_I = s_0 m_2 \tag{5.12}$$

where $m_2 = \sum_i z_i^2 / n$ is the second moment for Z (Anselin 1995). Note that for a row-standardized spatial weights matrix, $s_0 = n$, so that $\lambda_I = \sum_i z_i^2$, and for standardized Z, $m_2 = 1$, so that $\lambda_I = s_0$.

As discussed by Anselin (1995), the moments for I_i under the null hypothesis of no spatial association can be derived using the principles outlined by Cliff and Ord (1981, pp. 42–46). For example, for a randomization hypothesis, the expected value of I_i is

$$E[I_i] = -w_i/(n-1) \qquad (5.13a)$$

where w_i is the sum of the row elements, $\sum_j w_{ij}$, and the variance of I_i is calculated as

$$\text{var}[I_i] = w_{i(2)}(n-b_2)/(n-1) + 2w_{i(kh)}(2b_2-n)/[(n-1)(n-2)] + w_i^2/(n-1)^2 \qquad (5.13b)$$

where $b_2 = m_4/m_2^2$, with $m_4 = \sum_i z_i^4/n$ being the fourth moment of Z, $w_{i(2)} = \sum_{j \neq i} w_{ij}^2$, and $2w_{i(kh)} = \sum_{k \neq i}\sum_{h \neq i} w_{ik}w_{ih}$ (Anselin 1995; Getis and Ord 1996).

A test for significant local spatial association may be based on these moments, although the exact distribution of such a statistic is still unknown. For example, a test on I_i may be based on the normal distribution, as in Equation 5.7:

$$Z\_\text{test}(I_i) = [I_i - E[I_i]]/\sqrt{\text{var}[I_i]} \qquad (5.14)$$

as described by Getis and Ord (1996).

Similarly, a local Geary statistic for Z at location i may be defined as

$$c_i = \sum_j w_{ij}(z_i - z_j)^2 \qquad (5.15a)$$

or as

$$c_i = (1/m_2)\sum_j w_{ij}(z_i - z_j)^2, \qquad (5.15b)$$

where $m_2 = \sum_i z_i^2/n$ as in Equation 5.12. Using Equation 5.15b, we can sum the c_i over all locations:

$$\sum_i c_i = n\left[\sum_i\sum_j w_{ij}(z_i - z_j)^2 \Big/ \sum_i z_i^2\right] \qquad (5.16a)$$

In comparison, Geary's c statistic (global) is rewritten as

$$c = [(n-1)/2s_0]\left[\sum_i\sum_j w_{ij}(z_i - z_j)^2 \Big/ \sum_i z_i^2\right] \qquad (5.16b)$$

Thus, the factor of proportionality between the sum of the local and the global Geary statistics is

$$\lambda_c = 2ns_0/(n-1) \tag{5.17}$$

as described by Anselin (1995).

Consider a domain subdivided into n lattice cells, where each is identified with a point with known Cartesian coordinates, say, x_i $(i = 1, 2, ..., n)$, and associated with it is a value z_i (a weight) taken from a variable Z. The $G_i(d)$ statistic proposed by Getis and Ord (1992) measures the degree of association or the spatial concentration of the sum of z values associated with the j points within d of the ith point. The $G_i(d)$ statistic for an observation z_i is defined as

$$G_i(d) = \sum_j w_{ij}(d) z_j \Big/ \sum_j z_j \tag{5.18}$$

where $w_{ij}(d)$ is 1 for links defined as being within distance d of a given i and 0 if not (including the link of point i to itself). When used in conjunction with a statistic such as Moran's I, they provide better clues to the processes that give rise to the spatial patterns that have been observed and can detect local "pockets" of dependence that may not show up with global statistics, as shown by Getis and Ord (1992) in their empirical studies of sudden infant death syndrome by county in North Carolina and of house prices by ZIP-code district in the San Diego metropolitan area.

5.4   DISCUSSION

We have reviewed spatial autocorrelation in lattice data, which are sampled over finite cells partitioning continuous domains and, thus, can be considered as area aggregations of geostatistical data that are point referenced or can be treated as being so. Both global and local measures of spatial autocorrelation have been described and can be employed to examine the spatial variability of lattice-sampled fields throughout the problem domain or, locally, around individual cells, respectively. Proper use of correlograms provides valuable information about the ranges and strengths of spatial autocorrelation in the underlying fields and can thus be used to elucidate models of scale concerning the lattice processes and the data drawn from them.

In Section 5.1, a typology of lattice data was outlined such that there are spatially extensive and spatially intensive lattices corresponding to interval/ratio fields and categorical lattices recording class labels assuming homogeneity in individual cells. To frame further discussion about scale modeling, we classify lattice data in terms of their compatibility for geostatistical descriptions, as follows.

Lattice data, such as remotely sensed data (regularly shaped) and areal data sampled or aggregated over naturally regionalized zones (with *bona fide* boundaries), can be analyzed with respect to spatial dependence by using geostatistics through the use of finite blocks. These are called type I lattice data. On the other hand, type II lattice data, such as those collected over arbitrarily demarcated reporting zones (with *fiat* boundaries), are not directly amenable to geostatistical handling. This is because, in type II lattice data, lags over which spatial autocorrelation is quantified are not based on the Euclidean metric but on spatial adjacency, which is often not conformal to Euclidean distance. Therefore, for type II lattice data, spatial statistics, such as Moran's I and Geary's c indices, becomes particularly useful for analyzing the spatial autocorrelation and spatial scale inherent in the underlying fields, as spatial adjacency is effectively represented by spatial weight matrices. Even for type I lattice data, correlograms can be employed for scale modeling, not only complementing but also checking the appropriateness of the use of variograms for this purpose.

Although not elaborated in this chapter, the topic of scale change is extremely important for lattice data analysis and applications (Turner et al. 1989; Garcia-Gigorro and Saura 2005). As implied above, methods for regularization and deregularization of variograms, which are described in Chapter 4, can be applied to type I lattice data. For type II lattice data, regularization of spatial statistics should be based on proper weighting of spatial weight matrices in the aggregation of the underlying Z fields, leading to regularized correlograms. This process is similar to regularization of variograms, as described in Chapter 4. However, the process of de-regularization is far more complicated owing to the difficulty in deconvoluting lower-order spatial adjacency from its higher-order version (i.e., disaggregating already aggregated spatial weight matrices). Further research on de-regularization of correlograms will contribute to a better suite of tools for scale modeling and scale change for lattice data. Combined use and interactions of global and local indicators of spatial autocorrelation will add to the research agendas for scale description and scale change in lattice data (Jetz et al. 2005).

Scale may be changed on the basis of regularized and de-regularized correlograms for lattice data. For type I lattice data, the geostatistical methods for change of scale, to be described in Chapter 6, may be implemented. Methods for change of scale for gridded data (regular lattice data) will be the topic for Chapter 7, which may be usefully explored for type I lattice data. For type II lattice data, methods oriented to geostatistical data and gridded data are seldom applicable in a straightforward manner. There may be merits in enriching the semantics of type II lattice data by use of covariates that are correlated with the field under study and appear strongly regionalized. For instance, it is helpful to incorporate land-use information in the mapping and analysis of population distributions. Hierarchical models may also be found to be suitable for addressing scale issues in type II lattice data, which are often hierarchically structured (Wikle 2003; Johnson 2004). We need to develop specially devised methods for scale modeling and change of scale for such data.

In addition to geostatistical data and lattice data, there is yet another broad type of spatial data: point-process patterns. Point-process data concern locations of events, where the problem domain D is random, unlike the cases of geostatistical and lattice data where D is fixed. For example, the spatial locations of reported crimes or observed landslides are point patterns. There are two types of point patterns: (1) random points used to sample a field (geostatistics) and (2) points that may or may not be weighted, the result of a stochastic process, normally the entire population rather than a sample, and those conceptualized as discrete objects. There are metrics proposed for point-pattern analysis (Fortin et al. 2002), and some of the methods described in this and the previous chapters may shed light on how scale issues in point patterns should be addressed. However, because of the breadth of spatial statistics, a comprehensive description that covers the full spectrum of data types is beyond the scope of this book.

This and the previous chapters have reviewed geostatistical and autocorrelation-based modeling of scale, respectively. They provide the tools for modeling scale in geostatistical and lattice data. Scale models can be further exploited for scaling, which is essential for cross-scale information transfer and multi-scale spatial analyses and applications. The next two chapters discuss methods for scaling: Chapter 6 describes geostatistical methods, whereas Chapter 7 explores statistical and mechanistic approaches to scaling for gridded lattice data, such as remotely sensed images.

6 Geostatistical Methods for Scaling

Chapters 4 and 5 described variogram- and autocorrelation-based approaches to modeling scale, which are designed for geostatistical and lattice data, respectively. As mentioned toward the end of Chapter 5, this chapter seeks to describe geostatistical methods for scaling, whereas the next chapter addresses methods for scaling lattice data (e.g., various images or raster data).

Scaling, also known as scale transfer or scale transformation, usually refers to change of scale (spatial) in data or analysis concerning spatial processes and patterns, thus facilitating the translation of information between or across scales (Crow et al. 2005; Stoy et al. 2009; Brunsell and Anderson 2011). This definition of scaling goes far beyond scale invariance in power-law relations (e.g., functions) in which scaling by a constant leads to a proportionate scaling of the functions themselves (Blöschl and Sivapalan 1995; Bierkens et al. 2000). There are two kinds of scaling in terms of the relationships between the scale of the existing data (or analysis) and the required scale: (1) scaling up or upscaling, which is translating information from finer or higher resolution (smaller supports) to coarser or lower resolution (larger supports), and (2) scaling down or downscaling, which is translating information from coarser to finer resolution.

Upscaling is increasing the size of data support or decreasing the corresponding spatial resolution. In remote sensing, it is often necessary to relate point or quasi-point data (e.g., ground data with 1×1 m support) to area-support data (e.g., Landsat TM images with 30×30 m pixels). However, point data may not be compared directly to pixel values because of the scale mismatch, unless the underlying variation is homogeneous over a large area and can thus be assumed to be invariant across scales, or a sufficient number of point measurements can be made during the satellite overpass to enable point-to-area data aggregation and intercomparison (Liang et al. 2002). Therefore, upscaling from point measurements (or images of finer spatial resolution) to images at coarser spatial resolution is necessary and must be considered carefully (Liang 2003).

The choice of method to use for upscaling depends on the characteristics of the finer-resolution (smaller support) data. When finer-resolution data are on regular supports (i.e., regular lattice data, such as remote-sensing images) and abundant, upscaling to coarser-resolution counterparts may be performed by averaging finer-resolution data falling within coarser-resolution cells, although more sophisticated processing that takes care of sensor properties should be considered (Townshend and Justice 1988). On the other hand, consider the situation where the smaller-support data are irregular, but sparsely distributed. We may employ certain smart approaches to predict the values of a variable Z over larger supports (denoted Z_v) within which the point data fall, given smaller-support data. This kind of method for transferring information from finer to coarser resolution amounts to predicting the variable under study at an unsampled location of a larger support $Z_v(v)$ given samples of smaller supports at nearby locations; specifically, this is actually interpolation over supports larger than those of the samples and, because of the change of small to large supports, implies effects of spatial averaging. In this case, there will be uncertainty involved in the interpolated values, where the sampled data do not cover the large support completely, and uncertainty should be viewed as spatially varying depending on the local configuration of the samples, as interpolation at locations in the vicinity of dense sampled data will be more accurate than that with sparse local data. Therefore, interpolation-based methods for upscaling need to be designed carefully and evaluated with respect to accuracy.

It is known widely that spatial prediction is possible because of *spatial dependence* (i.e., the phenomenon that values of properties closer together tend to be more interdependent than those further away). As the unknown values at the unsampled location are dependent on the known values at sampled locations, the former can be predicted from the latter. Geostatistics provides a coherent framework for spatial prediction and uncertainty quantification, with the latter generated as a byproduct of the former, and both being important for research on scaling.

Geostatistical approaches for spatial prediction are known as kriging. For upscaling, we can apply block kriging, as will be described later in this chapter. Block kriging is chosen because it is optimal in the sense of minimum prediction variance and unbiasedness, and because it enables prediction over the larger supports automatically. Block-kriging-based upscaling is likely to be used where point sample data exist, and we would like to use the sample data (both within and outside each pixel) to predict values over larger supports (i.e., blocks). For example, block kriging may be used to predict the pixel values of remote-sensing images (e.g., Landsat TM images of 30 × 30 m resolution) given point samples (e.g., those with 1 × 1 m support), as mentioned previously, although the choice of image resolution depends on the spatial structure of the underlying field property concerned among other things (Woodcock and Strahler 1987; Townshend and Justice 1988). We obtain not only the kriging prediction, but also the pixel-wise kriging variance, which provides useful information on prediction uncertainty based on the spatial dependence of the point samples (or quasi-point supports in relative terms, to be more accurate) in relation to the individual pixels being predicted.

The other type of scaling concerns downscaling, which seeks to increase the spatial resolution of an existing dataset and, thus, has a goal opposite to that of upscaling. As for upscaling, kriging can also be usefully explored for downscaling. Block-to-point kriging, or area-to-point kriging in two-dimensional cases, provides a geostatistical solution to downscaling. The solution is general and can be applied to blocks of any size, shape, and orientation, which provides a potential solution to downscaling in both geostatistical and lattice data settings (the latter case will be discussed in more detail in Chapter 7). This kind of method depends on estimation of the point-to-point spatial covariance (i.e., variogram deconvolution to point support as described in Chapter 4), which may be difficult to implement in practice because regionalized variables are rarely measured over point supports and with adequate density.

Geostatistical inverse modeling is an adaptation of least-squares methods to spatially correlated quantities, such as those in groundwater systems, and has been used for estimating spatial patterns based on measurements (Kitanidis and Vomvoris 1983). In atmospheric remote sensing, for instance, geostatistical inversion can be applied to deconvolute the effects of atmospheric transport and recover source fluxes on the basis of atmospheric measurements (Michalak et al. 2004). Geostatistical inverse modeling is a Bayesian approach, as it is based on the principle of combining prior information with information supplied by available measurements. In the geostatistical approach, however, the prior information is not an initial estimate of variables for certain sampled locations. Instead, the prior information is in the form of a spatial or temporal correlation, which is prescribed as a prior covariance matrix for deviations from the mean, with the mean of the underlying distribution specified either as a constant or as a function of auxiliary variables (Kitanidis and Vomvoris 1983).

This chapter provides an overview of kriging, a branch of geostatistical methods for spatial prediction. Firstly, the kriging techniques to be discussed include simple, ordinary, and universal kriging (Journel and Huijbregts 1978; Isaaks and Srivastava 1989; Cressie 1993; Goovaerts 1997). The underlying principles of variants of kriging will be seen as unified in terms of implementing the criteria of best linear unbiased prediction and manipulating basic linear algebra. This unified description about kriging variants is important for various scaling purposes in which adaptive geostatistical implementation is required to accommodate data and scale heterogeneity. Secondly, we will describe indicator kriging, which is useful for spatial prediction of probabilities of class occurrences, whether the variables are originally categorical or continuous. Indicator kriging requires indicator-transform and indicator-variogram modeling before solving kriging equations, which are similar to those of interval/ratio fields. Indicator approaches can also be based on transition probability models, as described by Carle and Fogg (1996). Thirdly, block kriging will be described as a geostatistical strategy for upscaling. Then, this chapter will describe area-to-point kriging for downscaling. This is followed by geostatistical inverse modeling, which exists as a set of modified Bayesian approaches that utilize spatial dependence in deriving downscaled spatial distributions (Kitanidis 1996). Lastly, this chapter concludes with some discussion.

6.1 KRIGING

Spatial correlation allows the prediction of a property at unsampled locations from sample data. If samples were not correlated in space, the best linear unbiased estimate of $Z(x)$ would be the global mean $m_Z = E[Z]$. Kriging is a name for a spatial prediction technique that uses the variogram as a model of geospatial continuity and estimates unsampled locations on that basis (Journel and Huijbregts 1978; Goovaerts 1999). In the following descriptions about kriging variants, point or quasi-point supports are assumed.

The simplest form of kriging is termed simple kriging and consists of estimating the unsampled location say x as a linear combination of the neighboring data values, which is denoted as the set $\{z(x_\alpha), \alpha = 1, \ldots, n\}$, where x_α $(\alpha = 1, \ldots, n)$ stands for sampled locations (subscript α instead of i is used to avoid confusion with indicator data). Prediction with known stationary mean m_Z can be described as

$$\hat{z}(x) - m_Z = \sum_{\alpha=1}^{n} \lambda_\alpha (z(x_\alpha) - m_Z) \qquad (6.1)$$

where λ_α are termed the kriging weights, as described by Zhang and Yao (2008a). Kriging provides a single estimate $\hat{z}(x)$ of the true unknown value of $Z(x)$ at location x, and kriging weights are derived on the criteria of unbiasedness (in the mean) and minimum variance in kriging prediction $\hat{z}(x)$, as shown below where the prediction error $\hat{z}(x) - z(x)$ can be analyzed in terms of the mean and variance to ensure unbiasedness and minimum dispersion of the estimator (Journel and Huijbregts 1978).

The average prediction error is bound to be zero:

$$E[\hat{z}(x) - z(x)] = m_Z + \sum_{\alpha=1}^{n} \lambda_\alpha (E[z(x_\alpha)] - m_Z) - E[z(x_\alpha)] = 0 \qquad (6.2)$$

and the variance of the prediction error (i.e., the kriging variance) can be written as

$$\sigma_Z^2(x) = \text{var}[\hat{z}(x) - z(x)] = \text{var}\left[m_Z + \sum_{\alpha=1}^{n} \lambda_\alpha (z(x_\alpha) - m_Z) - z(x_\alpha) \right]$$

$$= \text{var}\left[\sum_{\alpha=1}^{n} \lambda_\alpha (z(x_\alpha) - m_Z) - (z(x) - m_Z) \right] \qquad (6.3)$$

To derive a reduced form for the kriging variance, some linear algebra is required. As shown by Zhang and Yao (2008a), denote an extended weight vector $\boldsymbol{\lambda}^*$ and an extended vector of regionalized variables at locations $\{x_1, \ldots, x_n, x\}$ $\mathbf{Z}^*$:

$$\boldsymbol{\lambda}^* = \begin{bmatrix} \boldsymbol{\lambda} \\ -1 \end{bmatrix} = \begin{bmatrix} \lambda_1 \\ \vdots \\ \lambda_n \\ -1 \end{bmatrix} \qquad (6.4)$$

$$\mathbf{Z}^* = \begin{bmatrix} \mathbf{Z}(\mathbf{x}_{(n)}) - \mathbf{m}_Z \\ z(x) - m_Z \end{bmatrix} = \begin{bmatrix} z(x_1) - m_Z \\ \vdots \\ z(x_n) - m_Z \\ z(x) - m_Z \end{bmatrix} \qquad (6.5)$$

Thus, the kriging variance formulated in Equation 6.3 can be rewritten as

$$\text{var}[\boldsymbol{\lambda}^{*\text{T}}\mathbf{Z}^{*}] = \begin{bmatrix} \boldsymbol{\lambda}^{\text{T}} & -1 \end{bmatrix} \begin{bmatrix} \mathbf{cov}(\mathbf{x}_{(n)}) & \mathbf{cov}(\mathbf{x}_{(n)}, x) \\ \mathbf{cov}(\mathbf{x}_{(n)}, x)^{\text{T}} & \text{var}[z(x)] \end{bmatrix} \begin{bmatrix} \boldsymbol{\lambda} \\ -1 \end{bmatrix} \tag{6.6}$$

$$= \boldsymbol{\lambda}^{\text{T}} \mathbf{cov}(\mathbf{x}_{(n)}) \boldsymbol{\lambda} - 2\boldsymbol{\lambda}^{\text{T}} \mathbf{cov}(\mathbf{x}_{(n)}, x) + \text{var}[z(x)]$$

where $\mathbf{cov}(\mathbf{x}_{(n)})$ and $\mathbf{cov}(\mathbf{x}_{(n)}, x)$ are the covariance matrix of the sample locations (i.e., $\mathbf{cov}(\mathbf{x}_{(n)}) = [c(x_\alpha, x_{\alpha'})]$ ($\alpha, \alpha' = 1, \ldots, n$)) and the covariance vector between sampled locations and the unsampled location x (i.e., $\mathbf{cov}(\mathbf{x}_{(n)}, x) = [c(x_\alpha, x)]$ ($\alpha = 1, \ldots, n$)), respectively, and $\mathbf{x}_{(n)}$ denotes the vector of sampled locations (i.e., $\mathbf{x}_{(n)} = [x_1, \ldots, x_n]^{\text{T}}$ with T being the transpose operator), as deduced also by Zhang and Yao (2008a).

To minimize the prediction variance, it is necessary to set n partial derivatives of the prediction variance to zero:

$$\frac{\partial \text{var}\left[\sigma_Z^2(x)\right]}{\partial \lambda} = 0 \tag{6.7}$$

which gives rise to the following simple kriging equation:

$$\mathbf{cov}(\mathbf{x}_{(n)})\boldsymbol{\lambda} - \mathbf{cov}(\mathbf{x}_{(n)}, x) = 0 \tag{6.8}$$

and hence the weight vector (with a length of n as shown in Equation 6.4):

$$\boldsymbol{\lambda} = \mathbf{cov}(\mathbf{x}_{(n)})^{-1}\mathbf{cov}(\mathbf{x}_{(n)}, x) \tag{6.9}$$

which constitutes the solution (unbiased and with minimum variance) to Equation 6.1 (Zhang and Yao 2008a).

Based on the weight vector, it is possible to obtain the simple kriging estimator:

$$\hat{z}(x) = m_Z + \sum_{\alpha=1}^{n} \lambda_\alpha(z(x_\alpha) - m_Z) = m_Z + \boldsymbol{\lambda}^{\text{T}}(\mathbf{Z}(\mathbf{x}_{(n)}) - m_Z) \tag{6.10}$$

and the minimized kriging variance:

$$\sigma_Z^2(x) = \text{var}[z(x)] - \boldsymbol{\lambda}^{\text{T}}\mathbf{cov}(\mathbf{x}_{(n)}, x) \tag{6.11}$$

When the stationary mean is unknown, it is necessary to perform estimation of $Z(x)$ through a linear combination of the sampled values $\{z(x_\alpha), \alpha = 1, ..., n\}$. Thus, the simple kriging estimation formulated by Equation 6.1 should be reformulated as

$$\hat{z}(x) = \sum_{\alpha=1}^{n} \lambda_\alpha z(x_\alpha) + \left(1 - \sum_{\alpha=1}^{n} \lambda_\alpha\right) m_Z \tag{6.12}$$

where the factor for m_Z should be imposed with a constraint of zero:

$$\sum_{\alpha=1}^{n} \lambda_\alpha - 1 = 0 \tag{6.13}$$

so that the effect of the unknown m_Z is cancelled out, resulting in the so-called ordinary kriging procedure. To streamline the algebraic treatment, it is helpful to construct a column vector of n 1s (Zhang and Yao 2008a):

$$\mathbf{1} = \begin{bmatrix} 1 \\ \vdots \\ 1 \end{bmatrix}_{n \times 1} \tag{6.14}$$

so that the constraint in Equation 6.13 is rewritten as

$$\boldsymbol{\lambda}^{\mathrm{T}} \mathbf{1} - 1 = 0 \tag{6.15}$$

The objective function for minimization of the ordinary kriging variance becomes

$$\phi = \boldsymbol{\lambda}^{\mathrm{T}} \mathbf{cov}(\mathbf{x}_{(n)})\boldsymbol{\lambda} - 2\boldsymbol{\lambda}^{\mathrm{T}} \mathbf{cov}(\mathbf{x}_{(n)}, x) + \mathrm{var}[z(x)] - 2v(\boldsymbol{\lambda}^{\mathrm{T}}\mathbf{1} - 1) \tag{6.16}$$

where v is the Lagrange multiplier. To solve for minimum variance, the objective function is differentiated against the vector of n weights $\{\lambda_\alpha, \alpha = 1, ..., n\}$ and the derivatives are set to zero:

$$\frac{\partial \phi}{\partial \lambda} = \mathbf{0} \tag{6.17}$$

where $\mathbf{0}$ is an $n \times 1$ vector of zeros (Zhang and Yao 2008a). Equation 6.17 can be written as

$$\mathbf{cov}(\mathbf{x}_{(n)})\boldsymbol{\lambda} - 1v = \mathbf{cov}(\mathbf{x}_{(n)}, x) \tag{6.18}$$

Furthermore, setting the partial derivative of ϕ with respect to the Lagrange multiplier v leads to

$$\frac{\partial \phi}{\partial v} = 0 \tag{6.19}$$

which gives rise to

$$\mathbf{1}^T \boldsymbol{\lambda} = 1. \tag{6.20}$$

Equations 6.18 and 6.20 can be combined to produce the following ordinary kriging equation system (Zhang and Yao 2008a):

$$\begin{bmatrix} \mathbf{cov}(\mathbf{x}_{(n)}) & \mathbf{1} \\ \mathbf{1}^T & 0 \end{bmatrix} \begin{bmatrix} \boldsymbol{\lambda} \\ -v \end{bmatrix} = \begin{bmatrix} \mathbf{cov}(\mathbf{x}_{(n)}, x) \\ 1 \end{bmatrix} \tag{6.21}$$

from which n weights $\{\lambda_\alpha, \alpha = 1, \ldots, n\}$ can be solved, so the estimator for $Z(x)$ is

$$\hat{z}(x) = \sum_{\alpha=1}^{n} \lambda_\alpha z(x_\alpha) \tag{6.22}$$

and the kriging variance is

$$\sigma_Z^2(x) = \text{var}[z(x)] - \boldsymbol{\lambda}^T \mathbf{cov}(\mathbf{x}_{(n)}, x) + v \tag{6.23}$$

Both simple and ordinary kriging assume stationary means for the problem domains. However, there may be evidence that the regionalized variable is better described as the sum of a nonstationary mean $m_Z(x)$ and a zero mean, but spatially correlated error $R(x)$ having covariance model $C_R(h) = \text{cov}(R(x), R(x + h))$. This prompts the development of nonstationary kriging or universal kriging, in which the residual at x is written as a linear combination of residuals evaluated at n sampled locations in the neighborhood:

$$\hat{z}(x) - m_Z(x) = \sum_{\alpha=1}^{n} \lambda_\alpha (z(x_\alpha) - m_Z(x_\alpha)) \tag{6.24}$$

where the local means are modeled as a trend surface:

$$m_Z(x) = \sum_{b=0}^{B} \beta_b f_b(x) = \mathbf{f}(x)^T \boldsymbol{\beta} \tag{6.25}$$

where the functions $\{f_b(x), b = 0, B\}$ of known forms constitute the following column vector:

$$\mathbf{f}(x) = \begin{bmatrix} f_0(x) \\ \vdots \\ f_B(x) \end{bmatrix} \quad (6.26)$$

for which a quadratic trend surface example may be written as

$$f_0(x) = 1 \quad f_1(x) = X \quad f_2(x) = Y \quad f_3(x) = X^2 \quad f_4(x) = Y^2 \quad f_5(x) = XY$$

where X and Y represent the coordinates of location x, as described by Zhang and Yao (2008a).

The basis functions $\mathbf{f}(x)$ defined in Equation 6.26 for local mean surfaces can be based on not only coordinates (X, Y) but also covariates $Z_{ED}(x)$ known as external drifts (Deutsch and Journel 1998; Bourennane and King 2003). As reviewed by Hudson and Wackernagel (1994), the external drift method, in which $\mathbf{f}(x)$ is substituted by covariates $Z_{ED}(x)$, originated in petroleum and gas exploration where a few accurately measured (expensive) borehole data $Z(x)$ needed to be combined with many fairly imprecise (but easily obtainable) seismic data $Z_{ED}(x)$ in order to map the top of a reservoir. This method has been applied in many subject areas, such as for mapping the log of transmissivity $Z(x)$ using the log of specific capacity as $Z_{ED}(x)$ (Ahmed and de Marsily 1987).

To develop universal kriging equations, we denote the trend coefficients by a vector $\boldsymbol{\beta}$

$$\boldsymbol{\beta} = \begin{bmatrix} \beta_0 \\ \vdots \\ \beta_B \end{bmatrix} \quad (6.27)$$

and a matrix of trend basis functions:

$$\mathbf{F}(\mathbf{x}_{(n)}) = \begin{bmatrix} f_0(x_1) & \cdots & f_0(x_n) \\ \vdots & \ddots & \vdots \\ f_B(x_1) & \cdots & f_B(x_n) \end{bmatrix} \quad (6.28)$$

By expressing $m_Z(x)$ and $m_Z(x_\alpha)$ $(\alpha = 1, \ldots, n)$ in terms of $\boldsymbol{\beta}$ and $\mathbf{F}(\mathbf{x}_{(n)})$, substituting them in Equation 6.24, and re-arranging, we get

$$\hat{z}(x) = \boldsymbol{\lambda}^T \mathbf{z}(\mathbf{x}_{(n)}) + (\mathbf{f}(x)^T - \boldsymbol{\lambda}^T \mathbf{F}(\mathbf{x}_{(n)})^T)\boldsymbol{\beta} \quad (6.29)$$

To account for unknown $\boldsymbol{\beta}$, we need to set a constraint:

$$\mathbf{f}(x)^{\mathrm{T}} - \boldsymbol{\lambda}^{\mathrm{T}} \mathbf{F}(\mathbf{x}_{(n)})^{\mathrm{T}} = 0 \tag{6.30}$$

Following the logic in deriving the ordinary kriging equations, we can write the equation for universal kriging below:

$$\begin{bmatrix} \mathbf{cov}_R(\mathbf{x}_{(n)}) & \mathbf{F}(\mathbf{x}_{(n)})^{\mathrm{T}} \\ \mathbf{F}(\mathbf{x}_{(n)}) & 0 \end{bmatrix} \begin{bmatrix} \boldsymbol{\lambda} \\ -\psi \end{bmatrix} = \begin{bmatrix} \mathbf{cov}_R(\mathbf{x}_{(n)}, x) \\ \mathbf{f}(x) \end{bmatrix} \tag{6.31}$$

We can thus compute the predicted value $\hat{z}(x)$ for $Z(x)$ at x:

$$\hat{z}(x) = \sum_{\alpha=1}^{n} \lambda_\alpha z(x_\alpha) \tag{6.32}$$

and the universal kriging variance:

$$\sigma_Z^2(x) = \text{var}[R(x)] - \boldsymbol{\lambda}^{\mathrm{T}} \mathbf{cov}_R(\mathbf{x}_{(n)}, x) + \mathbf{F}(\mathbf{x}_{(n)})^{\mathrm{T}} \psi \tag{6.33}$$

as described by Zhang and Yao (2008a).

Hudson and Wackernagel (1994) applied universal kriging for incorporating the information about elevation into the mapping of temperature. Their experiments confirmed the increased accuracy of the map of January mean temperatures in Scotland, as compared with kriging based on temperature data alone. However, some exaggerated extrapolation effects were observed for a region in the Highlands, where stations are scarce and located at low altitudes. In addition to linear models for incorporating internal (geospatial coordinates) and external drifts, mean surface modeling may also be based on nonlinear methods. To reconstruct missing data in pixel values in MODIS image time series due to cloud cover or other image noise, Poggio et al. (2012) presented and tested a hybrid generalized additive model–geostatistical space–time model, which includes the fitting of a smoother spatio-temporal trend and a spatial component to account for local details supported by information in covariates. The proposed method is not limited by the type of noise or degradation of pixel values, latitude, vegetation dynamics, and land uses, and is able to reproduce spatial patterns and local features of the measured products in the reconstructed images, even with substantial numbers of missing pixels.

Variants of kriging have been described neatly by applying the optimum criteria of best linear unbiased estimation and linear algebra. Their applications for spatial prediction in a few subject areas, such as hydrogeology, climatology, and environmental remote sensing, have also been reviewed, although only concisely. Indicator geostatistics for predicting the occurrences of categorical variables, such as certain soil types or classes of originally continuous variables, will be discussed in the next section.

6.2 INDICATOR APPROACHES

There are various occasions where simple, ordinary, and universal kriging techniques may be incapable of performing the desired spatial prediction task. One type of application concerns quantification of cumulative distribution functions (CDFs), or the probabilities of occurrences of categorical variables over space (Zhang and Yao 2008b). For instance, it is often necessary in environmental studies to evaluate probabilistically the risk of contamination exceeding certain thresholds (Goovaerts 1999) and the risk of plant (e.g., cotton fields) exposures to pathogens (e.g., southern root-knot nematode) (Ortiz et al. 2010). Such events can be conceived of as indicators of success or failure, positive or negative, when tested against some criteria. To perform probabilistic mapping, conventional kriging assumes a normal distribution at each prediction location. On the contrary, indicator kriging can be used to predict probabilities of occurrences of events (categorical variables); it makes no assumption of normality and is thus a nonparametric technique (Journel 1983; Brown et al. 2002; Zhang and Yao 2008b; Figueira et al. 2009). This section extends geostatistics from the realm of kriging with continuous variables to indicator kriging, which provides a nonparametric solution to probability mapping.

As for conventional kriging, indicator kriging performs prediction of probabilities based on spatial structure information in indicator-transformed data in the neighborhood of an unsampled location (Goodchild et al. 2009). For variables $C(x)$, originally of discrete nature, an indicator transform can be performed as in Equation 4.13a. There will be K indicator variables for K classes. When working with continuous variables, indicator kriging requires a series of threshold values between the smallest and largest data values in the dataset. These threshold values, referred to here as *cutoffs* (say there are $K-1$ of them for K interval classes), are used to build CDFs. For each cutoff, data in the neighborhood are transformed into 0s and 1s: 0s if the data are greater than the threshold and 1s otherwise, that is, $I_k(x) = 1$ when $Z(x) \le z_k$ ($I_k(x) = 0$ otherwise), as shown in Equation 4.13b.

A spatial measure of connectivity between any two points in space separated by a lag distance h is the indicator variogram (Journel 1983; Western et al. 1998), which is defined as $C_k(h) = E[I_k(x) - p_k(x))(I_k(x + h) - p_k(x + h))]$, where $p_k(x)$ and $p_k(x + h)$ are the probabilities of class occurrences for class k at locations x and $x + h$, respectively, as in Equation 4.14b. $C_k(h)$ can be modeled from the sample data by transforming each observation into a vector of size K of indicator data (zeros and ones).

Once the K indicator variograms (for K classes) are determined, one can perform indicator kriging to determine local probability, as it can be shown that

$$E[I_k(x)] = \Pr(I_k(x) = 1) \times 1 + \Pr(I_k(x) = 0) \times 0$$

$$= \Pr(I_k(x) = 1) \tag{6.34}$$

$$= \Pr(Z(x) \le z_k)$$

where class k (a total of K classes) is considered. Hence, any estimation of $I_k(x)$ is also an estimation of $\Pr(Z(x) \le z_k)$, which means that kriging of an indicator variable is able to determine the local class probability for the categorical variable, or the uncertainty about the original Z variable (Zhang and Goodchild 2002), as also described by Zhang and Yao (2008b).

For example, to determine the probability of porosity at unsampled location x to be above z_k, given the local well data, one only needs to krige the indicator variable using indicator kriging; then that kriged value is also a probability. The indicator kriging estimate is written as

$$[I_k(x)]^* = [\Pr(Z(x) \le z_k \mid (n))]^* = p_k + \sum_{\alpha=1}^{n} \lambda_a(x)(i_k(x_\alpha) - p_k) \qquad (6.35)$$

where p_k is the estimated proportion of data above the cut-off z_k, $i_k(x_\alpha)$ is the indicator-transformed sample data, $\lambda_a(x)$ is the kriging weight assigned for the sample location x_α, and (n) is the set of conditional data (n sampled locations in the neighborhood for kriging). The indicator kriging weights are determined by solving the indicator kriging (simple mode) equations (a set for each of the K classes)

$$\sum_{a'=1}^{n} \lambda_a(x) C_k(x_a, x_{a'}) = C_k(x_a, x) \qquad \forall a = 1, \dots, n \qquad (6.36a)$$

which, according to Carle and Fogg (1996), can also be rewritten in terms of transition probability models as described in Equation 4.16b

$$\sum_{a'=1}^{n} \lambda_a(x) [tp_{kk}(x_a, x_{a'}) - p_k] = tp_{kk}(x_a, x) - p_k \qquad \forall a = 1, \dots, n \qquad (6.36b)$$

which needs to be solved for kriging weights for each of the K classes under consideration, as for Equation 6.36a. As transition probability models have some advantages over traditional indicator kriging methods (e.g., being a simpler and more intuitive framework for interpreting categorical variables), they will be valuable extensions for indicator geostatistics (Jones et al. 2005).

By estimating $\Pr(Z(x) \le z_k \mid (n))$ for various thresholds z_k, indicator kriging provides the uncertainty about the property at an unsampled location x given any sample data (n). Note that no Gaussian or any parametric distribution model is assumed; the uncertainty is quantified through a series of cumulative probabilities at given thresholds z_k (Goovaerts 1997; Zhang and Yao 2008b). As for continuous variables, in addition to simple kriging, ordinary kriging and simple kriging with local prior means may also be applied for modeling local probabilities of classes (Oberthür et al 1999; Ortiz et al. 2010), although this is not elaborated here.

This section has discussed indicator geostatistics as providing a nonparametric solution to probabilistic mapping and CDF modeling for categorical variables and continuous variables, respectively. Indicator geostatistics comes with costs. This concerns the difficulties in modeling and using indicator variograms for all classes. Conditional simulation may be better pursued for probabilistic mapping of categorical occurrences or discretized classes of intervals of continuous variables.

6.3   UPSCALING BY BLOCK KRIGING

With regularly sampled points or lattice data, upscaling may proceed with a simple process of averaging all the multiple data points or finer-resolution cell values falling within the individual cells of coarser resolution, as mentioned in the introductory section. For example, leaf area index (LAI) is one of the most sensitive determinants of plant production and can vary abruptly over short distances. The landscape distribution of LAI is generally derived from remotely sensed surface reflectance (e.g., normalized difference vegetation index (NDVI)). However, LAI upscaling is complicated by the mismatch in scales between ground and satellite sensor measurements (Williams et al. 2008). Working from a fine-scale harvest LAI-NDVI relationship, Williams et al. (2008) collected NDVI data over a 500 × 500 m catchment in the Swedish Arctic, at resolutions from 0.2 to 9.0 m in a nested sampling design; performed linear upscaling of NDVI (i.e., NDVI at any scale was a simple average of multiple NDVI measurements taken at finer scales); and furnished a single exponential LAI-NDVI relationship from 1.5 to 9.0 m of resolution.

With irregular and sparsely sampled data (of finer resolution), upscaling needs to explore spatial dependence in the underlying distributions and to use such structural information, in addition to that contained in the sampled data, in making spatial predictions over blocks (of coarser resolution). Block kriging is a method for estimation of the average value of a variable Z within a prescribed local area or volume (i.e., block) (Mason et al. 1994). One method is to discretize the block into many points and then average the individual point estimates to estimate the average over the block. This method is computationally expensive. A computationally more efficient strategy is based on constructing and solving only one kriging system for each block estimate, as we will see below.

Block kriging is similar to point kriging, as elaborated below. First, we make the assertion that the mean value of a random variable (RV) over a block v_x is simply the average (a linear combination) of all the n_p point RVs discretizing the block. So, we can conceptualize a RV $Z(v_x)$ as being the average of n_p point RVs $Z(x_\beta)$ ($\beta = 1, \ldots, n_p$), i.e.,

$$Z(v_x) = \frac{1}{n_p} \sum_{\beta=1}^{n_p} Z(x_\beta) \tag{6.37}$$

As in the derivation of simple kriging previously, we can derive simple block kriging by applying two constraints, that is, unbiasedness and minimum variance of the estimator (Journel and Huijbregts 1978). These two constraints need to be examined with respect to block supports in the context of block kriging, as follows.

Unbiasedness is met as in point simple kriging, as we assumed stationarity of the mean and covariance in the problem domain. The kriging variance to be minimized in Equation 6.6 is, however, adapted to the block support of the location v_x to be predicted:

$$\sigma_Z^2(v_x) = \lambda^T \mathbf{cov}(\mathbf{x}_{(n)}) \lambda - 2\lambda^T \mathbf{cov}(\mathbf{x}_{(n)}, v_x) + \text{var}[z(v_x)] \tag{6.38}$$

Point-to-block covariance vector $\mathbf{cov}(\mathbf{x}_{(n)}, v_x)$, which is required for block kriging, is evaluated as

$$\mathbf{cov}(\mathbf{x}_{(n)}, v_x) = \frac{1}{n_p} \sum_{\beta=1}^{n_p} \mathbf{cov}(\mathbf{x}_{(n)}, x_\beta) \tag{6.39}$$

Substituting $\mathbf{cov}(\mathbf{x}_{(n)}, v_x)$ for $\mathbf{cov}(\mathbf{x}_{(n)}, x)$ in Equation 6.8, we have a block kriging weight vector solved as

$$\lambda = \mathbf{cov}(\mathbf{x}_{(n)})^{-1} \mathbf{cov}(\mathbf{x}_{(n)}, v_x) \tag{6.40}$$

and the simple block kriging prediction:

$$\hat{z}(v_x) = m_Z + \lambda^T (\mathbf{Z}(\mathbf{x}_{(n)}) - m_Z) \tag{6.41}$$

and the block kriging variance:

$$\sigma_Z^2(x) = \text{var}[z(v_x)] - \lambda^T \mathbf{cov}(\mathbf{x}_{(n)}, v_x) \tag{6.42}$$

where $\text{var}[z(v_x)]$ represents the average covariance within the block being predicted, which is evaluated as

$$\text{var}[z(v_x)] = \frac{1}{n_p n_p} \sum_{\beta=1}^{n_p} \sum_{\beta'=1}^{n_p} \text{cov}(z(x_\beta), z(x_{\beta'})) \tag{6.43}$$

It is interesting to note the similarity between Equations 6.9 and 6.40, between Equations 6.10 and 6.41, and between Equations 6.11 and 6.42, if we generalize the point support x to block support v_x.

The notation of blocks above does not imply regular grid cells. Thus, block kriging can be applied to aggregate data from finer resolution to versions of coarser resolution, regardless of whether the source and target data supports are regular or irregular. This means that block kriging provides a flexible solution to upscaling.

Adaptations similar to those described above for simple kriging can be made to ordinary, universal, and indicator kriging to arrive at block implementations of these kriging variants (see Tercan and Dowd [1995] for change-of-support issues concerning indicator approaches). This will require a suitable evaluation of point-to-block covariances in relevant equations and is not elaborated here.

6.4   DOWNSCALING BY AREA-TO-POINT KRIGING

As discussed by Kyriakidis (2004), geostatistics provides a framework for area-to-point prediction, which is formulated as point prediction of a regionalized variable using available area-support data of the same or a different variable (Journel and Huijbregts 1978; Gotway and Young 2002). This is in contrast to block kriging (i.e., point-to-area interpolation, as described in the preceding section). In this section, we describe a general geostatistical framework for area-to-point interpolation (univariate), following Kyriakidis (2004).

Let $z(x)$ denote the true (unknown) point-support value at a location x within a problem domain D. In geostatistics, the set of all point-support values $\{z(x), x \in D\}$ is viewed as a joint realization of a collection of spatially correlated random variables $\{Z(x), x \in D\}$ (i.e., a RF).

Suppose there are N area-support data denoted by $z(v_a)$, $a = 1, \ldots, N$, where $v_a = v(x_a)$ denotes the support of the ath areal datum with centroid at location x_a. We can use $\mathbf{z}$, an $N \times 1$ vector, to denote the set of area-support data (i.e., $\mathbf{z} = [z(v_1), \ldots, z(v_N)]^T$). The measure (length in one dimension, area in two dimensions, volume in three dimensions) of that support is denoted as $|v_a|$. The ath observed areal datum $z(v_a)$ is defined as the average of all point values within support v_a:

$$z(v_a) = \frac{1}{|v_a|} \int_{x \in v_a} z(x)\,dx \approx \frac{1}{n_a} \sum_{j=1}^{n_a} z(x_j) \qquad (6.44)$$

where n_a denotes the number of points discretizing v_a, although other kernels may be prescribed for sampling areal data (Gotway and Young 2002; Kyriakidis 2004).

As in the case of point-support data, the N observed areal data $\mathbf{z} = [z(v_1), \ldots, z(v_N)]^T$ are also viewed as a joint realization of N RFs $\{Z(v_a), a = 1, \ldots, N\}$. The mean and covariance of the area-support RF $Z(v)$ can be estimated readily from those of the point-support $Z(x)$, provided the latter are known, as shown in Chapter 4. In what follows, as in Kyriakidis (2004), we assume knowledge of the stationary point covariance $C_Z(h)$ in order to derive all area-to-area and area-to-point covariance terms required by area-to-point kriging to be developed below.

The objective of area-to-point kriging is to predict a point value $z(x)$ at any unsampled location x, using the areal data $\mathbf{z}$ mentioned previously. The predicted point value $\hat{z}(x)$ is expressed as a weighted linear combination of the available areal data:

$$\hat{z}(x) = \sum_{a=1}^{N} \lambda(v_a) z(v_a) = \boldsymbol{\lambda}^T \mathbf{z} \qquad (6.45)$$

where $\lambda = [\lambda(v_1), \ldots \lambda(v_N)]^T$ denotes a location-specific $N \times 1$ vector of weights, whose ath entry $\lambda(v_a)$ is the weight assigned to the ath areal datum $z(v_a)$ for prediction at location x (Kyriakidis 2004). Since the stationary point mean m_z is unknown, one has to introduce a constraint on the weights to ensure unbiasedness in prediction $\hat{z}(x)$:

$$\sum_{a=1}^{N} \lambda(v_a) = 1 \qquad (6.46)$$

as was also the case for ordinary kriging described in Section 6.1.

As described by Kyriakidis (2004), the N weights $\lambda = [\lambda(v_1), \ldots, \lambda(v_N)]^T$ that yield the minimum prediction error variance with the constraint in Equation 6.46 are obtained through solution of the system of normal equations:

$$\begin{cases} \sum_{\alpha'=1}^{N} \lambda(v_{\alpha'}) C_Z(v_a, v_{\alpha'}) - \psi = C_Z(v_a, x) \\ \qquad\qquad\qquad\qquad\qquad\qquad\qquad \forall \quad a = 1, \ldots, N \\ \sum_{\alpha'=1}^{N} \lambda(v_{\alpha'}) = 1 \end{cases} \qquad (6.47)$$

where ψ denotes the Lagrange multiplier that accounts for the unit sum constraint on the weights. Equation 6.47 is similar in format to Equation 6.21, except for the need to accommodate block support in the former: covariance matrix $C_Z(v_a, v_{\alpha'})$ (between blocks) and vector $C_Z(v_a, x)$ (between blocks and unsampled location), as explained below.

$C_Z(v_a, v_b)$ in Equation 6.47 denotes the covariance between two areal RFs $Z(v_a)$ and $Z(v_b)$,

$$C_Z(v_a, v_b) = \frac{1}{|v_a \| v_b|} \int_{x \in v_a} \int_{x' \in v_b} \mathrm{cov}(Z(x), Z(x')) \, dx' \, dx \approx \frac{1}{n_a n_b} \sum_{j=1}^{n_a} \sum_{j'=1}^{n_b} \mathrm{cov}(x_j - x_{j'})$$

$$(6.48)$$

where $\mathrm{cov}(x_j - x_{j'})$ is the covariance between two points (n_a and n_b being their total numbers) discretizing supports v_a and v_b, respectively. Similarly, $C_Z(x, v_a)$ in Equation 6.47 denotes the covariance between any point RF $Z(x)$ and any areal RF $z(v_a)$:

$$C_Z(v_a, x) = \frac{1}{|v_a|} \int_{x' \in v_a} \mathrm{cov}(Z(x'), Z(x)) \, dx' \approx \frac{1}{n_a} \sum_{j=1}^{n_a} \mathrm{cov}(x_j - x) \qquad (6.49)$$

where $\mathrm{cov}(x_j - x)$ is the point covariance between the prediction location x and any location discretizing v_a (i.e., $x_j \in v_a$) (see Kyriakidis 2004 for details).

The resulting ordinary kriging variance (or minimum prediction error) at location x is

$$\sigma^2(x) = \text{cov}(0) - \sum_{a=1}^{N} \lambda(v_a) C_Z(v_a, x) + \psi \tag{6.50}$$

In the following paragraphs, we discuss the coherence property of the area-to-point predictions, as described by Kyriakidis (2004). In matrix format, the solution to Equation 6.47 can be written in vector and matrix format as

$$\begin{bmatrix} \lambda \\ -\psi \end{bmatrix} = \begin{bmatrix} \mathbf{C(V,V)} & \mathbf{1}_N \\ \mathbf{1}_N^T & 0 \end{bmatrix}^{-1} \begin{bmatrix} \mathbf{C(V,}x) \\ 1 \end{bmatrix} \tag{6.51}$$

where $C_Z(\mathbf{V}, \mathbf{V}) = [C_Z(v_a, v_{a'})]$ $(a, a' = 1, \ldots, N)$, $\mathbf{C(V}, x) = [C_Z(v_a, x)]$ $(a = 1, \ldots N)$, $\mathbf{V}$ is the set of N blocks $v_a(a = 1, \ldots, N)$, and $\mathbf{1}_N = [1, \ldots, 1]^T$ denotes an $N \times 1$ vector of unit entries.

The kriging prediction previously formulated in Equation 6.45 can be equivalently written as (Kyriakidis 2004):

$$\hat{z}(x) = \begin{bmatrix} \mathbf{z}^T & 0 \end{bmatrix} \begin{bmatrix} \lambda \\ -\psi \end{bmatrix} = \begin{bmatrix} \mathbf{z}^T & 0 \end{bmatrix} \begin{bmatrix} \mathbf{C(V,V)} & \mathbf{1}_N \\ \mathbf{1}_N^T & 0 \end{bmatrix}^{-1} \begin{bmatrix} \mathbf{C(V,}x) \\ 1 \end{bmatrix} \tag{6.52}$$

where $\mathbf{z}$ is an $N \times 1$ vector of area-support data and Equation 6.51 is employed for $\begin{bmatrix} \lambda \\ -\psi \end{bmatrix}$.

We can define an augmented weight vector after transposing the right-hand side of Equation 6.52 and extracting the product of the last two parts:

$$\begin{bmatrix} \mathbf{w} \\ \varphi \end{bmatrix} = \begin{bmatrix} \mathbf{C(V,V)} & \mathbf{1}_N \\ \mathbf{1}_N^T & 0 \end{bmatrix}^{-1} \begin{bmatrix} \mathbf{z} \\ 0 \end{bmatrix} \tag{6.53}$$

where $\mathbf{w}$ is an $N \times 1$ vector of weights and φ denotes a Lagrange multiplier (specific to location x). Thus, the downscaling prediction $\hat{z}(x)$ can be derived as

$$\hat{z}(x) = \begin{bmatrix} \mathbf{w}^T & \varphi \end{bmatrix} \begin{bmatrix} \mathbf{C(V,}x) \\ 1 \end{bmatrix} \tag{6.54}$$

Equation 6.54 indicates that the ath entry of $\mathbf{w}$ is the weight assigned to the area-to-point covariance between the ath support v_a and the prediction location x (Kyriakidis 2004). More importantly, the kriging system (Equation 6.54) allows an easy proof of the coherence of area-to-point predictions, as follows.

Consider now the areal average of the point predictions $\hat{z}(x_j)$ $(x_j \in v_a)$ within the support v_a of the ath areal datum $z(v_a)$. Applying Equation 6.54, and averaging predictions $\hat{z}(x_j)$ over the points discretizing v_a, we get

$$\frac{1}{n_a}\sum_{j=1}^{n_a}\hat{z}(x_j)=\left[\begin{array}{cc} \mathbf{w}^{\mathrm{T}} & \varphi \end{array}\right]\left[\begin{array}{c} \dfrac{1}{n_a}\sum_{j=1}^{n_a}\mathbf{C}(\mathbf{V},x_j) \\ 1 \end{array}\right]=\left[\begin{array}{cc} \mathbf{w}^{\mathrm{T}} & \varphi \end{array}\right]\left[\begin{array}{c} \mathbf{C}(\mathbf{V},v_a) \\ 1 \end{array}\right] \quad (6.55)$$

By left-multiplying $\left[\begin{array}{cc} \mathbf{C}(\mathbf{V},\mathbf{V}) & \mathbf{1}_N \\ \mathbf{1}_N^{\mathrm{T}} & 0 \end{array}\right]$ and transposing both sides of Equation 6.53, we get

$$\left[\begin{array}{c} \mathbf{z} \\ 0 \end{array}\right]^{\mathrm{T}}=\left[\begin{array}{cc} \mathbf{w}^{\mathrm{T}} & \varphi \end{array}\right]\left[\begin{array}{cc} \mathbf{C}(\mathbf{V},\mathbf{V}) & \mathbf{1}_N \\ \mathbf{1}_N^{\mathrm{T}} & 0 \end{array}\right] \quad (6.56)$$

Comparing Equation 6.55 to Equation 6.56, and noting that $\mathbf{C}(\mathbf{V}, v_a)$ is the ath column of the covariance matrix $\mathbf{C}(\mathbf{V}, \mathbf{V})$ and that $z(v_a)$ is the ath element of the areal data vector $\mathbf{z}$, we can confirm the equivalence between the $\hat{z}(x_i)$ averaged over v_a and the ath datum value $z(v_a)$:

$$\frac{1}{n_a}\sum_{j=1}^{n_a}\hat{z}(x_j)=z(v_a) \quad (6.57)$$

as discussed by Kyriakidis (2004).

This proves the coherence characteristic of area-to-point kriging: the mean of point kriging predictions within any areal datum support identifies the corresponding areal-average datum. Thus, for predicting point density values from areal density data, the area-to-point kriging predictions satisfy the pycnophylactic constraint of Tobler (1979), as discussed by Kyriakidis (2004). The coherence property of area-to-point kriging is independent of the point covariance model adopted and will hold provided that the area-to-area and area-to-point covariance values are derived *consistently* from the point covariance model (Kyriakidis 2004), as by Equations 6.48 and 6.49, respectively.

In addition to the coherence property, the area-to-point kriging technique described above possesses some attractive advantages over conventional area-to-point interpolation methods. As discussed by Kyriakidis (2004), the former explicitly accounts for the different data supports involved in downscaling with areal and point data, unlike the latter in which areal data are simply collapsed into their respective polygon centroids in interpolation algorithms. Also, in the former, uncertainty concerning the area-to-point predictions is quantified (Equation 6.50). There are many interesting developments in the literature on areal interpolation, such as incorporating ancillary data in areal interpolation (Hawley and Moellering 2005; Zhang and Qiu 2011), which may be usefully incorporated into area-to-point interpolation.

6.5 GEOSTATISTICAL INVERSE MODELING

Samples are typically sparse or inadequate for downscaling the observed and coarse-resolution RFs. Thus, downscaling is often underdetermined, an inverse problem. It is essential to explore structural information about the spatial dependence of unknown RFs to obtain stable solutions to this kind of inverse problem. A stochastic method for the solution of the inverse problem in geophysics was proposed, which is named geostatistical inversion (Kitanidis and Vomvoris 1983; Kitanidis 1996). Geostatistical inversion seeks to combine measurements and spatial structural information for a solution to the inverse problem, as described below.

Let **s** be an $n \times 1$ column vector of the variable values at finer resolution that need to be estimated over a domain. The moments of **s** are

$$E[\mathbf{s}] = \mathbf{X}\beta \tag{6.58}$$

$$E[(\mathbf{s} - \mathbf{X}\beta)(\mathbf{s} - \mathbf{X}\beta)^{\mathrm{T}}] = \mathbf{C}_\mathbf{s}(\boldsymbol{\theta}) \tag{6.59}$$

where **X** is a known $n \times p$ matrix, β is a $p \times 1$ vector of unknown coefficients, $E[\cdot]$ denotes an expected value, and $\mathbf{C}_\mathbf{s}(\boldsymbol{\theta})$ is a covariance function parameterized by $\boldsymbol{\theta}$ (a vector of parameters), as described by Kitanidis (1996).

Let **z** be an $N \times 1$ vector of observations. As in Kitanidis (1996), we will neglect uncertainty in the drift coefficients so that the drift can be subtracted from the spatial process. This is equivalent to setting $\mathbf{X} = 0$. Second, although nonlinear models can be taken care of by geostatistical inversion (Kitanidis 1995), we consider that the relationship between the observations and the underlying spatial processes is linear:

$$\mathbf{z} = \mathbf{H} * \mathbf{s} + \boldsymbol{\varepsilon} \tag{6.60}$$

where **H** is an $N \times n$ convolution kernel matrix and $\boldsymbol{\varepsilon}$ is an error vector of the same dimension as that of **z** (Kitanidis 1996), as discussed previously in the context of remote-sensing imaging in Chapter 3.

It is common to model the observational errors as Gaussian, so we have

$$p(\mathbf{z} \mid \mathbf{s}) \propto \left| \mathbf{C}_\varepsilon \right|^{-\frac{1}{2}} \exp\left[-\frac{1}{2}(\mathbf{z} - \mathbf{H} * \mathbf{s})^{\mathrm{T}} \mathbf{C}_\varepsilon^{-1}(\mathbf{z} - \mathbf{H} * \mathbf{s}) \right] \tag{6.61}$$

where $\mathbf{C}_\varepsilon$ stands for the variance and covariance matrix of error $\boldsymbol{\varepsilon}$ and $|\cdot|$ denotes the determinant of a square matrix. The prior probability distribution of **s** is considered Gaussian:

$$p(\mathbf{s}) \propto \left|\mathbf{C}_\mathbf{s}\right|^{-\frac{1}{2}} \exp\left[-\frac{1}{2}(\mathbf{s}-\mathbf{X}\beta)^\mathrm{T}\mathbf{C}_\mathbf{s}^{-1}(\mathbf{s}-\mathbf{X}\beta)\right] \tag{6.62}$$

Then, the joint probability density function (PDF) of $\mathbf{z}$ and $\mathbf{s}$ is

$$p(\mathbf{z},\mathbf{s}\,|\,\varepsilon) \propto \left|\mathbf{C}_\varepsilon\right|^{-\frac{1}{2}}\left|\mathbf{C}_\mathbf{s}\right|^{-\frac{1}{2}} \exp\left[-\frac{1}{2}(\mathbf{z}-\mathbf{Hs})^\mathrm{T}\mathbf{C}_\varepsilon^{-1}(\mathbf{z}-\mathbf{Hs})+\mathbf{s}^\mathrm{T}\mathbf{C}_\mathbf{s}^{-1}\mathbf{s}\right] \tag{6.63}$$

and the PDF of the observations $\mathbf{z}$ is

$$p(\mathbf{z}\,|\,\varepsilon) \propto \left|\mathbf{C}_\varepsilon\right|^{-\frac{1}{2}}\left|\mathbf{C}_\mathbf{s}\right|^{-\frac{1}{2}} \int_\mathbf{s} \exp\left[-\frac{1}{2}\left\{(\mathbf{z}-\mathbf{Hs})^\mathrm{T}\mathbf{C}_\varepsilon^{-1}(\mathbf{z}-\mathbf{Hs})+\mathbf{s}^\mathrm{T}\mathbf{C}_\mathbf{s}^{-1}\mathbf{s}\right\}\right]d\mathbf{s} \tag{6.64}$$

as described by Kitanidis (1996).

Maximum *a posteriori* probability estimation consists of maximizing Equation 6.63 with respect to both $\mathbf{s}$ and $\boldsymbol{\theta}$, which is thus equivalent to the minimization of the negative logarithm of Equation 6.63,

$$\frac{1}{2}\ln\left|\mathbf{C}_\varepsilon\right| + \frac{1}{2}\ln\left|\mathbf{C}_\mathbf{s}\right| + \frac{1}{2}\left[(\mathbf{z}-\mathbf{Hs})^\mathrm{T}\mathbf{C}_\varepsilon^{-1}(\mathbf{z}-\mathbf{Hs})+\mathbf{s}^\mathrm{T}\mathbf{C}_\mathbf{s}^{-1}\mathbf{s}\right] \tag{6.65}$$

Setting the derivative with respect to vector $\mathbf{s}$ equal to zero and solving for $\mathbf{s}$ (as a function of $\boldsymbol{\theta}$):

$$\hat{\mathbf{s}} = \left(\mathbf{H}^\mathrm{T}\mathbf{C}_\varepsilon^{-1}\mathbf{H}+\mathbf{C}_\mathbf{s}^{-1}\right)^{-1}\mathbf{H}^\mathrm{T}\mathbf{C}_\varepsilon^{-1}\mathbf{z} \tag{6.66}$$

which is the best estimate of $\mathbf{s}$ (if $\boldsymbol{\theta}$ were given) and has the mean square error matrix

$$\mathbf{V}_{\hat{\mathbf{s}}} = \left(\mathbf{H}^\mathrm{T}\mathbf{C}_\varepsilon^{-1}\mathbf{H}+\mathbf{C}_\mathbf{s}^{-1}\right)^{-1} \tag{6.67}$$

The estimation of $\mathbf{s}$ follows the formula in Equation 6.66, along with estimation of error variance from Equation 6.67. Further details were provided by Kitanidis (1996).

Kitanidis (1996) also described a geostatistical approach to structural parameter estimation by devising algorithms for maximizing Equation 6.64. The geostatistical solution to structural parameters is found by minimizing, with respect to θ, the function

$$\frac{1}{2}\ln\left|\mathbf{C}_\varepsilon+\mathbf{HC}_\mathbf{s}\mathbf{H}^\mathrm{T}\right| + \frac{1}{2}\mathbf{z}^\mathrm{T}(\mathbf{C}_\varepsilon+\mathbf{HC}_\mathbf{s}\mathbf{H}^\mathrm{T})^{-1}\mathbf{z} \tag{6.68}$$

For applications of geostatistical inverse modeling, we review a few published research results in the literature. The geostatistical approach is often applied to estimate the sources and sinks of greenhouse gases at finer spatial and temporal discretizations (i.e., resolutions), using atmospheric mass fraction measurements combined with a numerical atmospheric transport model, as it takes advantage of the spatial and/or temporal correlation in fluxes and does not require prior flux (of CO_2) estimates. In the batch setup, the geostatistical approach requires the inversion of a matrix with dimensions of either the number of observations or the number of fluxes to be estimated, implying that the cost becomes prohibitive as inversions are performed using more data, at finer resolutions, and over longer periods, as discussed by Michalak (2008). To deal with the computational cost, Michalak (2008) adapted a computationally efficient Kalman smoother (we will describe a bit more about Kalman filtering in the context of multiresolution data fusion in Chapter 8) for integration with the geostatistical approach to facilitate inversions of global grid-scale CO_2 fluxes at fine resolutions, over multiple years (1997 to 2001), and with large quantity of data, yielding results consistent with those from a batch geostatistical inversion. Gourdji et al. (2009) tested geostatistical inverse modeling for mapping North American CO_2 fluxes using synthetic data. Their experimental results indicated that monthly ecoregion-scale CO_2 surface fluxes over North America can be obtained quite accurately, based on even a fairly sparse network of nine towers collecting continuous CO_2 measurements across the continent, without using auxiliary information or prior estimates of the flux distribution in time or space. It is anticipated that geostatistical inverse modeling will have a greater role to play in scaling (Göckede et al. 2010), especially when considering its adaptability for nonlinearity in measurement modeling and its extensibility for spatial–temporal interdependence.

6.6 DISCUSSION

This chapter discussed geostatistical methods for both upscaling and downscaling, based on a concise introduction to the kriging methodology. In this section, we summarize these methods for scaling, hoping to reinforce the view that geostatistics has a major role to play in scaling, while admitting that there is room for improvement and that alternative and complementary strategies are worth exploring for flexibility in scaling when the relationships between the source data and end products are complex in terms of data supports and data-generating mechanisms.

Block kriging was described as a geostatistical solution to upscaling where the data of smaller support are irregularly spaced and their density is low, relative to the number of locations of larger supports. If the data are densely distributed, averaging-based solutions may be implemented for upscaling so that values may be predicted for the larger supports by either simple or weighted averaging of the finer-resolution data. The weighting can be done more cheaply than required of a geostatistical approach where variogram modeling is a prerequisite. More importantly, in the context of upscaling finer-resolution images to versions of coarser resolution, sensor characteristics, such as point spread functions, can be incorporated in algorithmic developments, as will be described in the next chapter.

Block kriging-based upscaling, although preferred over conventional approaches such as weighting, suffers from the effects of smoothing, because data from outside the larger supports (e.g., coarser-resolution pixels) are used in prediction, leading to an effective support that is artificially extended beyond the original larger pixel edges (Atkinson 2006). This kind of smoothing alters the characteristics of the upscaled variable, leading to biases in the simulated versions of the underlying fields discretized at coarser resolution. Again, this may be addressed by taking advantage of the detailed information about the convolutions desired of an upscaling processor, such as a simulated sensor with larger pixel size.

Gotway and Young (2002) provided a comprehensive review of approaches to change of support. For downscaling, a change of scale in the direction (coarse-to-fine resolution) opposite to that of upscaling, the geostatistical technique, known as area-to-point kriging, can be used, where areal data are used to predict point values. These points to be predicted need not lie on a regular grid or comprise a surface (Kyriakidis 2004). Area-to-point interpolation can also be viewed as a special case of areal interpolation (Lam 1983; Haining 2003), whereby both source data and target values pertain to the same property and are defined over areal units and points, respectively. Existing literature on choropleth mapping and kernel smoothing may be revisited usefully for insights on downscaling techniques, which can handle complexity with respect to data support (hence variogram modeling and regularization) and data lineage (Martin 1996; Hay et al. 2005). For example, it would be helpful to conceptualize downscaling not as a simple area-to-point interpolation but, in a broader sense, as a larger blocks-to-smaller blocks interpolation, bearing in mind that the source data and target variables are often defined on finite supports.

The geostatistical framework was promoted for scaling for several good reasons. One is that a measure of uncertainty associated with each prediction can be calculated along with the prediction itself. As will be studied in detail in Chapter 11, information theory asserts that one cannot increase the information content of a dataset by manipulation alone (Cover and Thomas 2006). Thus, downscaling techniques make use of the available information (e.g., remote-sensing images) to provide maps or images at a finer spatial resolution. The implications of kriging variance for downscaling performance evaluation should be further studied.

Through geostatistical inverse modeling, the solution to the inverse problem is defined as the set of parameter values that represent an optimal balance between two requirements. First, the optimized, or *a posteriori*, $\hat{s}$ should be as close as possible to the *a priori* s. Second, the measurement values z^* that would result from the inversion-derived (*a posteriori*) $\hat{s}$ should agree as closely as possible with the actual measurements z (Kitanidis and Vomvoris 1983; Michalak et al. 2004). Thus, geostatistical inverse modeling forms a basis to get closer to the data-generation processes and provides a mechanism for multisource data conflation, such as the combination of ground measurements and satellite sensor data and multisensor image merging (Chatterjee et al. 2010).

7 Methods for Scaling Gridded Data

Scaling or change of scale is important for information transfer across scales, wherever analysis, modeling, or applications are performed on scales different from those of the given datasets, models, or problem domains (Gelfand et al. 2001; Hufkens et al. 2008). As in previous chapters, we focus on methods for (spatial) scaling of spatial data. Chapter 6 described geostatistical approaches to change of scale: upscaling can be performed through block kriging, whereas downscaling can be implemented using block-to-point (or larger support-to-smaller support) kriging, respectively. For downscaling, geostatistical inverse modeling approaches may be usefully explored and provide a motivation for this chapter, as both are explicitly oriented to gridded data, such as images and raster data. Below, we review briefly the geostatistical methods for scaling before moving to the cases of gridded lattice data, highlighting the complementarity of methods covered in Chapters 6 and 7.

Geostatistics provides a useful framework for scaling where the data to be scaled may be based on regular or irregular sampling, and are often sparse with respect to the density of locations to be predicted, as shown in Chapter 6. This is because geostatistical scaling makes use of spatial structural information (spatial dependence) latent in the limited samples in addition to the measurements themselves to ensure accuracy in scaling. Besides, uncertainty related to spatial interpolation and averaging in scaling can be quantified and used as valuable information for balancing between accuracy and resolution in down- or upscaled data products.

The geostatistical methods described in Chapter 6 can be employed for scaling not only geostatistical data but also lattice data, in particular type I lattice data that include regularly gridded data and areal data partitioned with *bona fide* boundaries, as discussed in Chapter 5. Although regularly gridded data can be suitably taken care of by geostatistics, they merit special treatment for the following reasons. Firstly, there are growing numbers of gridded data involved in remote-sensing and geographic information systems, such as multisource images and raster maps (Irons et al. 1985; Holben 1986; Masek et al. 2006; Zhang 2008; Wulder et al. 2010). Research on their scaling will be essential for enriching the methodology of change of scale in a fuller spectrum of spatial data. Without causing confusion, gridded data are used interchangeably with regular lattice data or simply image data in this chapter where such techniques as image convolution and image deconvolution are discussed. Secondly, the characteristics of gridded data should be considered in developing methods that are well suited for scaling. For instance, it is sensible to take advantage of fine-grained data in the case of upscaling and to handle the complex convolutions implied in coarse-grained data in the case of downscaling.

Upscaling may be used for image coarsening (i.e., to obtain coarse [low]-resolution images from fine [high]-resolution images). This is known as spatial degradation of images (Justice et al. 1989). In addition to image coarsening, upscaling can also be used as a means for compressing gridded data to reduce data volume, thus becoming an important technique for on-board processing of satellite data (Justice et al. 1989). Upscaling can be considered as a process of convolution, for which proper convolution kernels may need to be formulated, as elaborated below.

The approaches to image upscaling, as mentioned above, are often statistical in the sense that the aggregation of fine-grained data into coarse-grained data is performed through statistical operators, such as the mean for continuous data and the mode for categorical data. However, statistical approaches may be limited because the statistical operators employed may not be based on the imaging mechanisms, resulting in upscaled images that do not necessarily emulate what would be obtained from a sensor with coarser spatial resolution (Justice et al. 1989). For this, we need to prescribe convolution kernels based on proper parameterization of sensor point spread functions (PSFs). Justice et al. (1989) described and implemented a mechanistic approach to simulating coarse-resolution images (e.g., the National Oceanic and Atmospheric Administration [NOAA] Advanced Very High Resolution Radiometer [AVHRR]) from relatively fine-resolution images (e.g., Landsat multispectral scanner system). Such mechanistic approaches to upscaling will be a valuable extension to statistical ones.

Just as upscaling can be formulated as convolution, downscaling can be seen as a process of deconvolution, as it aims to reverse the effects of convolution on data, such as blurring, image motion, and noise in imaging, as described in Chapter 3. In other words, deconvolution is a technique for recovering the original image that would have been recorded without degradation. Different approaches to deconvolution have been developed (Castleman 1995; Jackett et al. 2011), catering for different situations (e.g., PSFs known or unknown *a priori*). Related to deconvolution is super-resolution mapping, which seeks to increase or refine the spatial resolution of the input data through various techniques, such as the combined use of multiple observed images of coarser spatial resolution (Park et al. 2003).

This chapter will first describe how gridded data may be upscaled via statistical operators. This is followed by an introduction to convolution, which is considered as a generic form for upscaling gridded data, providing a clue to associating upscaling and downscaling via convolution and deconvolution, respectively. A mechanistic strategy for upscaling is then described to take advantage of the characteristics of the sensors for simulation of coarser-resolution images based on the original finer-resolution image. We will also discuss alternative approaches to upscaling data products that are derived from gridded sources.

As the other focus for this chapter, downscaling of gridded data will be described as a process inverse to that of upscaling. Deconvolution will be described with emphasis on Wiener filtering, least squares, and blind approaches. We will wrap up downscaling by examining super-resolution techniques. This is done by looking at restoring band-limited images beyond their cutoff frequencies and then by reviewing some techniques for super-resolution mapping.

In the final section, we give some overviews of approaches to scaling of gridded data in terms of their possible links to geostatistical methods as well as their special aspects. Although gridded data may be scaled upward or downward employing geo-statistics to some extent, we should give due consideration to the physical models of imaging, characteristics of image data, and practicality of computing with gridded data. Further developments are likely to be seen in the integration of sensor/imaging models, statistical image analysis, and geostatistics.

7.1 UPSCALING

As mentioned earlier in this chapter, there seem to be merits in exploring alternatives to upscaling, especially those dealing with gridded data, such as remotely sensed images and other raster data that are information products derived from image analysis and geo-processing. In this section, we first describe statistical methods for upscaling, which work by deriving, from data values of the smaller supports falling within the bounds of larger supports, values to be assigned to the latter. Upscaling can be usefully formulated as convolution, to emulate the processes of imaging that are often characterized through convolution, as mentioned previously. For this, proper convolution kernels should be prescribed to facilitate implementation of convolution-based upscaling. This is followed by a description of upscaling techniques that are based on physical understanding about the imaging or data-generating processes, rather than merely statistical data manipulation for resolution coarsening.

7.1.1 STATISTICAL APPROACHES

Upscaling is used to aggregate fine-grained gridded datasets into coarse-grained ones. For example, to reduce the problems of data volume, the NOAA AVHRR system performs on-board data compression: the large area coverage (LAC) data at a nominal resolution of 1.1 km are sampled to provide the daily global area coverage (GAC) data, whose pixel represents a nominal 3.3×4-km area on the ground at nadir (Justice et al. 1989). For upscaling, a straightforward approach would be to assign values to the grid cells of larger support from the set of values of smaller grid cells bounded within the larger-support grid cells. Statistics provides a framework for value assignments to upscaled grid cells from the pool of values of smaller-support grid cells.

Upscaling implies averaging, for which a common statistical operator is to esti-mate the mean or arithmetic average of a sample of values. We sum all the values of smaller-support grid cells within the larger-support cell and then divide by the num-ber of these smaller-support cells. Secondly, we can derive, from a list of numbers (i.e., the sample of values of smaller-support grid cells), the median, which is the middle value for the list. Medians are robust to extreme values, as only the ordering of values in the list counts, not the values themselves. For this reason, median-based statistics can be usefully implemented for anomaly detection (Zhang et al. 2007). To find the median, we list the sample of values in numerical order and locate the mid-dle value of the list. Thirdly, there is the mode for a sample, which is the value that occurs most frequently among the set of sample values. The mode is the only statistic appropriate for categorical data, and majority votes in image classifications (e.g., the *k*-nearest-neighbor [*k*-NN] classification method; Duda et al. 2001) are based on this statistic. Fourthly, selecting randomly a value from the sample of smaller-support cell values for the larger-support cell would also be a candidate operator. To add to the list of statistical operators for upscaling, there are the options for maximum and minimum, which are the largest and smallest values in a sample of values, respectively.

The statistical sampling approaches described above can be applied for upscaling gridded data. For example, upscaling of the AVHRR LAC data based on individual blocks of 5 × 3 grid cells (each at LAC 1-km resolution) was necessary owing to the limited storage capacity of the on-board tape recorder (Justice et al. 1989). Five sampling approaches were applied for this purpose: NOA (mean of the first four values, as shown below), AVE (mean of all 15 values), DIS (mean of four dispersed values with a constant pattern), ONE (single pixel value from the center of the block), and MED (median of all 15 values; the values were ranked in ascending order and the eighth value in rank was selected), as shown by Justice et al. (1989). Clearly, GAC data are derived from LAC data by a selection of sampling approaches.

The NOA approach mentioned is as follows. For a given scan line, the first four pixels are averaged and the next pixel is skipped. This sequence is continued along the line. The next two lines of data are skipped entirely, and the averaging procedure is resumed for the following line. Thus, a 5 × 3-pixel block is sampled by taking the mean of the first four pixels of the first line, as described by Justice et al. (1989). This procedure was selected based primarily on ease of on-board data processing. Since GAC data contain only one of three original AVHRR lines and the data resolution is further reduced by averaging every four adjacent samples and skipping the fifth sample along the scan line, their effective resolution element is 1.1 × 4 km with a 3-km gap between pixels across the scan line (Justice et al. 1989).

With alternative sampling approaches, which may produce upscaled datasets with different correlations with the original data and varied dispersion, it is important to assess their suitability for particular applications (Justice et al. 1989). The experiment by Justice et al. (1989) found that sampling by single pixel selection (i.e., the ONE approach) gave the smallest correlations for all but one study area and the largest variance (thus being most similar to the random sample collected for the experiment), whereas a mean of all 15 pixels (i.e., the AVE approach) gave the largest correlations of all and the smallest variance (thus being most dissimilar from the random sample).

The statistical sampling approaches to upscaling can be usefully formulated as convolution, although some statistics, such as the median, are difficult to implement via convolution. For instance, sampling by means will correspond to an equally weighted convolution kernel, where each element in the kernel indicates a weight equaling the reciprocal of the total number of smaller-support grid cells within the larger-support cell.

Formally, assume that the set of samples representing the fine-resolution data are $\{s(i,j)\}$, where $i, j = 0, 1, ..., N_h - 1$. When the same scene is sampled at a coarser resolution, the data are denoted $\{z(k, l)\}$, where $k, l = 0, 1, ..., N_l - 1$. Here, $N_l = N_h/mag$, where *mag* is the ratio of the sampling rates of $s(i, j)$ over that of $z(k, l)$ and assumed to be an integer greater than 1. For example, if $N_h = 4$ and $N_l = 2$, then $mag = 2$.

The relationship between the fine- and coarse-resolution data can be expressed through a convolution process (Leung et al. 2001):

$$\mathbf{z} = \mathbf{Hs} + \mathbf{n} \tag{7.1}$$

where $\mathbf{s}$, $\mathbf{z}$, and $\mathbf{n}$ are lexicographically ordered vectors of the fine-resolution data, coarse-resolution data, and additive noise, respectively; $\mathbf{H}$ is the matrix describing the transformation from fine-resolution to coarse-resolution data.

There are two kinds of sampling for grid data acquisition: area sampling and point sampling. Consider grid data matrices that consist of regular grids of cells of uniform sensitivity (weighting) across their spatial extents and with no spaces between them. By area sampling, upscaling a fine-resolution grid to a coarse-resolution one is performed by averaging mag^2 samples of smaller-support cells within a cell of larger support; in point sampling, upscaling works by taking only every magth sample in each direction, as described by Leung et al. (2001).

Following Leung et al. (2001), we can write the convolution matrix $\mathbf{H}$ for a specific ratio of grid sampling rates. Consider the simple example mentioned above with $N_h = 4$ and $N_l = 2$. For area sampling, assuming uniform sensitivity in grid data acquisition, we may have an $\mathbf{H}$ matrix as follows:

$$\mathbf{H} = \begin{pmatrix} \mathbf{H}_1 & \mathbf{0} \\ \mathbf{0} & \mathbf{H}_1 \end{pmatrix}$$

where $\mathbf{0}$ is a matrix of dimension 2×8 with all elements being zeros, and block $\mathbf{H}_1$ is

$$\mathbf{H}_1 = \begin{pmatrix} 1/4 & 1/4 & 0 & 0 & 1/4 & 1/4 & 0 & 0 \\ 0 & 0 & 1/4 & 1/4 & 0 & 0 & 1/4 & 1/4 \end{pmatrix}$$

For point sampling, on the other hand, the convolution matrix $\mathbf{H}$ will be

$$\mathbf{H} = \begin{pmatrix} \mathbf{H}_2 & \mathbf{0} & \mathbf{0} & \mathbf{0} \\ \mathbf{0} & \mathbf{0} & \mathbf{H}_2 & \mathbf{0} \end{pmatrix}$$

where $\mathbf{0}$ is a matrix of dimension 2×4, again, with all elements being zeros, and block $\mathbf{H}_2$ is

$$\mathbf{H}_2 = \begin{pmatrix} 1 & 0 & 0 & 0 \\ 0 & 0 & 1 & 0 \end{pmatrix}$$

The convolution kernels illustrated above are based on the assumption of equal weighting (in scaling from fine to coarse resolution) or uniformity in a pixel's sensitivity to light during imaging, which is unlikely to hold true in reality. The convolution kernels should be quantified on the basis of sensor PSFs. This leads to the other school of upscaling for gridded data, as examined in the next subsection. Before moving on there, we briefly hint at the effects of upscaling on derivatives of gridded data.

As discussed by Justice et al. (1989), the relationship between the original dataset and the upscaled versions will be dependent on the spatial heterogeneity of the problem domain. Unless the sampling block for upscaling is uniform or homogeneous, there will be differences between them. The more heterogeneous the distributions under consideration, the greater the differences. Such differences will become complicated when upscaling derivative datasets.

For derivative gridded datasets, there are two alternatives for their upscaling from finer to coarser-resolution versions. One is to obtain finer-resolution derivatives (i.e., at the scales conformal to that of the input finer-resolution grid datasets) and to perform upscaling on the finer-resolution derivatives. The other is to upscale input finer-resolution grid datasets to coarser-resolution versions, from which coarser-resolution derivatives are then generated.

Suppose that the derived variable Z' (e.g., normalized difference vegetation index [NDVI]; Townshend and Justice 1986) is functionally related to the original variables $\{Z_1, ..., Z_b\}$, which are organized as a vector $\mathbf{z} = [Z_1, ..., Z_b]^T$ (e.g., reflectances) at the fine scale, via an algorithm tf that converts $\mathbf{z}$ to Z' (i.e., $Z' = tf(\mathbf{z})$). For instance, the calculation of NDVI, for a pixel with reflectance values in the infrared (IR) and the red (R) bands, respectively, is done via

$$VI = \frac{IR - R}{IR + R} \tag{7.2}$$

The question is whether the following is valid: $\overline{tf}(\mathbf{z}) = tf(\overline{\mathbf{z}})$, where the bars indicate averaging. Differences in the resultant upscaled derivative datasets should be analyzed to provide guidance for the upscaling of derivatives. Justice et al. (1989) demonstrated mathematically that only in the case where the sums of the radiances in the R and IR channels are equal for each of the smaller-support cells used in upscaling will the two approaches to upscaling yield the same upscaled NDVI values.

7.1.2 MECHANISTIC APPROACHES

Upscaling can be applied to simulate images (and the sensors) with different spatial resolutions on the basis of fine-resolution image data. The upscaling strategy applied by Justice et al. (1989) for image simulation is different from the simple averaging technique used by Woodcock and Strahler (1987), as the former involves using a spatial filter similar to that derived by Sadowski and Sarno (1976), which was designed via proper sensor PSFs. This kind of approach is thus termed mechanistic, as it is based on the physical understanding and modeling of the relevant imaging processes.

To develop upscaling methods based on the imaging mechanisms, we need knowledge of sensor PSFs (i.e., spatial responsivity) or transfer functions (TFs) in the frequency domain (Schowengerdt 2007), as described in Chapter 3. This requires some specific information about the sensor designs and parameters concerning the imaging processes. The sensor instantaneous field of view (IFOV) is a measure of image spatial resolution. For example, the Landsat Multispectral Scanner System (MSS) sensors had an effective IFOV of 68 m in the cross-track direction by 83 m in the along-track direction, with images often resampled at a spatial resolution of 57 × 57 m (Justice et al. 1989; Richards and Jia 2006). Alternatively, the measure of spatial resolution can be based on the PSF, as reviewed in Chapter 3. For instance, the nominal 30-m spatial resolution of Landsat-4 and Landsat-5 thematic mapper (TM) is based on the PSF definition of the IFOV (Mather and Koch 2011). The optical PSF describes the distribution of energy in the image plane that results from a single point source. It should be noted that the PSF is sometimes represented by its Fourier transform, referred to as the modulation transfer function (MTF) (i.e., the modulus of the system optical transfer function). Below, we review some examples of MTF analysis of remote sensors.

Park et al. (1984) demonstrated that a meaningful system response can be calculated by averaging over an ensemble of point-source system inputs to yield an MTF that accounts for the combined effects of image formation, sampling, and image reconstruction. This random sample-scene phase approach was used by them for the MTF analysis of Landsat MSS. It was found that the combined effects of the image-forming optics, scanning aperture, electronic filter, two-dimensional sampling, and bilinear image reconstruction produce an effective IFOV of 104 × 148 m. In the moderate resolution-imaging spectroradiometer (MODIS), for example, the geometric spatial response is the ground instantaneous field of view (GIFOV) blurred cross-track by time integration during scanning. The spatial response profile of this effective GIFOV is not a rectangle function, but approximately a triangle, in the cross-track direction. In addition, the MODIS optics causes some additional blurring in-track and cross-track (Rojas et al. 2002). Rojas et al. (2002) undertook a project to characterize the MODIS spatial response using preflight MTF validation data and on-orbit imagery. The preflight validation data were used to estimate the optical blurring parameters to complete a system MTF model for MODIS bands 1 and 2.

Proper parameterization of sensor PSFs allows for the design of spatial filters for image or sensor simulation. Justice et al. (1989) designed a spatial filter to simulate Landsat MSS sensor response, simulating datasets with a range of spatial resolutions. These simulated coarser-resolution datasets were identical in terms of spectral band pass, radiometric sensitivity and noise, viewing geometry, and all other properties with the exception of spatial resolution. The approach by Justice et al. (1989) is to model the transfer function between the initial data and the desired data so that a spatial filter and sampling mechanism are derived to simulate the coarser-resolution data from the finer-resolution data. Justice et al. (1989) selected Landsat MSS data as the finest-resolution data.

According to Markham (1985), the Landsat MSS TF is assumed to be a rectangularly (X, Y) separable function:

$$\text{TF}_X(f_X) = \exp\left(-2\pi^2\sigma^2 f_X^2\right)\text{sinc}(f_X d)\frac{1}{1 + 2\text{I}(f_X/f_c) - 2(f_X/f_c)^2 - \text{I}(f_X/f_c)^3} \quad (7.3a)$$

$$\mathrm{TF}_Y(f_Y) = \exp\left(-2\pi^2\sigma^2 f_Y^2\right)\mathrm{sinc}(f_Y d) \tag{7.3b}$$

where $\sigma = 12$ m (1.7×10^{-5} rad) ground projected sensor blur at nadir, $d = 78.3$ m (1.11×10^{-4} rad) ground projection detector dimension at nadir, $f_c = 7.454 \times 10^{-3}$ cycles/m (5255 cycles/rad) (i.e., Butterworth filter cut-off frequency at nadir), and $I^2 = -1$, as described also by Justice et al. (1989). When we say a TF is separable with respect to the (X, Y) coordinate system, we have $\mathrm{TF}(f_X, f_Y) = \mathrm{TF}_X(f_X)\,\mathrm{TF}_Y(f_Y)$ and $\mathrm{PSF}(X, Y) = \mathrm{LSF}_X(X)\,\mathrm{LSF}_Y(Y)$. Note that, by definition, $\mathrm{TF}_X(f_X) = \int_{-\infty}^{\infty} \mathrm{LSF}_X(X)\,e^{-12\pi X f_X}\,dX$, where I is the imaginary unit $I^2 = -1$, as in Equation 3.1.

According to Justice et al. (1989), to simulate Landsat MSS data with spatial resolutions of 125, 250, 500, 1000, 2000, and 4000 m, the Landsat MSS data at 83-m resolution were used as the source data, which were degraded by a factor of 1.513 in the upscaling from 83 to 125 m at the first step and by a factor of 2 for successive upscalings. Thus, for the 125-m upscaling, the desired transfer function (TF_d) is the same as that for the original data without resampling, with the parameters changed to $\sigma = 18$ m, $d = 118.5$ m, and $f_c = 4.927 \times 10^{-3}$ cycles/m. As an example for successive upscalings, a second filter similar to the first was used for upscaling to 250-m resolution with a degradation factor of 2: $\sigma = 36$ m, $d = 237$ m, and $f_c = 2.464 \times 10^{-3}$ cycles/m.

Assume the sensor (e.g., Landsat MSS) is an analogue system, with aliasing effects ignored. After the TF of the source data (TF_s) and the TF desired (TF_d) of the output data are derived, we can determine the TF for the filter (TF_f) required to generate the output upscaled data

$$TF_f = \frac{TF_d}{TF_s} \tag{7.4}$$

according to the linear system theory, as described by Justice et al. (1989). Based on frequency sampling, Justice et al. (1989) generated two one-dimensional filters along the X (scan) and Y (track) directions from corresponding TFs via inverse discrete Fourier transforms. The two-dimensional filter is created as the product of two filters using the scan and track filters for the row and column, respectively (i.e., the first column of the two-dimensional filter is the product of the first column of the scan filter and all the rows of the track filter). Sensor noise needs to be restored to approximately the original level to make the simulated coarser-resolution images more realistic.

Indeed, as discussed by Kavzoglu (2004), simulation is often used to generate images that will be acquired by a future sensor such as using aircraft imagery to simulate a new orbital sensor. Simulation has usually been carried out as a spatial degradation process, as shown previously. However, to make a realistic simulation, it is necessary to take into consideration both PSF and the spectral response of the sensor. In a study by Kavzoglu (2004), Digital Airborne Imaging Spectrometer (DAIS 7915) data for a test site from the La Mancha Alta region of Spain were used to simulate a Landsat enhanced thematic mapper plus (ETM+) image. The results were promising in that comparable images were produced by considering the PSF and spectral response of the Landsat ETM+. Further research along this line is important for developments in mechanistically based upscaling.

The most common way to simulate new instruments is to use data of existing sensors with finer spatial and spectral resolution. Valor et al. (2000) proposed a procedure to simulate a thermal instrument with coarser spatial resolution (as is usually done), but finer spectral resolution, than the original dataset. The method was applied to the simulation of a medium-scale-surface-temperature instrument. It focused on land observation, the land surface temperature being the primary physical quantity to be estimated, with main applications related to agriculture, forests, hydrology, and environment. The sensor is planned with two thermal bands in the wavelength range of 10.0–11.0 and 11.5–12.5 μm, a spatial resolution of 240 m at nadir, a 1- to 3-day revisit, and a swath of 1400 km with incident angles between ±39°. The input data for the simulation were provided by Landsat TM band 6, which is placed at 10.5–12.5 mm and has a spatial resolution of 120 m. The methodology is general and could be applied in other cases.

7.2 DOWNSCALING

Downscaling is the process of deriving finer-resolution data from coarser-resolution ones, or inversion of the original signals **s** (e.g., finer-resolution data) from their spatially degraded versions **z**, which is represented in Equation 7.1, where **H** is a convolution kernel. As convolution was described as a process for upscaling in the preceding section, deconvolution is taken here to refer to techniques for downscaling, although deconvolution usually means deblurring. In this section, we describe various approaches to deconvolution, such as Wiener filters, least squares methods, and blind deconvolution (when **H** is not known in advance as is often the case in practice). This is followed by some introductory description of super-resolution techniques, which seek to refine the spatial resolution of input data.

7.2.1 DECONVOLUTION

As shown previously, the process of imaging is subject to blurring. When the blur is linear and shift invariant, the imaging process amounts to a convolution of the signal **s** with the PSF, say, **H**, as shown in Chapter 3. Also, an additive noise term **n** is often assumed, resulting in an imaging (convolution) equation as in Equation 7.1. Deconvolution seeks to derive an estimated signal $\hat{s}$ (Castleman 1995).

We start with *Wiener deconvolution*, which works in the frequency domain and seeks to minimize the impact of deconvoluted noise at frequencies that have a poor signal-to-noise ratio, as reviewed by Castleman (1995). Given a system as shown by $z = h*s + \varepsilon$, where z is measurement, h is convolution kernel, s is signal, * means convolution, and ε is noise, the goal is to find some function $g(t)$ so that $s(t)$ can be estimated:

$$\hat{s}(t) = g(t)*\, z(t) \tag{7.5a}$$

where $\hat{s}(t)$ is an estimate of $s(t)$ that minimizes the mean square error. Equation 7.5a may be rewritten in the frequency domain:

$$\hat{S}(\omega) = G(\omega)Z(\omega) \tag{7.5b}$$

where $\hat{S}(\omega)$, $G(\omega)$, and $Z(\omega)$ are the Fourier transforms of $\hat{s}(t)$, $g(t)$, and $z(t)$, respectively (Castleman 1995).

$G(\omega)$ is actually the Wiener deconvolution filter described in the frequency domain and is written as

$$G(\omega) = \frac{H^*(\omega)PSD_S(\omega)}{|H(\omega)|^2 PSD_S(\omega) + PSD_N(\omega)} \tag{7.6a}$$

where $H(\omega)$ is the Fourier transform (at frequency ω) of h; $PSD_S(\omega)$ is the mean power spectral density of the input signal $s(t)$ $(PSD_S(\omega) = E|S(\omega)|^2)$; $PSD_N(\omega)$ is the mean power spectral density of the noise $\varepsilon(t)$ $(PSD_N(\omega) = E|NOISE(\omega)|^2$, where $NOISE(\omega)$ is the Fourier transform (at frequency (ω)) of $\varepsilon(t)$); and the superscript "*" denotes complex conjugation. The Wiener filter can also be written as

$$G(\omega) = \frac{1}{H(\omega)}\left[\frac{|H(\omega)|^2}{|H(\omega)|^2 + \frac{PSD_N(\omega)}{PSD_S(\omega)}}\right] = \frac{1}{H(\omega)}\left[\frac{|H(\omega)|^2}{|H(\omega)|^2 + \frac{1}{SNR(\omega)}}\right] \tag{7.6b}$$

Here, $1/H(\omega)$ is the inverse of the original system convolution kernel's Fourier transform and $SNR(\omega) = PSD_S(\omega)/PSD_N(\omega)$ is the signal-to-noise ratio (Castleman 1995). Wiener filters are derived from minimizing the mean squared error:

$$err\,(\omega) = E\left|S(\omega) - \hat{S}(\omega)\right|^2 \tag{7.7}$$

where E denotes expectation. To find the minimum error value, we differentiate $err\,(\omega)$ with respect to $G(\omega)$ and set the differential equal to zero. This will provide the solution $G(\omega)$ (i.e., the Wiener filter), which has $err\,(\omega)$ minimized (Gonzalez et al. 2003).

The filter discussed above involves the assumption of stationary signals and noise. However, most images are highly nonstationary. Nevertheless, they can be assumed locally stationary, as their power spectrum computed over a small window changes slowly as one moves the window within the images (Castleman 1995). Equation 7.6b can be adapted for locally modeled signal and noise power spectra $PSD_S(\omega, x)$ and $PSD_N(\omega, x)$.

Deconvolution can be set up as a *least squares* minimization problem (Castleman 1995). Consider first the so-called unrestrained deconvolution when **n** = **0** or we know nothing about the noise. Let **e**($\hat{s}$) be a vector of residual errors that result from using $\hat{s}$ as an approximation to **s**. Equation 7.1 then becomes

$$\mathbf{z} = \mathbf{Hs} = \mathbf{H\hat{s}} + \mathbf{e}(\hat{s}) \text{ or } \mathbf{e}(\hat{s}) = \mathbf{z} - \mathbf{H\hat{s}} \tag{7.8}$$

and we seek to minimize the objective function

$$\left\| \mathbf{e}(\hat{\mathbf{s}}) \right\|^2 = \left\| \mathbf{z} - \mathbf{H}\hat{\mathbf{s}} \right\|^2 = (\mathbf{z} - \mathbf{H}\hat{\mathbf{s}})^{\mathrm{T}} (\mathbf{z} - \mathbf{H}\hat{\mathbf{s}}) \tag{7.9}$$

Setting to zero the derivative of $\left\| \mathbf{e}(\hat{\mathbf{s}}) \right\|^2$ with respect to $\hat{\mathbf{s}}$, we have

$$\frac{\partial \left\| \mathbf{e}(\hat{\mathbf{s}}) \right\|^2}{\partial \hat{\mathbf{s}}} = -2\mathbf{H}^{\mathrm{T}} (\mathbf{z} - \mathbf{H}\hat{\mathbf{s}}) = 0 \tag{7.10}$$

and solving for $\hat{\mathbf{s}}$ yields

$$\hat{\mathbf{s}} = (\mathbf{H}^{\mathrm{T}}\mathbf{H})^{-1}\mathbf{H}^{\mathrm{T}}\mathbf{z} \tag{7.11}$$

which is reduced to $\mathbf{H}^{-1}\mathbf{z}$ when $\mathbf{H}$ is invertible (Castleman 1995).

Then, consider constrained least squares deconvolution (Hunt 1973), for which we introduce the constraint $\left\| \mathbf{z} - \mathbf{H}\hat{\mathbf{s}} \right\|^2 - \left\| \mathbf{z} - \mathbf{H}\mathbf{s} \right\|^2 = \left\| \mathbf{z} - \mathbf{H}\hat{\mathbf{s}} \right\|^2 - \left\| \mathbf{n} \right\|^2 = 0$ into the minimization. The problem of deconvolution is now the minimization of the objective function *objective* ($\hat{\mathbf{s}}$) below:

$$objective(\hat{\mathbf{s}}) = \left\| \mathbf{Q}\hat{\mathbf{s}} \right\|^2 + \lambda \left(\left\| \mathbf{z} - \mathbf{H}\hat{\mathbf{s}} \right\|^2 - \left\| \mathbf{n} \right\|^2 \right) \tag{7.12}$$

where $\mathbf{Q}$ is a matrix to define some linear operator on $\hat{\mathbf{s}}$ (for setting the goal of deconvolution) and λ is a constant called a *Lagrange multiplier* (Castleman 1995).

As before, a solution for $\hat{\mathbf{s}}$ is obtained by setting to zero the derivative of *objective* ($\hat{\mathbf{s}}$) with respect to $\hat{\mathbf{s}}$. This yields

$$\hat{\mathbf{s}} = (\mathbf{H}^{\mathrm{T}}\mathbf{H} + \lambda^{-1}\mathbf{Q}^{\mathrm{T}}\mathbf{Q})^{-1}\, \mathbf{H}^{\mathrm{T}}\mathbf{z} \tag{7.13}$$

where λ^{-1} is a constant that must be adjusted so that the constraint $\left\| \mathbf{z} - \mathbf{H}\hat{\mathbf{s}} \right\|^2 = \left\| \mathbf{n} \right\|^2$ is satisfied. If we set $\lambda^{-1} = 0$, this reduces to the filter of Equation 7.11. $\mathbf{Q}$ may be set to be the noise-to-signal ratio

$$\mathbf{Q} = \mathbf{R}_s^{-1/2}\mathbf{R}_n^{1/2} \tag{7.14}$$

where $\mathbf{R}_s$ and $\mathbf{R}_n$ are the signal and noise covariance matrices, respectively. Then, Equation 7.13 becomes

$$\hat{\mathbf{s}} = \left(\mathbf{H}^{\mathrm{T}}\mathbf{H} + \lambda^{-1}\mathbf{R}_f^{-1}\mathbf{R}_n \right)^{-1} \mathbf{H}^{\mathrm{T}}\mathbf{z} \tag{7.15}$$

which, with $\lambda^{-1} = 1$, is the classical Wiener filter that minimizes the mean square difference between the original and the restored images, as described by Castleman (1995).

In many situations, the blurring (i.e., PSFs; Chalmond 1991) is often unknown, and little information is available about the original image **s**. Therefore, the original image **s** must be identified directly from observed image **z** by using partial or no information about the blurring process and **s**. Such an estimation problem, assuming the linear degradation model of Equation 7.1, is called blind deconvolution (Kundur and Hatzinakos 1996; Carasso 2001; Levin et al. 2011).

Kundur and Hatzinakos (1996) described a maximum-likelihood (ML) estimation method for blind deconvolution, in which the true image is modeled as a two-dimensional autoregressive (AR) process and the PSF as a two-dimensional moving average process. Specifically, the true image **s** (formed by lexicographically ordering the two-dimensional signals $\{s(i, j)\}$) is modeled as an AR process:

$$\mathbf{s} = \mathbf{As} + \mathbf{e} \tag{7.16}$$

where $\mathbf{A} = \{a(l, m)\}$ $(a(0, 0) = 1)$ are the AR model coefficients and **e** (a vector formed by lexicographically ordering matrix $\{e(i, j)\}$) is the modeling error, which is a zero-mean noise process with covariance matrix $\mathbf{cov}_e$ and is statistically independent of **s**. From the linear degradation model of Equation 7.1, the degraded image **z** (as for **s**, formed by lexicographically ordering the two-dimensional image $\{z(i, j)\}$) can be expressed as

$$\mathbf{z} = \mathbf{Hs} + \mathbf{n} \tag{7.17a}$$

where $\mathbf{H} = \{h(l, m)\}$ over the finite support (l, m) and **n** (lexicographically ordered matrix $\{n(i, j)\}$) is the additive noise of the imaging system assumed to be zero-mean Gaussian with covariance $\mathbf{cov}_n$. Rearranging Equation 7.16, substituting into Equation 7.17a, and rearranging yields

$$\mathbf{z} = \mathbf{H}(\mathbf{I} - \mathbf{A})^{-1}\mathbf{e} + \mathbf{n} \tag{7.17b}$$

where **I** is the identity matrix (Kundur and Hatzinakos 1996).

Deconvolution is about estimating the PSF, the AR model coefficients of the original image **s**, and the variance of additive noise and the error term in Equation 7.17b. These are denoted by the parameter set $\boldsymbol{\theta} = \left\{\{h(l, m)\}, \{a(l, m)\}, \sigma_n^2, \sigma_e^2\right\}$. By the ML method, parameter estimation is made such that the probability or likelihood of receiving the observed image **z** given $\boldsymbol{\theta}$ is maximized:

$$\hat{\boldsymbol{\theta}}_{ml} = \arg\left\{\max_{\boldsymbol{\theta}\in\Theta} L(\boldsymbol{\theta})\right\} = \arg\left\{\max_{\boldsymbol{\theta}\in\Theta} \log pdf(\mathbf{z}; \boldsymbol{\theta})\right\} \tag{7.18}$$

where $L(\boldsymbol{\theta})$ denotes the log-likelihood function of $\boldsymbol{\theta}$, Θ specifies the range of elements of $\boldsymbol{\theta}$, and $pdf(\mathbf{z}; \boldsymbol{\theta})$ is the probability density function of **z** for a given $\boldsymbol{\theta}$, as described by Kundur and Hatzinakos (1996).

Assume both **n** and **e** are zero-mean Gaussian processes so that **z** is also zero-mean and Gaussian. Lagendijk et al. (1990) derived the following equivalent likelihood function:

$$L(\boldsymbol{\theta}_{ml}) = \log\left(\det|\mathbf{cov}_{\mathbf{z}}|\right) + \mathbf{z}^{\mathrm{T}}\mathbf{cov}_{\mathbf{z}}^{-1}\mathbf{z} \tag{7.19}$$

where $\mathbf{cov}_z$ is the covariance matrix of $\mathbf{z}$ given by

$$\mathbf{cov}_z = \sigma_e^2 \mathbf{H}(\mathbf{I}-\mathbf{A})^{-1}(\mathbf{I}-\mathbf{A})^{-T}\mathbf{H}^T + \sigma_n^2 \mathbf{I} \tag{7.20}$$

and $(\cdot)^{-T}$ and $(\cdot)^T$ represent inverse transpose and transpose matrix operations, respectively. The ML solution is obtained by minimizing Equation 7.19 with respect to the set of parameters $\boldsymbol{\theta}$:

$$\hat{\boldsymbol{\theta}}_{ml} = \arg\left\{ \min_{\boldsymbol{\theta}\in\Theta} \log\left(\det\left|\mathbf{cov}_z\right|\right) + \mathbf{z}^T \mathbf{cov}_z^{-1}\mathbf{z} \right\} \tag{7.21}$$

Further details can be found in Kundur and Hatzinakos (1996).

There is increasing research on blind image deconvolution (Levin et al. 2011; Shen et al. 2012). Levin et al. (2011) showed that it is important to consider the overall shape of the posterior distribution $p(\mathbf{s}, \mathbf{H}|\mathbf{z})$ (not only its mode) in the maximum *a posteriori* probability (MAP) deconvolution, where $\mathbf{s}$, $\mathbf{H}$, $\mathbf{z}$ stand for the original image, the blur kernel (a matrix), and the blurred image, respectively. For this, there are two approaches to MAP-based deconvolution: $\text{MAP}_{\mathbf{s},\mathbf{H}}$ that estimates the mode pair $\mathbf{s}$, $\mathbf{H}$, and $\text{MAP}_{\mathbf{H}}$ that selects the best $\mathbf{H}$ while marginalizing over all possible $\mathbf{s}$ images. The latter is significantly more robust than the former, yet has not been applied widely as it involves a process of marginalization over latent images. Levin et al. (2011) developed a simple approximate $\text{MAP}_{\mathbf{H}}$ algorithm based on modest modification of common $\text{MAP}_{\mathbf{s},\mathbf{H}}$ algorithms, which is computationally efficient.

7.2.2 SUPER-RESOLUTION MAPPING

We have seen various techniques for image deconvolution, which may be employed for downscaling. Theoretically, however, deconvolution can only restore the spectrum of an object up to, but not beyond, the diffraction limit for images. Techniques that seek to recover information beyond the diffraction limit are referred to as super-resolution mapping, which increases or refines the spatial resolution of images where the pixel size to be predicted is smaller than the pixel size of the images or data available (Starck et al. 2002). Super-resolution mapping has attracted growing research efforts, although challenges remain concerning robustness, treatment of color, and dynamic operation modes (Farsiu et al. 2004). We review frequency domain and multiframe methods below.

First, super-resolution mapping can be based on the extrapolation of band-limited functions. As explained by Harris (1964), under noiseless conditions, any recorded image can correspond to one and only one object. Thus, it should be possible to reconstruct that object in infinite detail from its diffraction-limited image. As described by Castleman (1995), Harris's super-resolution technique works by applying the sampling theorem, in the frequency domain, to obtain a system of linear equations that can be resolved for values of the signal spectrum outside the diffraction-limited passband, as follows.

Suppose $z(x)$ is spatially bounded, with T being the bandwidth. According to the sampling theorem, with the time and frequency domains reversed, $\mathbb{F}_Z(\omega)$ (the Fourier transform of $z(x)$) can be completely reconstructed from a series of equally spaced samples provided they are taken no more than $1/2T$ apart. The reconstruction is expressed as

$$\mathbb{F}_Z(\omega) = \left[\sum_{n=-\infty}^{\infty} \delta(\omega - 2nT)\mathbb{F}_Z(\omega) \right] * \frac{\sin(2\pi\omega T)}{2\pi\omega T}$$

$$= \sum_{n=-\infty}^{\infty} \mathbb{F}_Z(2nT) \frac{\sin(2\pi\omega T - 2nT)}{2\pi\omega T - 2nT} \tag{7.22}$$

If $z(x)$ is passed through a linear system that is band limited to frequency s_m, deconvolution can recover the signal such that the spectrum is known exactly for frequencies out to s_m (Harris 1964; Castleman 1995).

Super-resolution mapping aims to determine $\mathbb{F}_Z(\omega)$ over a range $[-s_n, s_n]$, which is wider than $[-s_m, s_m]$. Suppose that $\mathbb{F}_Z(\omega)$ is sampled so that M sample points fall within the passband $[-s_m, s_m]$. As described by Castleman (1995), an estimate of the spectrum band limited at s_n (denoted by $\hat{\mathbb{F}}_Z(\omega)$) can be computed from

$$\mathbb{F}_Z(\omega) \approx \hat{\mathbb{F}}_Z(\omega) = \sum_{n=-N}^{N} \mathbb{F}_Z(2nT) \frac{\sin(2\pi\omega T - 2nT)}{2\pi\omega T - 2nT} \tag{7.23}$$

Equation 7.23 may be viewed as a linear equation in $2N + 1$ unknowns, which are the values of $\mathbb{F}_Z(2nT)$ at the $2N + 1$ sample points. We can generate a system of $2N + 1$ linear equations in $2N + 1$ unknowns by selecting $2N + 1$ frequencies $\mathbb{F}_Z(\omega)$ within the passband and substituting the known values of $\mathbb{F}_Z(\omega)$ into Equation 7.23. After solving the system of linear equations (Equation 7.23) for the unknown values of the spectrum, we substitute them into Equation 7.23 to generate an estimate of the spectrum (i.e., $\hat{\mathbb{F}}_Z(\omega)$), which is band limited at a frequency higher than the diffraction limit of the imaging system.

Theoretical simplicity is a major advantage of the frequency domain approaches, because the relationship between finer- and coarser-resolution images is evidently demonstrated in the frequency domain. Kim et al. (1990) described a recursive approach to reconstructing the original finer-resolution image from observed noisy coarser-resolution images. However, the observation models are restricted to only global shifting and linear space-invariant blurring. Spatial domain knowledge incorporation and spatially adaptive regularization would be valuable extensions to frequency domain-based super-resolution techniques.

As shown above, super-resolution techniques break the diffraction limit of systems and, in general, improve over the resolution of digital imaging sensors. There are multiple-frame variants of super-resolution techniques (Park et al. 2003), which use the subpixel shifts between multiple coarse-resolution images of the same scene to create an improved-resolution image. Most approaches to super-resolution mapping with multiple coarser-resolution images consist of three stages: registration (i.e., to align multiple coarser-resolution images with the finer-resolution source image), interpolation (i.e., to obtain a uniformly spaced finer-resolution image from coarser-resolution images), and restoration (Park et al. 2003).

First, we need to formulate an observation model that relates the original finer-resolution image to the observed coarser-resolution images, as described by Park et al. (2003). Let the desired finer-resolution image of size $mag_1 N_1 \times mag_2 N_2$ be written as a vector $\mathbf{s} = [s_1, \ldots, s_{N'}]^T$ ($N' = mag_1 N_1 \times mag_2 N_2$), in lexicographical notation, which is sampled at or above the Nyquist rate from a continuous scene that is assumed to be band limited. The parameters mag_1 and mag_2 represent the magnification (or down sampling) factors in the observation model for the X and Y directions, respectively, with each of the observed coarser-resolution images being of size $N_1 \times N_2$. Let the lth coarser-resolution image be denoted in lexicographic notation as $\mathbf{z}_l = [z_{l,1}, \ldots, s_{l,N}]^T$ ($l = 1, \ldots, nl$; $N = N_1 \times N_2$). Assume $\mathbf{s}$ remains constant during the acquisition of the multiple coarser-resolution images. Following Park et al. (2003), the observation model is represented as

$$\mathbf{z}_l = \mathbf{H}_l \mathbf{s} + \mathbf{n}_l \quad \text{for } l = 1, \ldots, nl \tag{7.24}$$

where matrix $\mathbf{H}_l$ of dimension $(N_1 \times N_2) \times (mag_1 N_1 \times mag_2 N_2)$ is the convolution kernel for the lth observed image, representing the contribution of pixels in $\mathbf{s}$ to pixels in $\mathbf{z}_l$, and $\mathbf{n}_l$ is the additive noise term.

Super-resolution mapping seeks to estimate $\mathbf{s}$ from nl images $\mathbf{z}_l$ ($l = 1, \ldots, nl$), a process of image restoration, which is often an ill-posed problem owing to an insufficient nl and ill-conditioned $\mathbf{H}_l$. Thus, regularization (a technique for introducing additional information in solving ill-posed problems, not to be interpreted in the sense of variogram regularization over finite supports) is often applied to stabilize the inversion. There are deterministic and stochastic approaches to regularization, exemplified by constrained least squares and maximum *a posteriori* methods, respectively, as reviewed by Park et al. (2003).

For constrained least squares, the solution is found by choosing an $\mathbf{s}$ to minimize the Lagrangian:

$$\sum_{l=1}^{nl} \left\| \mathbf{z}_l - \mathbf{H}_l \mathbf{s} \right\|^2 + \beta \left\| \mathbf{C} \mathbf{s} \right\|^2 \tag{7.25}$$

where $\| \ \|^2$ represents an l_2 norm, the matrix $\mathbf{C}$ is often a high-pass filter, β is the regularization parameter that controls the tradeoff between fidelity to the data (the first part) and the smoothness of the solution (as expressed by $\left\| \mathbf{C} \mathbf{s} \right\|^2$), as shown by Park et al. (2003). Stochastic super-resolution reconstruction is typically Bayesian in the sense that the *a posteriori* probability density function of the original image $\mathbf{s}$, $p(\mathbf{s} \mid \mathbf{z}_1, \ldots, \mathbf{z}_{nl})$, can be established. The maximum *a posteriori* probability with respect to $\mathbf{s}$ is the solution to the so-called MAP estimator of $p(\mathbf{s} \mid \mathbf{z}_1, \ldots, \mathbf{z}_{nl})$

$$\mathbf{s} = \arg \max p(\mathbf{s} \mid \mathbf{z}_1, \ldots, \mathbf{z}_{nl}) \tag{7.26a}$$

which, after applying Bayes's theorem, is equivalent to

$$\mathbf{s} = \arg\max\ [\ln p(\mathbf{z}_1, \ldots, \mathbf{z}_{nl}|\mathbf{s}) + \ln p(\mathbf{s})] \tag{7.26b}$$

where $p(\mathbf{s})$ is the *a priori* image model and $p(\mathbf{z}_1, \ldots, \mathbf{z}_{nl}|\mathbf{s})$ is the conditional density conveying statistical information about noise. Markov random field models are often adopted for modeling the *a priori* image $\mathbf{s}$ (Park et al. 2003).

7.2.3 SUBPIXEL MAPPING

Super-resolution mapping is also a valuable technique for deriving data products from image data. For instance, because of surface heterogeneity, there are always some pixels at any spatial resolution that contain multiple cover types, especially when the spatial resolution becomes coarse. Effective methods for extracting subpixel information about multiple cover types within individual pixels would be valuable in this situation. Subpixel classification is a kind of super-resolution mapping that refines the spatial resolution of a classified remotely sensed image compared to that of the input images by predicting the spatial distribution of land-cover classes at subpixel scale (Atkinson 2006; Ling et al. 2010; Jin et al. 2012).

Subpixel distributions of land cover are often determined based on the principle of spatial dependence and from land cover fraction images derived from spectral mixture modeling or soft classification (DeFries et al. 1997; Hansen et al. 2003). The purpose of unmixing algorithms (Settle and Drake 1993; Lu et al. 2011) is to estimate fractional abundances from a group of pixels at multiple wavebands, as reviewed by Liang (2003). Assume that the signal on a pixel level is a linear combination of the signals of all components that are called end members. The pixel xs reflectance for a specified band b can be written as

$$RFL_b(x) = \sum_{k=1}^{K} p_{k,b}(x) rfl_{k,b}(x) + \varepsilon_b(x) \tag{7.27}$$

where $RFL_b(x)$ is the average reflectance at band b, $p_{k,b}(x)$ is the proportion of end-member k at band b (also called fractional abundances, $p_{k,b}$ are positive and sum to 1 over k for each b), $rfl_{k,b}(x)$ is the average reflectance of end-member k, and $\varepsilon_b(x)$ is a random error term. Settle and Drake (1993) provided further details about unmixing. Subpixel classification can be performed once the linear mixture model has been calibrated through solving the linear system as prescribed in Equation 7.27.

Spectral unmixing remains a challenging topic after two decades of research. Regardless of the available spatial resolution, the spectral signals collected in natural environments are invariably a mixture of the spectra of the various materials found within the spatial extent of the ground IFOV of the remote-sensing imaging instrument. The availability of hyperspectral imaging spectrometers, whose number of spectral bands exceeds those of spectral mixture components, has attracted increasing research efforts directed toward such issues as linear vs. nonlinear unmixing, end-member determination with or without pure pixels, and unmixing for moderate spatial resolution data or multitemporal time-series images (Plaza et al. 2011). Application areas of spectral unmixing are set to widen.

Ling et al. (2010) proposed a novel method for super-resolution land cover mapping, which uses multiple subpixel-shifted remotely sensed images taken by observation satellites. These satellites take images over the same area once every several days, but with slight orbit translations. Coarse-resolution pixels in these remotely sensed images therefore contain different land-cover fractions that can provide useful information for super-resolution land cover mapping. We will discuss the issue of downscaling area-class maps in more detail in Chapter 10.

7.3 DISCUSSION

This chapter has reviewed some of the methods for upscaling and downscaling of gridded data, respectively. For upscaling, both statistical and mechanistic approaches were described, with the former proceeding as aggregation by sampling from smaller-support cell values within the bounds of larger-support cells and the latter requiring knowledge of imaging system parameters for coarser-resolution image or sensor simulations. Mechanistic approaches are becoming increasingly popular in research and applications, although statistical ones remain dominant (Bian and Butler 1999; Tan et al. 2006). Statistical and mechanistic upscaling of gridded data can be formulated in a unified framework as convolution. Both approaches to upscaling are useful and should be combined for design and implementation of relevant algorithms.

For downscaling, on the other hand, statistical methods have been developed for image deconvolution and super-resolution restoration (Baker and Kanade 2002). Deconvolution aims to reverse the effects of convolution on recorded data, which may or may not refine the spatial resolution of the input datasets. Super-resolution mapping means refining the spatial resolution of an input dataset based on input data of coarser resolution. Downscaling of gridded data can be performed in both the spatial and the frequency domains. The geostatistical inverse modeling approaches described in Chapter 6 can be extended to mechanistic downscaling of gridded data. Thus, both statistical and mechanistic approaches to downscaling of gridded data are useful and should be explored in combination for technical enhancement and practical utility, as in the case of upscaling. The magnification factors in super-resolution mapping are by no means infinite, as there are fundamental limits to them (Baker and Kanade 2002). Assuming local translation among coarse-resolution images, Lin and Shum (2004) determined the explicit limits of reconstruction-based algorithms for super-resolution mapping, under both real and synthetic conditions, based on the perturbation theory of linear systems.

The statistical and mechanistic methods for scaling gridded data should be viewed as being complementary to the geostatistical approaches described in Chapter 6. Thus, we should consider geostatistical adaptation of the methods described in this chapter for scaling gridded data. In other words, geostatistics should be incorporated into relevant procedures concerning scaling of gridded data (Tarnavsky et al. 2008). For example, spatial dependence in error terms, such as those in Equations 7.1 and 7.24, may be quantified via variograms and accommodated properly in the computation for upscaling and downscaling. It is worth examining the comparative performances of methods designed originally for images and gridded data vs. those built on geostatistics.

In parallel, for geostatistical scaling of gridded data, we should consider integrating mechanisms of imaging or data generation in the relevant computing procedures (Huang et al. 2002). This is to use the physical models of imaging and sensor parameters for geostatistical scaling of gridded data, as in geostatistical inverse modeling. Equally important is to make use of the characteristics of image data in geostatistically modified procedures for scaling. This is to exploit the characteristics of image data and other gridded data in geostatistical scaling and to enhance the efficiency of computing with gridded data. Further developments should be directed toward integration of sensor/imaging models, statistical image analysis, and geostatistics. Such integral strategies will be valuable for conflation of multiscale spatial data (Gotway and Young 2002; Alparone et al. 2004; Zhang 2008; Chatterjee et al. 2010), for which spatial dependence, imaging mechanisms, and data semantics are key elements in algorithm design, as will be discussed in Chapter 8.

Although this chapter has focused on methods for scaling of both image data and their derivatives, such as NDVI, it is important to consider their extensions to more complex applications, such as terrain analysis and area-class mapping. Although some terrain derivatives, such as slope, may be discussed with respect to scaling in a way similar to simple reflectance derivatives, such as NDVI, other terrain derivatives, such as flow paths, should be examined carefully in terms of data semantics and the procedures for their derivation to ensure meaningfulness and accuracy in the results of data scaling. This is also true of area-class mapping, for which a process-based strategy is essential to scaling of area classes as well as their scale modeling. We will review scale issues in terrain analysis and area-class mapping in Chapters 9 and 10, respectively.

8 Multiscale Data Conflation

As described in Chapters 6 and 7, methods for scaling are based on geostatistics, spatial statistics, and characterization of imaging processes. These methods can handle change of scale for geostatistical and lattice data. Spatial problem solving often involves multisource data, which are usually multiscale. It is thus important to devise methods for integration of multiscale data, the subject for this chapter, extending the methods described in Chapters 6 and 7 to multivariate scenarios.

It is often necessary in geo-processing applications to integrate spatial data derived from different sources and, hence, existing at different scales, so that they can be collectively employed for spatial problem solving. Methods should thus be developed to facilitate multisource and multiscale data integration. Image fusion that merges images of coarse spatial resolution with fine-resolution ones is a process of spatial data integration. For example, in many remote-sensing systems, coarser-resolution multispectral bands are often complemented by a single finer-resolution panchromatic band, keeping the acquired data volume manageable (Aiazzi et al. 2002). We can fuse the fine-resolution panchromatic band data with the coarse-resolution multispectral data, generating fused images that contain the characteristics of both the spectral information in multispectral bands and the spatial information in the panchromatic band. Often, the objective of image fusion is to minimize redundancy in the output while maximizing information relevant to an application. In other words, successful image fusion will reduce the amount of data in storage or processing without significant loss of relevant information (Simone et al. 2002; Goshtasby and Nikolov 2007).

We use the term conflation as a superset of various approaches to integrating heterogeneous geospatial data, such as multisource images and maps, with the aim of increasing the quality and applicability of information products. There are four scenarios for data conflation, depending on the data types involved and specific objectives: raster–raster, raster–vector, vector–raster, and vector–vector, exemplified by image fusion, image classification utilizing ancillary vector data, map revision based on remote-sensing images, and conflation of digital maps, respectively (Zhang et al. 2007, 2009). The raster–vector and vector–raster scenarios are differentiated as they are situations for which the sources for data conflation are considered primary or secondary depending on the order of the sources (so for the raster–vector scenario, the raster and vector data sources are primary and secondary, respectively); the order of the data sources does not matter (i.e., they will not be treated differently as primary or secondary) if the output information remains in the same data model or can considered to be so. Although image fusion and map conflation have been addressed in remote-sensing and geographic information systems, respectively (Saalfeld 1988; Pohl and van Genderen 1998; Doytsher et al. 2001; Stein 2005; Hastings 2008), a coherent framework for multiscale data conflation remains underdeveloped. Thus, the availability of such a framework is crucial to further research on spatial data conflation for integral processing of massive geospatial data.

This chapter synthesizes geostatistical methods and other multiscale methods for conflating multiscale data, which often result from measurement and sampling employing differing data supports and sampling density. As discussed by Zhang et al. (2009), a coherent framework for data conflation should be built upon geospatial conceptualization, for which there are two complementary models, objects and fields. By the former, the real world is conceived of as populated by discrete entities $ID(x, attr)$ geo-referenced by position x and associated with attributes $attr$, whereas the latter views the world as consisting of layers of single-valued functions defined everywhere over the problem domain, as described in Chapter 2. The field variables may be interval/ratio $Z(x)$, such as terrain variables and population density, or discrete $C(x)$, such as vegetation and soil types. A prerequisite for spatial data conflation is alignment of multiple datasets. Depending on the objectives, conflation may be directed toward averaging, correction, or updating, in which different routines are involved. For instance, averaging of positional data needs to be implemented differently from correction or updating where detection and handling of inconsistency due to positional errors or arising from truly different objects are required. The same is also true of fields where disagreements signaling significant changes have to be detected and properly handled before averaging can proceed (Zhang et al. 2007, 2009).

As geospatial measurements are subject to various errors, the resultant data can be considered to be the sums of deterministic means and stochastic residuals. Thus, geostatistics provides a framework for conflation of spatial data (Journel and Huijbregts 1978; Cressie 1993; Zhang et al. 2007, 2009). Specifically, cokriging can be applied to fields of positional displacements to derive estimates of most probable positions for spatial objects. For interval/ratio fields, such as elevation, conflation may proceed to derive predictions about the variables under study by combining multisource data (Kyriakidis et al. 1999; Zhang et al. 2007). Zimmerman and Holland (2005) developed the so-called complementary cokriging method incorporating spatial dependence and measurement errors, in which observation network-specific biases and variances were used as the basis for combined analysis of data from all networks, outperforming naïve combined predictors.

As will be described, to better utilize cokriging for data conflation, cross-scale cokriging techniques are required. Atkinson et al. (2008) discussed cokriging for so-called super-resolution mapping, taking into account image correlations, data supports, sensors' point spread functions, and ancillary data. Liu and Journel (2009) discussed geostatistical methods for combining data with point and block supports.

The increasing availability of combined-resolution data stimulated the development of image fusion algorithms (Hill et al. 1999), which aim for visual enhancement and simulation of the multispectral images at finer spatial resolution. A common image fusion scheme for multispectral (coarser spatial resolution) and panchromatic (finer spatial resolution) images is to inject the spatial information from the panchromatic image into the multispectral image without overly modifying the spectral information of the latter. One method of achieving this is by transforming the image from the red–green–blue (RGB) color space to the intensity–hue–saturation (IHS) space

(Carper et al. 1990; Amolins et al. 2007). Wavelet-based approaches to image fusion were also developed (Aiazzi et al. 2002; Pajares and de la Cruz 2004), as they perform more accurately than standard schemes, particularly in terms of minimizing color distortion (Tu et al. 2001; Li et al. 2002; Thomas et al. 2008). As its second focus, this chapter will describe approaches to image fusion, with emphasis on wavelets.

Lastly, this chapter will describe multiresolution modeling and processing based on pyramidally organized trees (Basseville et al. 1992; Willsky 2002). The multi-resolution models and methods facilitate efficient algorithms for multiresolution data fusion and have broad applicability. There are various reasons for multiresolution modeling. Firstly, the phenomenon to be modeled can exhibit distinctive behavior over a range of scales or resolutions. Secondly, the available data are often at multiple resolutions, either as a result of transforming the data (e.g., using wavelet transforms) or because the data measure directly the quantities of interest at multiple resolutions. Thirdly, the objectives of users may be at multiple resolutions, as different studies focus on behavior over different ranges of scales. Lastly, it is important to develop algorithms at multiple resolutions for computational efficiency, as discussed by Willsky (2002). It is anticipated that multiresolution methods coupled with geostatistics and wavelets will have much to offer to efficient conflation of multiscale data.

8.1 MULTIVARIATE GEOSTATISTICS

There are geostatistical approaches to integrating multisource data in spatial prediction. For a primary variable Z_1, a simple way to introduce secondary variable Z_2 in either kriging or a simulation algorithm is to use secondary information as a trend, as reviewed by Goovaerts (1999) in the context of soil science. In *kriging with locally varying means*, the primary variable is decomposed into a mean $m(x)$ and a residual $R(x)$ component:

$$Z_1(x) = m(x) + R(x) \tag{8.1a}$$

where the mean component $m(x)$ can be modeled as a function (trend) of $Z_2(x)$

$$m(x) = trnd(Z_2(x)) \tag{8.1b}$$

rather than a function of spatial coordinates as often implemented in universal kriging (see Chapter 5 for details). It is often required that the Z_2 data (i.e., secondary data) are available at locations where $Z_1(x)$ needs to be predicted (i.e., secondary data are exhaustive) (Journel and Huijbregts 1978; Zhang and Yao 2008b).

In kriging with locally varying means, the spatial variability of Z_2 data is largely neglected. Often, secondary data have their own particular spatial correlation or exhibit cross-correlation with the primary variable Z_1. As shown in a case study in a mountainous region of southern Italy by Diodato (2005), topographic covariables (i.e., terrain elevation and topographic index) were found to influence the spatial variability of precipitation over small regions of complex terrain. In such cases, one would like to use spatial correlation of the secondary data (e.g., topographic covariables) expressed in $\mathrm{cov}_{Z_2}(h)$ and intercorrelation between secondary and primary data (e.g., precipitation) expressed through the cross-covariance $\mathrm{cov}_{Z_1Z_2}(h)$, as in bivariate cokriging (Journel and Huijbregts 1978; Diodato 2005). Multivariate geostatistics can be utilized for analyzing and processing spatially indexed multivariate data where the indexing is continuous across space (Kyriakidis et al. 1999; Bailey and Krzanowski 2012). For example, cokriging can be used to incorporate exhaustively sampled secondary data for spatial prediction of the primary variable that is undersampled relative to the coverage and density required of a given application, and to combine different measurements of the same property (e.g., a few precise laboratory measurements of clay content and more numerous field data collected using cheaper measuring devices) (Goovaerts 1999). We describe multivariate geostatistics next.

Consider a problem domain characterized by a set of K spatially intercorrelated random functions (RFs) $\{Z_k(x), k = 1 \text{ to } K\}$. The first- and second-order moments of these variables, assuming stationarity, are

$$E(Z_k(x)) = m_k \forall x \tag{8.2a}$$

$$\text{cov}_{kk'}(h) = E((Z_k(x_{s1}) - m_k)(Z_{k'}(x_{s2}) - m_k)) \text{ for } x_{s1} - x_{s2} = h \tag{8.2b}$$

as also described by Journel and Huijbregts (1978).

Many spatial variables, such as reflectance and land cover, are actually defined and measured on supports of finite size, such as pixels and parcels, which are assumed to be generated from finite-support random variables that originate from point-support variables $Z(x)$ (Krajewski 1987). Denote such a block-support variable as $Z(v(x))$, where $v(x)$ stands for a finite support centered at x. This notation applies also to point-support variables, when a block $v(x)$ is reduced to a point x. Thus, block-support variables and their corresponding data are used as a general setting in the following text.

As described in Chapter 6, it is usually assumed that the value over a block support (coarse-scale) is a linear average of all the point-support (fine-scale) values within that block. For instance, we may estimate variable Z_k over a block $v_k(x)$ centered at location x as the arithmetic average of the array of n_v points used to discretize $v_k(x)$:

$$z_k(v_k(x)) = \frac{1}{n_v} \sum_{k_v=1}^{n_v} z_k(x_{k_v}) \tag{8.3}$$

Suppose that there are K regionalized variables operating in the problem domain. The available data $\left\{ z_k\left(v_{\alpha_k}\right), \alpha_k = 1, ..., n_k, k = 1, ..., K \right\}$ are defined on the support $\{v_{\alpha_k}, \alpha_k = 1, ..., n_k, k = 1, ..., K\}$, which may be points or areas centered at points $\{x_{\alpha_k}, \alpha_k = 1, ..., n_k, k = 1, ..., K\}$ in a two-dimensional problem domain. Similar notations for block-support data (e.g., areal data in a two-dimensional setting) were used in Chapter 6 and by Kyriakidis (2004). Let k_0 be the descriptor for the variable being analyzed, $k_0 \in \{1, ..., K\}$. Prediction of Z_{k_0} over a block $v_{k_0}(x)$ centered at location x is provided by a linear combination:

$$\hat{z}_{k_0}(v_{k_0}(x)) = m_{k_0} + \sum_{k=1}^{K} \sum_{\alpha_k=1}^{n_k} \lambda_{\alpha_k}(z_k(v_{\alpha_k}) - m_k) \tag{8.4}$$

where $z_k(v_{\alpha_k})$ represents the Z_k data defined on the set of supports $\{v_{\alpha_k}\}$ $(\alpha_k = 1, 2, \ldots n_k)$, whereas λ_{α_k} represents the weight assigned to support v_{α_k}. These weights are derived by the solution of the following system of simple cokriging:

$$\sum_{k'=1}^{K}\sum_{\alpha_{k'}=1}^{n_{k'}} \lambda_{\alpha_{k'}} \overline{c}_{k'k}(v_{k'}(x_{\alpha k'}), v_k(x_{\alpha_k})) = \overline{c}_{k_0 k}(v_{k_0}(x), v_k(x_{\alpha_k})),$$

$$\forall \alpha_k = 1 \text{ to } n_k, \ k = 1, \ldots, K$$

(8.5)

where $\overline{c}_{k'k}$ and $\overline{c}_{k_0 k}$ represent the average covariance between pairs of locations within data supports $v_{k'}(x_{\alpha_{k'}})$ and $v_k(x_{\alpha_k})$, and the average covariance between locations discretizing the support to be estimated (i.e., $v_{k_0}(x)$) and those within data support $v_k(x_{\alpha_k})$, respectively (Journel and Huijbregts 1978; Zhang et al. 2007; Zhang and Yao 2008b). For instance, $\overline{c}_{k'k}$ is calculated by

$$\overline{c}_{k'k}(v_{k'}(x_{\alpha_{k'}}), v_k(x_{\alpha_k})) = \frac{1}{N_{\alpha_k} N_{\alpha_{k'}}} \sum_{n_{\alpha_{k'}}=1}^{N_{\alpha_{k'}}} \sum_{n_{\alpha_k}=1}^{N_{\alpha_k}} \text{cov}_{k'k}\left(x_{n_{\alpha_{k'}}} - x_{n_{\alpha_k}}\right)$$

(8.6)

where $x_{k'_v}$ and x_{k_v} denote locations falling within data supports $v_{k'}(x_{k'_\alpha})$ and $v_k(x_{k_\alpha})$, respectively (Zhang and Yao 2008b).

As described by Journel and Huijbregts (1978), to develop the ordinary cokriging equations, we can rewrite the simple cokriging estimator for Z_{k_0} over a block $v_{k_0}(x)$ (Equation 8.4) as

$$\hat{z}_{k_0}(v_{k_0}(x)) = \sum_{k=1}^{K}\sum_{\alpha_k=1}^{n_k} \lambda_{\alpha_k} z_k(v_{\alpha_k}) + m_{k_0}\left(1 - \sum_{\alpha_{k_0}=1}^{n_{k_0}} \lambda_{\alpha_{k_0}}\right) - \sum_{k \neq k_0}\left(\sum_{\alpha_k=1}^{n_k} \lambda_{\alpha_k}\right) m_k$$

(8.7)

To cancel out the effects of unknown means of coregionalized variables, the following constraints are imposed:

$$\begin{cases} \displaystyle\sum_{\alpha_{k_0}=1}^{n_{k_0}} \lambda_{\alpha_{k_0}} = 1 \\[2em] \displaystyle\sum_{\alpha_k=1}^{n_k} \lambda_{\alpha_k} = 0 \ \forall \ k \neq k_0 \end{cases}$$

(8.8)

which lead to the so-called ordinary cokriging estimator below:

$$\hat{z}_{k_0}(v_{k_0}(x)) = \sum_{k=1}^{K}\sum_{\alpha_k=1}^{n_k} \lambda_{\alpha_k} z_k(v_{\alpha_k})$$

(8.9)

where $z_k(v_{\alpha_k})$ $(\alpha_k = 1, 2, ..., n_k)$ represents the data defined on the set of supports $\{v_{\alpha_k}\}$, whereas λ_{α_k} represents the weight assigned to v_{α_k}, obtained by solving the following equations:

$$\sum_{k'=1}^{K} \sum_{\alpha_{k'}=1}^{n_{k'}} \lambda_{\alpha_{k'}} \bar{c}_{k'k}(v_{\alpha_{k'}}, v_{\alpha_k}) - v_k = \bar{c}_{k_0 k}(v_{k_0}, v_{\alpha_k}) \ \forall \ \alpha_k = 1, 2, ..., n_k \ \ k = 1, 2, ..., K$$

$$\sum_{\alpha_{k_0}=1}^{n_{k_0}} \lambda_{\alpha_{k_0}} = 1 \tag{8.10}$$

$$\sum_{\alpha_k=1}^{n_k} \lambda_{\alpha_k} = 0 \ \forall \ k \neq k_0$$

where the v_ks are Lagrange multipliers, with cokriging variance evaluated as

$$\sigma_{v_{k_0}}^2 = \bar{c}_{k_0 k_0}(v_{k_0}, v_{k_0}) + v_{k_0} - \sum_{k=1}^{K} \sum_{\alpha_k=1}^{n_k} \lambda_{\alpha_k} \bar{c}_{k_0 k}(v_{k_0}, v_{\alpha_k}) \tag{8.11}$$

as described by Journel and Huijbregts (1978). The ordinary cokriging equations were also reviewed by Zhang et al. (2007) and Zhang and Yao (2008b).

However, as warned by Goovaerts (1999), cokriging is much more computationally expensive than kriging because $K(K + 1)/2$ direct and cross-semivariograms must be jointly modeled for cokriging, with a larger cokriging system to solve. For joint modeling of direct and cross-semivariograms, if linear models of coregionalization are used, their permissibility can be checked through cumbersome procedures of checking whether basic variogram models are permissible and coregionalization matrices are positive semi-definite. Nevertheless, cokriging is well worth the efforts when the primary variable is undersampled with regard to the secondary variables and the secondary variables are well correlated with the primary variable.

As shown above, cokriging provides the best linear unbiased prediction of spatially correlated multivariate RFs and requires the solution of a large linear system based on the auto- and cross-covariance matrix of the data at sampled locations. It tends to be inefficient to solve the linear systems directly, especially with a large number of cross-correlated RFs. Furrer and Genton (2011) proposed an efficient linear unbiased predictor based on a linear aggregation of the covariables. The primary variable together with this single meta-covariable is used to perform cokriging. The proposed approach was used to create reanalysis-type, high-resolution historical temperature fields.

Cokriging based on matrix formulation (Myers 1982; Ver Hoef and Cressie 1993) may be pursued for applications where vector-valued spatial properties are concerned. For example, multiple variables, such as lead, zinc, and copper, are often involved in soil pollution studies, and their joint modeling and prediction are necessary (Majumdar and Gelfand 2007). Cokriging formulated in vector and matrix notations is obviously efficient for vector-valued fields, although this is not elaborated further here.

Multivariate geostatistics requires auto- and cross-variogram modeling of multivariate data. Bourgault and Marcotte (1991) presented a multivariable variogram-based method for finding the elementary variograms that characterize the different spatial scales contained in a set of data with K variables. A multivariable variogram is defined in a way similar to that of the traditional variogram, by the expected value of a distance, squared, in a space with K dimensions. Combined with the linear model of coregionalization, when the number of elementary components is less than or equal to the number of variables, it is possible, by means of nonlinear regression of variograms and cross-variograms, to estimate the coregionalization parameters directly in order to obtain the elementary variables themselves, either by cokriging or by direct matrix inversion. This new tool greatly simplifies the conventional procedure of multivariate variogram modeling (requiring fitting $K(K + 1)/2$ auto- and cross-variogram models), as it completes the search for the elementary variograms using only one variogram (multivariable).

There is continuing research on multivariate variogram modeling. Again, for linear models of coregionalization, Emery (2010) described three iterative algorithms to find coregionalization matrices that minimize a weighted sum of the squared deviations between sample and modeled variograms. These three algorithms are variogram fitting with no constraint other than mathematical consistency, simulated annealing for variogram fitting subject to constraints on the simple variogram sills, and a nonlinear least squares algorithm for fitting a plurigaussian model. In all three algorithms, the sample variogram matrices need not be known entirely for every lag vector, a situation of interest with heterotopic samplings.

As has been discussed previously, these methods for geostatistical data conflation feature cross-scale cokriging, drawing upon multiscale data for information fusion, which requires auto- and cross-covariance models defined on point supports. These models may be estimated from deconvoluting empirical variograms computed with finite-support data, as described in Chapter 4. Once the set of point-support models are inverted from empirical and finite-support variograms, any block-support covariance and cross covariance can be obtained easily by regularization. The techniques for derivation of point-support variogram models from empirical variograms computed over block supports are known as deconvolution, as shown in Chapter 4. Goovaerts (2008) applied deconvolution in area-to-point kriging to map the property of interest within regular or irregular data units (i.e., cancer rates aggregated at the county level). This method yields a more accurate prediction than a naïve point kriging of areal data that simply collapses each county into its geographic centroid, as also discussed via location and attribute error interactions by Cressie and Kornak (2003).

In addition to simple and ordinary cokriging for the solution of spatial prediction with coregionalized variables, as has been discussed previously, other implementations of cokriging should also be explored, such as cokriging variants that can handle nonstationarity in coregionalization. There is increasing research on multivariate geostatistics, as reviewed by Bailey and Krzanowski (2012). With further developments, there will be a systematic treatment for cokriging, just like simple, ordinary, and universal variants that have been discussed (Stein at al. 1991; Gertner et al. 2007). Moreover, cokriging with constraints, such as those ensuring consistency of estimates across scales and honoring geomorphologic features (e.g., ridges vs. valleys) in the case of terrain data conflation, should be explored.

It is anticipated that the method presented here will consolidate the basis for cross-scale spatial data handling. Cross-scale cokriging can be applied not only to continuous variables such as elevation, but also to categorical variables such as land cover and soil types, whereby improved discrimination of one area class from another is achievable by bringing together multisource covariates for classification. Data conflation can also be implemented for merging multisource map data depicting discrete points, lines, and areas.

8.2 IMAGE FUSION

8.2.1 CONVENTIONAL APPROACHES

Most algorithms for image fusion can be roughly divided into two groups: spectral substitute and spatial domain methods (Zhukov et al. 1999; Liang 2003). The former type of method is reviewed first, followed by a brief introduction to the latter type of method, for which wavelet-based methods are described in Sections 8.2.2 and 8.2.3.

Spectral component substitution is based on replacing a spectral component of the coarse-resolution multispectral image by a finer-resolution substitute, usually a panchromatic image. The most frequently used include the IHS method, principal component analysis (PCA) methods, and regression methods (Carper et al. 1990; Pohl and Van Genderen 1998).

In the *IHS* color space, intensity (*I*) is a measure of brightness. It is also sometimes called luminance (*L*). Hue (*H*) is the color, measured as the angle around a color wheel or color hexagon, whereas saturation (*S*) is the amount of color, with zero representing gray, or no color, and one representing full color. With three multispectral bands, we can produce *RGB* color composite imagery. We can then transfer three multispectral bands from the *RGB* color space to the *IHS* color space using the formula of Pratt (1991).

$$
\begin{bmatrix} I \\ vr_1 \\ vr_2 \end{bmatrix} = \begin{bmatrix} 1/3 & 1/3 & 1/3 \\ -\sqrt{2}/6 & -\sqrt{2}/6 & 2\sqrt{2}/6 \\ 1/\sqrt{2} & -1/\sqrt{2} & 0 \end{bmatrix} \begin{bmatrix} R \\ G \\ B \end{bmatrix} \tag{8.12}
$$

The variables vr_1 and vr_2 in Equation 8.12 can be considered as the X and Y axes in the Cartesian coordinate system, whereas intensity I indicates the Z axis. Thus, the hue (*H*) and saturation (*S*) can be represented by

$$
H = \tan^{-1}(vr_2 / vr_1) \text{ and } S = \sqrt{vr_1^2 + vr_2^2} \tag{8.13}
$$

as shown by Tu et al. (2001). The panchromatic band can then be used to replace component I after being linearly stretched to match the mean and variance of the intensity component (Liang 2003), which can be denoted as I'. The last step is to transform the revised IHS space into the RGB space:

$$\begin{bmatrix} R \\ G \\ B \end{bmatrix} = \begin{bmatrix} 1 & -1/\sqrt{2} & 1/\sqrt{2} \\ 1 & -1/\sqrt{2} & -1/\sqrt{2} \\ 1 & \sqrt{2} & 0 \end{bmatrix} \begin{bmatrix} I' \\ vr_1 \\ vr_2 \end{bmatrix} \tag{8.14}$$

as described by Carper et al. (1990) and Tu et al. (2001). Although there are several different formulae for the IHS transformation (Carper et al. 1990), Nunez et al. (1999) demonstrated that the best results can generally be obtained when the intensity is calculated as the average of the R, G, and B bands (Tu et al. 2001).

The PCA technique, also known as the Karhunen–Loeve transform, is a decorrelation scheme used for statistical data analysis. We can derive an orthogonal color coordinate system for PCA:

$$\begin{bmatrix} pc_1 \\ pc_2 \\ pc_3 \end{bmatrix} = \textbf{Trns} \begin{bmatrix} R \\ G \\ B \end{bmatrix} \tag{8.15a}$$

where **Trns** is the transformation matrix consisting of eigenvectors of the covariance matrix $\textbf{COV}_{RGB}$ of the RGB vectors. **Trns** satisfies the following relationship:

$$\textbf{Trns } \textbf{COV}_{RGB} \textbf{ Trns}^\mathrm{T} = \mathrm{diag}(\lambda_1, \lambda_2, \lambda_3) \tag{8.15b}$$

where λ_1, λ_2, λ_3 are eigenvalues corresponding to **Trns** with a descending order (Tu et al. 2001; Richards and Jia 2006).

The merging of the RGB and the pan images using the PCA method is similar to that of the IHS method. The first component, pc_1, of the PCA space is replaced by the pan image and transformed back into the original space (Tu et al. 2001).

To preserve the available radiometric information of the coarse-resolution image, spatial-domain techniques transfer excess high-spatial-frequency components to the coarse-resolution bands. This can be achieved by high-pass filters in the image domain or by using various multiresolution representations, such as the wavelet decomposition, which works on the basis of an image's local frequency content, as reviewed by Liang (2003). We introduce the wavelet-based method for image fusion after describing the wavelet transform and multiresolution representations.

8.2.2 Multiresolution Representations Based on Wavelets

The wavelet transform provides a framework to decompose an image into a series of images, each at different resolution. Therefore, it is a valuable method for multiresolution analysis and, thus, image fusion (Yocky 1996; Amolins et al. 2007). Through the wavelet transform, a function, say, $Z(x)$, is represented as a superposition of wavelets, from which we can also reconstruct the original function. In this subsection, we describe concepts and principles underlying wavelets and wavelet representations, based on Mallat (1989).

We start the description of wavelets with a multiresolution approximation of one-dimensional signals. Suppose the original signal $Z(x)$ is measurable and has a finite energy or variance (Lark and Webster 1999). Mathematically, this is denoted as $Z(x) \in L^2(R)$, with $L^2(R)$ being a vector space of measurable, square-integrable functions (i.e., the integrals of the square of their absolute values are finite). Let A_{2^j} be the operator that approximates a signal at a resolution 2^j, j being an integer, so $A_{2^j}Z(x)$ represents the approximation of $Z(x)$ at a resolution 2^j. For a given multi-resolution approximation U_{2^j}, there exists a unique scaling function $\phi(x) \in L^2(R)$ such that $\left(\sqrt{2^{-j}}\phi_{2^j}(x - 2^{-j}n)\right)$, $n \in \{\text{integers}\}$ is an orthonormal basis of U_{2^j}, where $\phi_{2^j}(x) = 2^j\phi(2^j x)$ (the dilation of $\phi(x)$ by 2^j) and the coefficient $\sqrt{2^{-j}}$ is for normalization of the functions in the $L^2(R)$ norm. $A_{2^j}Z(x)$ can be computed by decomposing it on the orthonormal basis above:

$$A_{2^j}Z(x) = 2^{-j}\sum_{n=-\infty}^{+\infty}\left\langle Z(x'),\phi_{2^j}(x' - 2^{-j}n)\right\rangle\phi_{2^j}(x - 2^{-j}n)$$

$$= 2^{-j}\sum_{n=-\infty}^{+\infty}A_{2^j}^d Z\,\phi_{2^j}(x - 2^{-j}n) \tag{8.16}$$

where $\left\langle Z(x'),\phi_{2^j}(x' - 2^{-j}n)\right\rangle$ stands for an inner product (i.e., $\int_{-\infty}^{+\infty}Z(x')\phi_{2^j}(x' - 2^{-j}n)$ $dx' = (Z(x') * \phi_{2^j}(-x'))(2^{-j}n))$ and is denoted as $A_{2^j}^d Z$, which stands for a discrete approximation of $Z(x)$ at resolution 2^j (Mallat 1989).

The discrete approximation $A_{2^j}^d Z$ is further analyzed below. As the orthonormal basis of U_{2^j} is included in that of $U_{2^{j+1}}$, the former can be expanded in the latter:

$$\phi_{2^j}(x - 2^{-j}n) = 2^{-j-1}\sum_{k=-\infty}^{+\infty}\left\langle \phi_{2^j}(x' - 2^{-j}n),\phi_{2^{j+1}}(x' - 2^{-j-1}k)\right\rangle\phi_{2^{j+1}}(x - 2^{-j-1}k)$$

$$= \sum_{k=-\infty}^{+\infty}\left\langle \phi_{2^{-1}}(x'),\phi(x' - (k - 2n))\right\rangle\phi_{2^{j+1}}(x - 2^{-j-1}k) \tag{8.17}$$

We can compute the inner products of $Z(x)$ with both sides of Equation 8.17 as

$$\left\langle Z(x'),\phi_{2^j}(x' - 2^{-j}n)\right\rangle = \sum_{k=-\infty}^{+\infty}\left\langle \phi_{2^{-1}}(x'),\phi_{2^{j+1}}(x' - (k - 2n))\right\rangle\left\langle Z(x'),\phi_{2^{j+1}}(x' - 2^{-j-1}k)\right\rangle$$

$$\tag{8.18a}$$

which can be rewritten as

$$A_{2^j}^d Z = \sum_{k=-\infty}^{+\infty} \tilde{h}(2n-k)A_{2^{j+1}}^d Z \tag{8.18b}$$

where $\tilde{h}(n) = h(-n)$ and $h(n) = \langle \phi_{2^{-1}}(x'), \phi(x'-n) \rangle$. Equation 8.18b shows that $A_{2^j}^d Z$ can be computed by convolving $A_{2^{j+1}}^d Z$ with $\hat{h}$ and keeping every other sample of the output. All the discrete approximations $A_{2^j}^d Z$ ($j < 0$) can thus be computed from $A_1^d Z$ by repeating this process (called pyramid transform) (Mallat 1989).

As shown above, the approximation at resolutions 2^{j+1} and 2^j of a signal is equal to its orthogonal projection on $U_{2^{j+1}}$ and U_{2^j}, respectively. The detail signal at resolution 2^j is the difference in information between the approximation of a function $Z(x)$ at resolutions 2^{j+1} and 2^j and is thus given by the orthogonal projection of the original signal on the orthogonal complement O_{2^j} of U_{2^j} in $U_{2^{j+1}}$ (Mallat 1989). To compute the detail signal of $Z(x)$, denoted by $D_{2^j}Z$, we need to find an orthonormal basis of O_{2^j}: $\sqrt{2^{-j}}\psi_{2^j}(x-2^{-j}n)$, where $\psi(x)$ is an orthogonal wavelet, with a dilation by 2^j denoted by $\psi_{2^j}(x) = 2^j\psi(2^j x)$, similar to that of $\phi(x)$ (Mallat 1989; Lark and Webster 1999). The detail signal can then be computed as $D_{2^j}Z = \langle Z(x), \psi_{2^j}(x-2^{-j}n) \rangle$.

We can show that $D_{2^j}Z$ can be calculated by convolving $A_{2^{j+1}}^d Z$ with a discrete filter. Because the function $\psi_{2^j}(x-2^{-j}n)$ is a member of $O_{2^j} \subset U_{2^{j+1}}$, as in Equation 8.17, this function can be expanded in an orthonormal basis of $U_{2^{j+1}}$. Then, we obtain

$$D_{2^j}Z = \sum_{k=-\infty}^{+\infty} \tilde{g}_{2^{-1}}(2n-k)A_{2^{j+1}}^d Z \tag{8.19}$$

where $\tilde{g}(n) = g(-n)$ and $g(n) = \langle \psi_{2^{-1}}(x'), \phi(x'-n) \rangle$, a high-pass filter. Equation 8.19 shows that we can compute the detail signal $D_{2^j}Z$ by convolving $A_{2^{j+1}}^d Z$ with the filter $\tilde{g}$ and retaining every other sample of the output. The orthogonal wavelet representation of a discrete signal $A_1^d Z$ can therefore be computed by successively decomposing $A_{2^{j+1}}^d Z$ into $A_{2^j}^d Z$ (Equation 8.18b) and $D_{2^j}Z$ (Equation 8.19) for $-J \le j \le -1$ (Mallat 1989; Lark and Webster 1999).

The original signal $Z(x)$ can also be reconstructed with a pyramid transform. According to Mallat (1989), $\left(\sqrt{2^{-j}}\phi_{2^j}(x-2^{-j}n), \sqrt{2^{-j}}\psi_{2^j}(x-2^{-j}n)\right), (n \in \{\text{integers}\})$ is an orthonormal basis of $U_{2^{j+1}}$. Thus, we can show that

$$A_{2^{j+1}}^d Z = 2\sum_{k=-\infty}^{+\infty} h(n-2k)\langle Z(x'), \phi_{2^j}(x'-2^{-j}k)\rangle + 2\sum_{k=-\infty}^{+\infty} g(n-2k)\langle Z(x'), \psi_{2^j}(x'-2^{-j}k)\rangle$$

$$= 2\sum_{k=-\infty}^{+\infty} h(n-2k)A_{2^j}^d Z + 2\sum_{k=-\infty}^{+\infty} g(n-2k)D_{2^j}Z \tag{8.20}$$

where h and g are filters defined previously. Equation 8.20 shows that $A^d_{2^{j+1}}Z$ can be reconstructed by putting zeros between each sample of $A^d_{2^j}Z$ and $D_{2^j}Z$ and convolving the resulting signals with the filters h and g, respectively. The original discrete signal $A^d_1 Z$ at resolution 1 is reconstructed by repeating this procedure for $-J \leq j < 0$ (Mallat 1989).

Consider the extension of the orthogonal wavelet representation to two-dimensional images $Z(X, Y) \in L^2(R^2)$. According to Mallat (1989), the approximation of $Z(X, Y)$ at a resolution 2^j is equal to its orthogonal projection on the vector space U_{2^j} $(-J \leq j < 0)$, whose orthonormal basis may be built upon a scaling function $\Phi(X,Y)$ through its dilation and translation:

$$\left(2^{-j}\Phi_{2^j}(X-2^{-j}n_X, Y-2^{-j}n_Y)\right), \quad \left(n_X, n_Y \in \{\text{integers}\}^2\right) \qquad (8.21a)$$

where $\Phi_{2^j}(X,Y) = 2^{2^j}\Phi(2^j X, 2^j Y)$ and the factor 2^{-j} normalizes each function in the $L^2(R^2)$ norm. For separable multiresolution approximations of $L^2(R^2)$, the scaling function $\Phi(X, Y)$ can be written as $\Phi(X, Y) = \phi(X)\phi(Y)$, where $\phi(X)$ is the one-dimensional scaling function of $\left(U^1_{2^j}\right)$. Equation 8.21a is then rewritten as

$$\left(2^{-j}\Phi_{2^j}(X-2^{-j}n_X, Y-2^{-j}n_Y)\right)$$
$$=\left(2^{-j}\phi_{2^j}(X-2^{-j}n_X)\phi_{2^j}(Y-2^{-j}n_Y)\right), \quad \left(n_X, n_Y \in \{\text{integers}\}^2\right) \qquad (8.21b)$$

The approximation of a signal $Z(X, Y)$ at a resolution 2^j is therefore characterized by the set of inner products $A^d_{2^j}Z = \left(\left\langle Z(X,Y)\phi_{2^j}(X - 2^{-j}n_X)\phi_{2^j}(Y - 2^{-j}n_Y)\right\rangle\right)$.

The detail signal at resolution 2^j is equal to the orthogonal projection of the signal on the orthogonal complement (say, O_{2^j}) of U_{2^j} in $U_{2^{j+1}}$. An orthonormal basis of O_{2^j} can be built by scaling and translating three wavelet functions, $\Psi^1(X, Y)$, $\Psi^2(X, Y)$, and $\Psi^3(X, Y)$:

$$2^{-j}\Psi^1_{2^j}(X-2^{-j}n_X, Y-2^{-j}n_Y),$$
$$2^{-j}\Psi^2_{2^j}(X-2^{-j}n_X, Y-2^{-j}n_Y), \quad \left(n_X, n_Y \in \{\text{integers}\}^2\right) \qquad (8.22)$$
$$2^{-j}\Psi^3_{2^j}(X-2^{-j}n_X, Y-2^{-j}n_Y)$$

where $\Psi^1(X, Y) = \phi(X)\psi(Y)$, $\Psi^2(X, Y) = \psi(X)\phi(Y)$, and $\Psi^3(X, Y) = \psi(X)\psi(Y)$, with $\psi(X)$ being the one-dimensional wavelet associated with the scaling function $\phi(X)$ (Mallat 1989). The difference images are characterized by the inner products of $Z(X, Y)$ with each vector of an orthonormal basis of O_{2^j} (Equation 8.22). The three detail images are

$$D^1_{2^j}Z = \left(\left\langle Z(X,Y), \Psi^1_{2^j}(X-2^{-j}n_X, Y-2^{-j}n_Y)\right\rangle\right)$$
$$= \left(\left(Z(X,Y)*\phi^j(-X)\psi^j(-Y)\right)(2^{-j}n_X, 2^{-j}n_Y)\right) \qquad (8.23a)$$

$$D_{2^j}^2 Z = \left(\left\langle Z(X,Y), \Psi_{2^j}^2 (X - 2^{-j} n_X, Y - 2^{-j} n_Y) \right\rangle \right)$$
$$= \left(\left(Z(X,Y) * \psi_{2^j}(-X) \phi_{2^j}(-Y) \right) (2^{-j} n_X, 2^{-j} n_Y) \right)$$

(8.23b)

$$D_{2^j}^3 Z = \left(\left\langle Z(X,Y), \Psi_{2^j}^3 (X - 2^{-j} n_X, Y - 2^{-j} n_Y) \right\rangle \right)$$
$$= \left(\left(Z(X,Y) * \psi_{2^j}(-X) \psi_{2^j}(-Y) \right) (2^{-j} n_X, 2^{-j} n_Y) \right)$$

(8.23c)

where the three wavelets $\Psi_1(X, Y)$, $\Psi_2(X, Y)$, and $\Psi_3(X, Y)$ are given by separable products of the functions ϕ and ψ, and n_X, $n_Y \in \{\text{integers}\}^2$, as shown by Mallat (1989).

As seen above, the image $A_{2^{j+1}}^d Z$ can be decomposed into four images: $A_{2^j}^d Z$, $D_{2^j}^1 Z$, $D_{2^j}^2 Z$, and $D_{2^j}^3 Z$, with $A_{2^j}^d Z$, $D_{2^j}^1 Z$, $D_{2^j}^2 Z$, and $D_{2^j}^3 Z$ corresponding to the lowest frequencies, the vertical high frequencies, the horizontal high frequencies, and the high frequencies in both directions, respectively. For any $J > 0$, an image $A_1^d Z$ is represented completely by the $3J + 1$ discrete images

$$\left(A_{2^{-J}}^d Z, \left(D_{2^j}^1 Z \right)_{-J \leq j \leq -1}, \left(D_{2^j}^2 Z \right)_{-J \leq j \leq -1}, \left(D_{2^j}^3 Z \right)_{-J \leq j \leq -1} \right)$$

which is called an orthogonal wavelet representation in two dimensions (Mallat 1989; Amolins et al. 2007).

Specifically, the filtering operations are conducted along the horizontal direction (row-wise), followed by row-wise downsampling. The two resulting rectangular images are further processed along the vertical direction (column-wise), followed by column-wise downsampling. After one stage of processing, one image at resolution level $j - 1$ is decomposed into four subimages: an image at coarser-resolution level j (low–low band), a horizontal orientation image (low–high band), a vertical orientation image (high–low band), and a diagonal orientation image (high–high band). The filtering is applied recursively to the outputs of the low–low band until the level of resolution desired is obtained (Mallat 1989; Pajares and de la Cruz 2004; Amolins et al. 2007). This is shown in Figure 8.1a, where $\tilde{G}$ and $\tilde{H}$ (corresponding to the aforementioned filters $\tilde{g}$ and $\tilde{h}$, respectively) represent the filters used for decomposition.

Figure 8.1b shows the procedures for reconstruction, where G and H (i.e., the filters g and h described previously) represent the filters used for reconstruction. At each step, the image $A_{2^{j+1}}^d Z$ is reconstructed from $A_{2^j}^d Z$, $D_{2^j}^1 Z$, $D_{2^j}^2 Z$, and $D_{2^j}^3 Z$. Between each column of the images $A_{2^j}^d Z$, $D_{2^j}^1 Z$, $D_{2^j}^2 Z$, and $D_{2^j}^3 Z$, we add a column of zeros and convolve the rows with a one-dimensional filter H; we then add a row of zeros between each row of the resulting image and convolve the columns with another one-dimensional filter G. The image $A_1^d Z$ is reconstructed from its wavelet transform by repeating this process for $-J \leq j \leq -1$ (Mallat 1989; Pajares and de la Cruz 2004; Amolins et al. 2007).

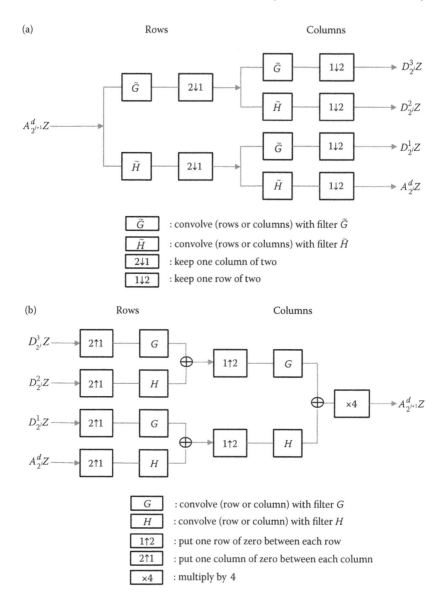

FIGURE 8.1 (a) One stage in a multiresolution image decomposition; (b) one stage in a multiresolution image reconstruction. (Based on Mallat, S.G., A theory for multiresolution signal decomposition: the wavelet representation. *IEEE Transactions on Pattern Analysis and Machine Intelligence* 11(7): 674–693, 1989.)

8.2.3 WAVELET-BASED IMAGE FUSION

As a two-dimensional discrete wavelet transform (DWT) decomposes images into multiresolution representations with both low-frequency coarse information and high-frequency detail information (Aiazzi et al. 2002), it is possible to make use of the multiresolution information to produce a fused image with increased visual interpretability and informational efficiency. As Zhou et al. (1998) demonstrated in the fusion of SPOT PAN and Landsat TM images, multisensor data fusion is a trade-off between the spectral information from a coarse-spatial-, fine-spectral-resolution sensor and the spatial structure from a fine-spatial-, coarse-spectral-resolution sensor, and is relatively easy to control with wavelet-based methods.

The wavelet transforms of images can be computed by the algorithms described by Pajares and de la Cruz (2004). In the decomposition phase of a two-dimensional DWT, each row of the image is separately filtered, as shown in Figure 8.1a. The resulting row-transformed image is then similarly filtered in the column direction, yielding four subbands at the first decomposition level 2^{j+1} and 2^j. The three detail subbands, $D_j LH$, $D_j HL$, and $D_j HH$, contain the vertical, horizontal, and diagonal high-frequency information, respectively, whereas the approximation subband $S_j LL$ is a low-pass filtered version of the original image. This approximation subband is subsequently passed to the next level for further subband decomposition (Li et al. 2002; Pajares and de la Cruz 2004).

It is necessary for images for fusion to be first registered (Toutin 2011). For instance, if a Landsat TM image and a SPOT PAN image are used for fusion, the Landsat TM image is resampled to the same resolution as the SPOT PAN image (10-m resolution), so that both images occupy the same geographic space and possess the same pixel size. After image registration, the images to be fused are then decomposed by DWT to the same resolution. Wavelet coefficients from TM's approximation subband and SPOT's detail subbands are then combined together, and the fused image is reconstructed by performing the inverse DWT. Since the distributions of coefficients in the detail subbands have a mean of zero, the fusion result does not change the radiometry of the original multispectral image (Li et al. 2002).

For illustration, assume that the number of decomposition levels is J, and SPOT PAN and Landsat TM images are to be fused. Denote the decomposed subbands of the SPOT PAN image by $\left(A_{2^{-i}}^d PAN, \left(D_{2^{-i}}^1 PAN \right)_{-J \leq j \leq -1}, \left(D_{2^{-i}}^2 PAN \right)_{-J \leq j \leq -1}, \left(D_{2^{-i}}^3 PAN \right)_{-J \leq j \leq -1} \right)$ and one band of the Landsat TM image by $\left(A_{2^{-i}}^d MS, \left(D_{2^{-i}}^1 MS \right)_{-J \leq j \leq -1}, \left(D_{2^{-i}}^2 MS \right)_{-J \leq j \leq -1}, \left(D_{2^{-i}}^3 MS \right)_{-J \leq j \leq -1} \right)$. It is straightforward to merge the approximation subband of the Landsat TM image and the detail subbands of the SPOT PAN image into a fused DWT representation. Denote the decomposed subbands in the fused DWT representation by $\left(A_{2^{-i}}^d F, \left(D_{2^{-i}}^1 F \right)_{-J \leq j \leq -1}, \left(D_{2^{-i}}^2 F \right)_{-J \leq j \leq -1}, \left(D_{2^{-i}}^3 F \right)_{-J \leq j \leq -1} \right)$. As shown by Li et al. (2002), we can write the fusion scheme as

$$A_{2^{-i}}^d F = A_{2^{-i}}^d MS \qquad (8.24a)$$

$$D^1_{2^{-i}}F = D^1_{2^{-i}}PAN, \quad -J \le j \le -1 \tag{8.24b}$$

$$D^2_{2^{-i}}F = D^2_{2^{-i}}PAN, \quad -J \le j \le -1 \tag{8.24c}$$

$$D^3_{2^{-i}}F = D^3_{2^{-i}}PAN, \quad -J \le j \le -1 \tag{8.24d}$$

The fused image is obtained by performing the inverse DWT on the fused DWT representations.

As shown previously, in the wavelet-based fusion of images, detail information is extracted from the SPOT PAN image using wavelet transforms and injected into the Landsat TM image. Distortion of the spectral information is minimized (Li et al. 2002). This kind of substitutive wavelet method is similar to the standard IHS and PCA schemes, as they all involve replacing information from one image with information from another image.

Amolins et al. (2007) described the additive wavelet method. Its main steps are as follows:

1. Generate one fine-resolution panchromatic image (PAN) for each band of the coarse-resolution multispectral image (MS), with the former histogram matched to the latter.
2. Apply one of the DWTs described above to both the MS and the new PAN images.
3. Add the detail images from the transformed PAN images to those of the transformed MS images. If multiple decompositions were applied, add the detail images at each resolution level.
4. Perform the inverse transform on the MS images with added PAN detail.

It is also possible to combine the standard methods and wavelet methods for image fusion. For example, it is sensible to incorporate the IHS transform into wavelet methods to make good use of spatial and spectral information in the source PAN and MS images. IHS and wavelet decomposition can be combined as follows (Amolins et al. 2007):

1. Convert the RGB of the MS image to an IHS image.
2. Generate a new PAN image histogram matched to the "I" component of the IHS image generated above.
3. Apply DWT to the new PAN and "I" images.
4. To obtain the set of approximation and detail images: the approximation of the I component can replace directly the approximation of the PAN image; the detail images of the I component and the PAN image can be added directly or through proper weighting according to the mean and standard deviation of the MS and PAN detail images, the approximation images, the original images, or other relevant criteria.
5. Perform inverse DWT to obtain a fused I component.
6. Convert the new IHS (image) to RGB.

As shown previously, a wavelet transform contains vertical (low–high band), horizontal (high–low band), and diagonal (high–high band) high-frequency information at different resolutions, plus the low (low–low band) frequency information at the coarsest resolution. Except for the low–low band, which has all positive transform values, all the other bands contain transform values that are fluctuating around zero. The larger absolute transform values in these bands correspond to sharper brightness changes and, thus, to the salient features in the image such as edges, lines, and region boundaries (Li et al. 1995). Therefore, a good integration rule is to select the larger absolute value of the two wavelet coefficients at each point (thus called the maximum fusion scheme) and to construct a composite image by performing an inverse wavelet transform based on the combined transform coefficients. As shown by Li et al. (1995), wavelet transforms can effectively extract salient features at different scales from source images, and, as a result, the maximum fusion scheme mentioned above can be used for image fusion.

Clearly, fusion of PAN and MS images amounts to the synthesis of multispectral images at finer spatial resolution, which should be as close as possible to those that would have been acquired by the corresponding sensor (e.g., Landsat TM) if it had this fine resolution. This synthesis is performed with the help of fine-spatial-, but coarse-spectral-, resolution imagery (PAN image). The fusion of the PAN and MS images is classically referred to as pan-sharpening. A fused product meets quality criteria only if the characteristics and differences between input images are taken into account. Dissimilarities existing between these two datasets originate from temporality in data acquisition times and discrepancies in the spectral bands of the sensors employed, which may lead to spectral distortion in the fused image (Wald et al. 1997). Remote-sensing physics should be carefully considered while developing the methods for image fusion (Thomas et al. 2008).

The fusion results are mostly evaluated visually. The problem with defining a quantitative measure lies in the difficulty of defining an ideal composite image based on multisensor images or images taken at different times. Nevertheless, there are numerous measures to assess the performance of image fusion, such as root mean squared errors, image-quality index (Wang and Bovik 2002), and information-theoretic measures (Qu et al. 2002). We will describe some informational measures in Chapter 11, which may be employed for performance evaluation of image fusion and spatial data conflation, since it is important to be able to quantify information transformation in an image's spatial detail and spectral bands beyond conventional measures for image fusion (Chavez and Bowell 1988; Wald et al. 1997).

8.3 OTHER MULTISCALE METHODS

It is of interest for many applications to analyze spatial distributions occurring at different scales. The wavelet transforms described in the preceding section provide a framework for multiresolution signal decomposition. There should be a corresponding statistical framework to support multiscale statistical signal processing (Benveniste et al. 1994). The dyadic tree-based models described by Chou et al. (1994) and Fieguth et al. (1995) furnish multiresolution image analyses and information fusion.

In this section, we introduce dynamic models on a tree structure in which stochastic processes are indexed by nodes in the tree and different depths in the tree correspond to different spatial scales in representing a signal or image (Basseville et al. 1992). Specifically, each node of the tree is denoted by an index i (i.e., $i = (resel, n)$) ($Tree = \{i\}$ the set of all nodes), with $resel(i)$ denoting the scale (resolution) of i (increasing $resel$ indicating the coarse-to-fine forward direction, 0 indicating the root node) and n being the translation. For a dyadic tree, we may define a backward shift $\bar{\gamma}$ and two forward shifts β_1 and β_2: for $i = (resel, n)$, $i\beta_1 = (resel + 1, 2n)$, $i\beta_2 = (resel + 1, 2n + 1)$, and $i\bar{\gamma} = (resel - 1, [n/2])$, as described by Chou et al. (1994) and Fieguth et al. (1995).

The structure of $Tree$ admits a class of scale-recursive linear dynamic models defined locally on $Tree$ and evolving from coarse to fine resolution:

$$\mathbf{s}(i) = \mathbf{A}_S(i)\mathbf{s}(i\bar{\gamma}) + \boldsymbol{\varepsilon}(i) \tag{8.25}$$

where $\mathbf{s}(i)$ is a zero-mean multiresolution process with covariance $\mathbf{cov}_S(i)$, $\mathbf{A}_S(i)$ is a matrix specified at each node other than 0 (i.e., the root node), $\boldsymbol{\varepsilon}(i)$ is a Gaussian white noise process with covariance $\mathbf{cov}_\varepsilon(i)$, and the term $\mathbf{A}_S(i)\mathbf{s}(i\bar{\gamma})$ represents a coarse-to-fine-resolution prediction (Chou et al. 1994; Fieguth et al. 1995; Willsky 2002). Thus, Equation 8.25 represents a generalization of the coarse-to-fine synthesis form of the wavelet transform. In Equation 8.25, the covariance of $\mathbf{s}(i)$ is quantified via the law of variance and covariance propagation:

$$\mathbf{cov}_S(i) = \mathbf{A}_S(i)\mathbf{cov}_S(i\bar{\gamma})\mathbf{A}_S^T(i) + \mathbf{cov}_\varepsilon(i) \tag{8.26}$$

Suppose also that there is a set of linear measurements (represented as a column vector $\mathbf{z}(i)$), which are related to the original process $\mathbf{s}(i)$:

$$\mathbf{z}(i) = \mathbf{B}(i)\,\mathbf{s}(i) + \mathbf{e}(i) \tag{8.27}$$

where $\mathbf{e}(i)$ is a zero-mean white noise process on the tree, independent of $\boldsymbol{\varepsilon}(i)$ and with covariance $\mathbf{cov}_{z_e}(i)$, and the matrix $\mathbf{B}(i)$ is the measurement matrix specifying what is measured at each node of the tree (Chou et al. 1994; Fieguth et al. 1995; Willsky 2002).

Let $\hat{\mathbf{s}}_i(i)$ denote the optimal estimate of $\mathbf{s}(i)$ given all the data in Equation 8.27 throughout the tree (i.e., given $\mathbf{z}(i)$ $\forall$ $i \in Tree$). Let $\mathbf{cov}_e(i)$ denote the covariance of the error in this estimate. The computation of these quantities throughout the entire tree can be accomplished using a two-sweep algorithm described by Willsky (2002) and Van de Vyver and Roulin (2009). This two-sweep algorithm includes an upward (fine-to-coarse) sweep and a downward (coarse-to-fine) sweep, as described below.

The first sweep is a generalization of the temporal Kalman filter. The objective of this sweep is the computation, at each node i of $\hat{\mathbf{s}}(i|i)$, of the optimal estimate of $\mathbf{s}(i)$ based on all of the data in $Tree_s$, the subtree rooted at node i (i.e., node i and all of its descendants), together with $\mathbf{cov}\,(i|i)$, the covariance of the error in this estimate (Chou et al. 1994; Fieguth et al. 1995). The following description of the three steps in this upward sweep is based on Willsky (2002).

Suppose that we have computed the best estimate $\hat{s}(i|i-)$ and the corresponding error covariance $\mathbf{cov}(i|i-)$ at node s, given all of the data in $Tree_s$ except for the measurement at node i itself. In the first step of the first fine-to-coarse sweep, the estimate (and associated error covariance) is updated by incorporating the measurement at node i in a way similar to measurement update in the usual Kalman filter. Given $\hat{s}(i|i-)$, $\mathbf{cov}(i|i-)$, and $\mathbf{z}(i)$, we have

$$\hat{s}(i|i) = \hat{s}(i|i-) + \mathbf{Gain}(i)\, \mathbf{d}_Z(i) \tag{8.28}$$

where $\mathbf{d}_Z(i)$ is the measurement innovation $\mathbf{d}_Z(i) = \mathbf{z}(i) - \mathbf{B}(i)\hat{s}(i|i-)$ (with zero mean and covariance $\mathbf{cov}_{d_z}(i) = \mathbf{B}(i)\mathbf{cov}(i|i-)\mathbf{B}^T(i) + \mathbf{cov}_{z_e}(i)$) and $\mathbf{Gain}(i)$ is the gain $\mathbf{Gain}(i) = \mathbf{cov}(i|i-)\mathbf{B}^T(i)\mathbf{cov}_{d_z}^{-1}(i)$. The updated error covariance $\mathbf{cov}(i|i)$ is quantified as $\mathbf{cov}(i|i) = [\mathbf{I} - \mathbf{Gain}(i)\, \mathbf{B}(i)]\, \mathbf{cov}\,(i|i-)$, where $\mathbf{I}$ is an identity matrix (Chou et al. 1994; Fieguth et al. 1995; Willsky 2002).

The second step of the first fine-to-coarse sweep involves the fusion of estimates that come from all of the immediate children of node i. Specifically, let $\hat{s}(i|i\beta_l)$ denote the optimal estimate for node i based on all of the data in $Tree_{i\beta_l}$ (i.e., the subtree rooted at node $i\beta_l$) and call this the estimate of the subtree $Tree_{i\beta_l}$, with $\mathbf{cov}(i|i\beta_l)$ being the corresponding error covariance. As described by Fieguth et al. (1995) and Willsky (2002), the estimate (and error covariance) at node i is based on all of the data at nodes descended from i, through *fusion of subtree estimates* $\hat{s}(i|i\beta_l)$, as follows:

$$\hat{s}(i|i-) = \mathbf{cov}(i|i-)\sum_{l=1}^{nd(i)}\mathbf{cov}^{-1}(i|i\beta_l)\hat{s}(i|i\beta_l) \tag{8.29a}$$

$$\mathbf{cov}^{-1}(i|i-) = \mathbf{cov}_S^{-1}(i) + \sum_{l=1}^{nd(i)}\left[\mathbf{cov}^{-1}(i|i\beta_l) - \mathbf{cov}_S^{-1}(i)\right] \tag{8.29b}$$

where $\mathbf{cov}_S(i)$ is the prior covariance at node i, computed from Equation 8.26, given $\hat{s}(i|i\beta_l)$ and $\mathbf{cov}(i|i\beta_l)$ for all l (where $nd(i)$ denotes the number of descendants of node i).

The third step in the first sweep involves the computation of the estimates $\hat{s}(i|i\beta_l)$ (and error covariance) for node i (refer to Verghese and Kailath (1979) for details). Given $\hat{s}(i\beta_l|i\beta_l)$ and $\mathbf{cov}(i\beta_l|i\beta_l)$, we have

$$\hat{s}(i|i\beta_l) = \mathbf{K}_1(i\beta_l)\hat{s}(i\beta_l|i\beta_l) \tag{8.30a}$$

$$\mathbf{cov}(i|i\beta_l) = \mathbf{K}_1(i\beta_l)\mathbf{cov}(i\beta_l|i\beta_l)\mathbf{K}_1^T(i\beta_l) + \mathbf{K}_2(i\beta_l) \tag{8.30b}$$

where $\mathbf{K}_1(i) = \mathbf{cov}_S(i\bar{\gamma})\mathbf{A}_S^T(i)\mathbf{cov}_S^{-1}(i)$ and $\mathbf{K}_2(i) = \mathbf{cov}_S(i\bar{\gamma}) - \mathbf{cov}_S(i\bar{\gamma})\mathbf{A}_S^T(i)\mathbf{cov}_S^{-1}(i)$ $\mathbf{A}_S(i)\mathbf{cov}_S(i\bar{\gamma})$, as described by Willsky (2002).

The initial estimates to the prior mean (here assumed to be 0) and the prior covariance, at each leaf node (the finest-scale), need to be set for the first sweep to proceed. For each finest-scale leaf node i, we have

$$\hat{s}(i|i-) = 0 \tag{8.31a}$$

$$\mathbf{cov}(i|i-) = \mathbf{cov}_s(i) \tag{8.31b}$$

as shown in Willsky (2002) and Van de Vyver and Roulin (2009).

When the fine-to-coarse sweep reaches the root node, the estimate and covariance computed at that node provide the initial conditions for the second coarse-to-fine sweep, as follows:

$$\hat{s}_i(0) = \hat{s}(0|0) \tag{8.32a}$$

$$\mathbf{cov}_e(0) = \mathbf{cov}(0|0) \tag{8.32b}$$

The computations in the second sweep are carried out in parallel at each of the children of the node $i\bar{\gamma}$:

$$\hat{s}_i(i) = \hat{s}(i|i) + \mathbf{K}_3(i)\left[\hat{s}_i(i\bar{\gamma}) - \hat{s}(i\bar{\gamma}|i)\right] \tag{8.33a}$$

$$\mathbf{cov}_e(i) = \mathbf{cov}(i|i) + [\mathbf{K}_3(i)]^2\left[\mathbf{cov}_e(i\bar{\gamma}) - \mathbf{cov}(i\bar{\gamma}|i)\right] \tag{8.33b}$$

where $\mathbf{K}_3(i) = \mathbf{cov}(i|i)\mathbf{K}_1^{\mathrm{T}}(i)\mathbf{cov}^{-1}(i\bar{\gamma}|i)$ (see Willsky [2002] and Van de Vyver and Roulin [2009] for details).

Slatton et al. (2001) applied a multiresolution Kalman filter to combine accurate yet sparse laser altimeter (LiDAR) observations with extended yet less accurate interferometric synthetic aperture radar (INSAR) inversion results to increase the accuracy of estimates of ground elevation and vegetation height. Such a multiresolution Kalman filter provides both the estimates (of ground elevation and vegetation heights) and a measure of their uncertainty at each location. Combining data from the two sensors provides estimates that are more accurate than those obtained from INSAR alone yet have dense, extensive coverage, which is difficult to obtain with LIDAR. Additionally, nonlinear measurement–state relationships were accommodated in their proposed method. The multiresolution methods reviewed above will have wide applicability in remote sensing where multisource data are often involved (Van de Vyver and Roulin 2009; Cheung et al. 2012).

8.4   DISCUSSION

Spatial data acquired from different sources are often available at different scales, and data originating from the same processes are likely to behave differently when observed, represented, and analyzed at different scales (Quattrochi and Goodchild 1997; Bierkens et al. 2000; Gertner et al. 2007). Therefore, scale poses some serious issues for data conflation, one of which is scale mismatch, which arises from the scale multiplicity that is often encountered in applications (Gotway and Young 2007). Scale mismatch hampers data conflation, affects the interpretability of the conflated data, and should be properly accommodated.

For multivariate geostatistics, scale mismatch can, in principle, be handled with properly regularized auto- and cross-variograms, assuming availability of point-support variogram models. However, it is often the case that point-support variogram models may not be able to be straightforwardly deconvolved from experimental variograms. Moreover, even if point-support variograms can be derived, there is the issue of how far we can go with downscaling (Merlin et al. 2009; we will discuss this issue in more detail in Chapter 9 when looking into the issue of terrain data disaggregation). For example, if cokriging is based on primary data (point-support, but sparse) and secondary data (block-support, but extended coverage) for prediction over point supports (thus, downscaling relative to the given data), and if there is a large discrepancy between the resolutions of the primary and secondary data, it remains to be seen what advantages cokriging holds over univariate kriging and its variants, such as simple kriging with local means, because the large discrepancy in data resolution (and sampling rates) is likely to render cokriging less effective than univariate kriging. Lastly, RFs underlying primary data are likely to have nonstationary variograms, making deconvolution of block-support experimental variograms for point-support variogram modeling extremely difficult. It is important to revisit the Shannon sampling theorem in the light of scale mismatch and to demarcate the limits to scale discrepancy in data conflation for value-added end products. Work by researchers such as Modis and Papaodysseus (2006) needs to gain more momentum. The issue of resolution and accuracy in downscaling, which is related to scale mismatch, will be readdressed in the context of terrain analysis in Chapter 9.

Wavelet transforms are useful tools for multiresolution data analysis. They are valuable not only for image fusion, as reviewed above, but also for scale description. As reviewed by Lark and Webster (1999), a wavelet is a compact analyzing kernel that can be used to measure variation locally. There are several families of wavelet, and, within any one family, wavelets of different smoothness and their corresponding scaling functions can be assembled into a collection of orthogonal functions, as shown in Section 8.2.2. Wavelets and associated scaling functions can be applied to filter spatial data into a series of independent components at varying scales (Lark and Webster 1999). Thus, multiresolution analysis based on wavelets is valuable for spatial data analysis as it can help to reveal multiscale properties in the dataset being considered. For example, Viscarra Rossel and Lark (2009) applied multiresolution analysis to identify local features in the soil diffuse reflectance spectra that contain information about soil composition. We will discuss some of the scale descriptors that are built on wavelets in Chapters 9 and 10 where terrain analysis and area-class mapping are emphasized, respectively.

We have reviewed applications of wavelet transforms for image fusion in Section 8.2.3. Further developments in image fusion may well be seen in the enrichment of data-generating mechanisms in multiresolution analysis for enhanced fusion of multisource images (Thomas et al. 2008). The other front is to explore scale mismatch in multisource data through wavelet-based multiresolution analysis, by which local variation in the data can be examined effectively.

The multiresolution estimation framework described in Section 8.3, which is built on pyramidally structured trees, is for efficient fusion of extremely large datasets, as it performs statistical analysis, interpolation, and smoothing at multiple resolutions directly (Chou et al. 1994; Fieguth et al. 1995). It is closely related to wavelets, image fusion, and geostatistics. It can be utilized for space–time data analysis. Johannesson et al. (2007) discussed a multiresolution spatial model that produces fast optimal estimates using a change-of-resolution Kalman filter, which incorporates temporal dependence. The ultimate goal is to synthesize developments in spatial and temporal data analysis, and to make multiscale data conflation scientifically rigorous and operationally efficient.

In reality, datasets vary on a continuum in terms of accuracy and scale. Also, because complex convolution functions and supports are usually involved in creating datasets, characterization of measurement scales and uncertainty, and differentiation between datasets on these bases, is rarely precise. Moreover, spatial variation in the effective measurement scales of data complicates the already vexing issue of scale mismatch. It is worth exploring the combinations of cokriging, wavelets, and multiscale methods as described in this section so that multiscale analysis can be performed more effectively.

9 Scale in Terrain Analysis

Up to this point, descriptions and discussions about issues of scale and scaling have been directed mostly toward data modeling, measurements, geostatistical data, lattice data, and multiscale data conflation. As important components in geographic information science, digital terrain analysis and area-class mapping have attracted substantial research into scale and scaling in spatial analysis and modeling. Thus, we introduce models of scale and methods for scaling with specific reference to terrain analysis in this chapter, whereas scale issues will be elaborated with respect to area-class mapping in Chapter 10.

Geographic forms and patterns are viewed as resulting from various physical and social processes operating at different scales (Wilson and Gallant 2000; Elith and Leathwick 2009). The meso- and topo-scales (a few to hundreds of kilometers) are important for environmental, hydrological, and ecological research (Hammond 1964; Pike 2000; Shary et al. 2002; Iwahashi and Pike 2007; Evans and Lindsay 2010). For example, finer-scale variation in climate results from meso- and topo-scale topography and rock type, thus highlighting the importance of topography in climatic regionalization (Bailey 1996). Many topographic properties are incorporated as controlling factors for hydrologic modeling (Beven and Kirkby 1979; Wood et al. 1988), as reviewed by Wilson and Gallant (2000). Soil development in many landscapes occurs in response to water movements in the landscape. Terrain properties can characterize these flow paths and, therefore, soil development, as demonstrated by a terrain–soil correlation study on a 5.4-ha topo-sequence in Colorado (Moore et al. 1993). Vegetation distribution in space and time, which responds to water, light, and nutrient availability (Poorter and Nagel 2000; Wilson and Gallant 2000), has been modeled using combinations of topographic properties, as shown by Guisan et al. (2002) and Van Niel et al. (2004).

Spatial analysis and modeling at meso- and topo-scales have been promoted by the growing availability of fine-resolution digital elevation data and the development of computerized terrain-analysis tools, as demonstrated by increasing research in this regard (Peucker and Douglas 1975; Wilson and Burrough 1999; Evans and Lindsay 2010). As reviewed by Gallant and Hutchinson (1996), terrain analysis is the quantitative analysis of topographic surfaces with the aim of studying surface and near-surface processes, and provides the basis for a wide range of landscape pattern analysis and environmental modeling. Several topographic properties can be calculated, including slope, aspect, curvature, and specific catchment area, as reviewed by Zhou and Liu (2006).

Scale dependence of topographic properties, such as slope and specific catchment area, has been studied intensively, for instance, in the context of hydrological modeling (Zhang and Montgomery 1994; Quinn et al. 1995; Brasington and Richards 1998). In these studies, analysts are concerned with the spatial scale (grid resolution) for analysis and modeling. Slope and aspect angles estimated from digital elevation models (DEMs) have been used commonly as if they depicted the surface orientation of only the single central cell (Hodgson 1995). Using DEM data of 30-m grid spacing, Hodgson (1995) showed that the slope and aspect angles derived from the neighboring elevation points best depict the surface orientation for a larger cell, either 1.6 or 2.0 times larger than the size of the central cell. This suggests that elevation data should be collected at a greater spatial sampling intensity than the other data layers in an application. When the elevation data must be generalized to a coarser sample size, it is more accurate to determine the slope and aspect angles for the larger cells from a vector mean of the smaller cell sizes.

As reviewed by Wilson and Gallant (2000), sensitivity analysis concerning scale effects highlights the importance of scale in terrain analysis. Many studies have examined the sensitivity of terrain surface properties to cell sizes in DEMs (Kienzle 2004). Upscaling with progressively larger grid spacings tends to filter out fine-detail or high-frequency information, leading to, for example, slope reduction because averaging through aggregation makes high-altitude areas lower and low-altitude areas higher (Zhang et al. 1999; Le Coz et al. 2009). Garbrecht and Martz (1994) examined the impact of DEM resolution on extracted drainage properties for an 84-km^2 study area in Oklahoma using hypothetical drainage network configurations and DEMs of varying spatial resolution. They derived various quantitative relationships and concluded that the grid spacing must be selected relative to the size of the smallest drainage features considered important for the work at hand. The effects of DEM source scale and DEM cell spacing on the topographic wetness index and TOPMODEL watershed model predictions were also examined by researchers, such as Beven and Kirkby (1979) and Zhang and Montgomery (1994).

Scaling is necessary for integrating data, models, and applications across scales. As described previously, there are two types of scaling operating in opposite directions: upscaling and downscaling, which are also known as aggregation and disaggregation, respectively (McBratney 1998). Scaling suffers from the fact that it is difficult to define *a priori* how the change of scale should proceed owing to nonlinear behaviors, especially when dealing with landscapes consisting of nonhomogeneous terrain patches. With a growing number of digital terrain data available, digital terrain analysis offers greater advantages for scale modeling and the development of methods for scaling. This is because terrain analysis can be built upon readily available databases (we will review some of them later in this chapter). Also, there are already many research contributions made by scientists in this area.

Firstly, we will discuss the key types of elevation datasets, including those derived from map digitizing, photogrammetric processing of aerial and space images, interferometry (Farr et al. 2007), and combinations of multisource topographic data. In the process, we will focus on such issues as measurement scales, accuracy, and semantics in measurements (i.e., what elevation means in certain situations). It will become apparent that there are disparate scales of measurement for multisource elevation data, which are complicated by issues of accuracy and semantics in these datasets. These will have implications for terrain analysis because of scale multiplicity and mismatch.

Secondly, we will review models of scale in terrain variables. Geostatistics is valuable for modeling scales in terrain variables and will be discussed with an aim to emphasize its possible extensions or links to alternative modeling approaches. For this, local geostatistical models of scale will be described, since they provide a non-stationary model of spatial dependence over a problem domain. Alternative models will then be reviewed, such as local variance-based models, fractals, and representative elementary areas. Various ruggedness indices that seek to quantify complexity in topography have been proposed in the literature; they are thus indicative of the scale inherent to the terrain properties. These ruggedness indices will be introduced as scale indicators.

Thirdly, we will review some approaches to scaling terrain variables. Scaling may be performed using geostatistics, as described in Chapter 6. In this chapter, the emphasis will be on alternative approaches. For upscaling, we may use various statistical methods reviewed in Chapter 7, such as those based on nearest neighbor, mean, median, mode, maximum, and minimum. Their functionality should be considered carefully in light of the terrain properties under study, as they are often more suited for certain terrain properties than others. Other approaches to upscaling will also be described. There are various approaches to downscaling, as described in Chapters 6 and 7 where scaling of geostatistical data and lattice data were discussed, respectively. Fractals are useful for downscaling slope within certain scale limits, but it is important to recognize these limits to avoid imputing spurious detail. There are criteria proposed for this, which seek to optimize the finest possible resolution for downscaling (Merlin et al. 2009).

Lastly, this chapter will discuss multiscale characterization of terrain properties. Future developments in scaling will also be discussed. This is followed by prospects with respect to models of scale and methods for scaling for discrete objects, such as flow paths extracted from DEMs. Landform classifications based on pattern analysis of terrain variables are reviewed briefly to make explicit the link to the topic of area-class mapping in Chapter 10.

9.1   DIGITAL ELEVATION DATA AND THEIR SCALES

Geographic properties such as elevation Z are often represented as surfaces approximated by single-valued functions of the form $Z = Z(X, Y)$, with X, Y being a coordinate pair. It is impossible, however, to collect data for every point on the surface. Instead, a sample of measurements is often collected, which implies a certain sampling density and data support, as shown in Chapter 3; surface $Z = Z(X, Y)$ may be reconstructed in grid format by interpolation.

Ground surveys are a traditional source of elevation data and continue to be important because of their high accuracy (Keim et al. 1999). The advent and widespread use of global positioning systems provides a cost-effective means for the collection of elevation data (Casas et al. 2006). Most of the currently available digital elevation datasets are products of photogrammetric mapping systems, which are manual or automatic stereoplotters taking in aerial photographs or satellite sensor images to extract topographic information (Klinkenberg and Goodchild 1992; Mikhail et al. 2001). Elevation datasets can also be acquired by digitizing the contour lines on topographic maps, by which we can often achieve higher accuracy. More recently, interferometric synthetic aperture radar (SAR) technology was developed for topographic mapping, such as Shuttle Radar Topography Mission (SRTM) (Farr et al. 2007). GTOPO30 is a global DEM dataset that was derived from various sources and techniques, with a mixed lineage and accuracy.

We review some main types of digital elevation datasets created from major techniques as mentioned above, such as the US Geological Survey (USGS) DEMs, SRTM DEMs based on the SAR technique, and GTOPO30 DEMs. In doing so, we discuss the issues of scale, accuracy, and semantics in these datasets. It will become clear that elevation data have their inherent resolution, accuracy, and semantics, which are spatially varying depending on the specific data sources and the algorithms involved in the data production processes.

First, we review methods employed by the USGS for DEM production. As reviewed by Hodgson et al. (2003), the USGS has used four primary methods, which are manual profiling, automatic correlation, contour-to-grid interpolation, and integrated contour-to-grid interpolation, for creating DEMs that are categorized into three levels of quality: levels 1, 2, and 3 (USGS 1986). Each production method uses a unique combination of source materials and processing methods, as shown below.

The manual profiling method requires that the operator view and pick samples from a reconstituted stereo model, with the DEMs created as byproducts of the orthophoto production process. The operator views the image through a narrow slit, which is moved systematically as the work proceeds. Photographic content in the slit is transferred to the orthophotoquad while removing the relief displacement, with vertical motions of the floating slit recorded as elevation heights, thus generating the DEM (Klinkenberg and Goodchild 1992; Hodgson et al. 2003).

An automated correlation technique known as the gestalt photomapper (GPM) was used extensively by the USGS for creating DEMs from black and white stereo pairs of aerial photographs (Klinkenberg and Goodchild 1992). A USGS 7.5-min DEM was created from two stereo models of the National High Altitude Photography program (USGS 1986). As reviewed by Hodgson et al. (2003), image correlation was performed on one small patch of the photographs at a time, with a small kernel of pixels centered at each point on one photograph matched to its corollary on the opposite photograph. The horizontal displacement of the kernel pairs (i.e., X-parallax) is a measure of the surface height. The set of mass points produced in this process was subject to interpolation over a regular lattice, and merging and resampling to a 30 × 30-m grid. The resultant DEM was edited for water bodies and corrected for minor errors identified visually.

USGS DEMs derived from both GPM and manual profiling have a horizontal spacing of 30 m and a vertical resolution of 1 m, referenced to mean sea level (Klinkenberg and Goodchild 1992). However, their elevation values are derived differently. The manual profiling system records elevations every 30 m or so along north–south running profiles, which are spaced approximately 90 m apart. The intervening elevations are estimated through interpolation. On the other hand, elevations produced using the GPM method are obtained with an initial ground spacing of approximately 14 m (both north–south and east–west), which is gridded to the final

DEM spacing of 30 m with average elevations registered over the 30 × 30-m grid cells (Hodgson et al. 2003).

DEMs derived from manual profiling and GPM are of level 1 quality, whereas gridded DEMs of levels 2 and 3 quality (USGS 1986) are derived from contour-to-grid interpolation algorithms, as reviewed by Hodgson et al. (2003). The contour lines typically come from 1:24,000 scale or 1:100,000 scale topographic maps or directly from the stereo models of aerial photographs. Similar to level 1 DEMs, these data are also smoothed for consistency and corrected for any obvious artifacts. For level 2 DEMs, although their vertical accuracy and representation of surface slope are typically improved over the level 1 DEMs, the interpolated grids may have a bias in the form of too-frequent DEM elevations coinciding with the contour-line elevations (i.e., overrepresented elevations along contours). Attempts to correct for this bias may lead to underrepresentation of contour-line elevations (Guth 1999). The level 3 DEMs from the USGS are created by contour-to-grid interpolation constrained by other planimetric or vertical information including hydrographic features and vertical or horizontal control networks (Hodgson et al. 2003).

All fine-scale DEMs produced by the USGS undergo a process of quality assessment (USGS 1986). As reviewed by Hodgson et al. (2003), a minimum of 28 reference points are used to determine the root mean squared error (RMSE) for each 7.5-min quad-based DEM. The outcome of the accuracy assessment is used to establish whether the DEM meets the accuracy standard required (Chapter 12 will discuss accuracy assessment in more detail). An RMSE of 15 m is the maximum allowed for level 1 DEMs produced directly from the National High Altitude Photography program/National Aerial Photography Program (GPM or manual profiler methods). Level 2 and 3 DEMs have a maximum RMSE of 1/2 and 1/3 of the contour interval, respectively. In addition, the RMSE of level 2 and 3 DEMs should not exceed 7 m, regardless of the contour intervals employed (Hodgson et al. 2003).

Secondly, we describe SAR-based topographic mapping. As described in Chapter 3, a SAR synthetically simulates a very long antenna using the forward movement of its platform (Hodgson et al. 2003). An interferometric synthetic aperture radar (InSAR) is used to derive topographic information by measuring the differences in the phase of signals reflected from a terrain when observed from slightly different locations and times. The observable terrain shift is of the order of the radar wavelength rather than of the resolution cell size, circumventing a major problem in stereo techniques (i.e., that of recognizing the conjugate resolution cells in the stereo image pair). This problem is particularly difficult in coherent radar images because of the speckle phenomenon (i.e., different views of the same resolution cell are only statistically related), as described by Zebker and Goldstein (1986). SRTM used C-band and X-band InSAR to acquire global topographic data (high latitudes, above 60° north and below 56° south, were omitted because of the Shuttle orbit) at a fine scale during an 11-day Shuttle mission in February 2000 (http://www.jpl.nasa.gov/srtm/). Although 1 arcsecond data (approximately 30 m of posting at the Equator) are available only for locations within the United States, data outside of the United States are publicly available (after July 2004) at 3 arcsecond sampling (approximately 90 m posting) covering the globe, with the 90 m derived from coarsening of the 30-m data (Slater et al. 2006). SRTM data coarsening works either by subsampling central pixel values over 3 × 3 pixel windows for the "finished" dataset or, for the research-quality dataset, by averaging elevation values within the 3 × 3 pixel windows (http://www.jpl.nasa.gov/srtm/).

The vertical precision of the SRTM data depends on the inherent phase noise in the SRTM radar, whereas the horizontal resolution depends on the signal-to-noise (SNR) ratio as a function of horizontal wavelength (Smith and Sandwell 2003). The absolute vertical and horizontal accuracies of the data collected were specified to be ±16 and ±20 m (linear error at 90% confidence level), respectively, for the 1 arc-second data release of the United States (Rabus et al. 2003). The absolute vertical accuracy of SRTM data was validated to be about 9 m (linear error at 90% confidence level), much better than its design goal of 16 m. The greatest errors are associated with steep terrain (Himalaya, Andes) and very smooth sandy surfaces with low SNR (Sahara Desert) (Falorni et al. 2005), as noted by Farr et al. (2007). As the 3 arc-second "finished" DEM was created by 3 × 3 pixel averaging of the 1 arcsecond data, elevation data errors (measured by standard deviation) are considered to be reduced by approximately a factor of 3 (i.e., $\sigma_{3\times3} \approx \sigma/3$) at the cost of coarsened resolution (USGS 2004). The interferometric processing of SRTM datasets included averaging of multiple data takes (if acquired) and adaptive smoothing to reduce noise (Pierce et al. 2006). As the adaptive smoothing algorithm applies more averaging to smooth regions, but less, if any, averaging to areas with abrupt relief, the effective horizontal resolution after applying such filters is expected to vary spatially (Pierce et al. 2006). (See Farr et al. [2007] for further details about the SRTM DEM dataset.)

The meanings of elevation values in SRTM data need to be examined. As SRTM heights were determined by the phase of the vector sum of all the returned signals from within the pixel being imaged, the resultant data might refer to a nonground surface, unless the pixels recorded signals entirely from bare ground (Farr et al. 2007). For instance, the pixel return over vegetated surfaces is influenced by the vegetation height, structure, and density. Radar waves can penetrate into frozen snow or ice or very dry soil, potentially up to several meters. Since SRTM flew in February, there was significant snow cover in the northern hemisphere, indicating that the C-band-derived heights might be from the top of the snowpack or from the ground surface buried underneath, depending on the state of the snow (Farr et al. 2007). Similarly, areas of very dry sand cover in the Sahara Desert, observed to be penetrated by L-band (25 cm) signals, may also be penetrated a short distance by the shorter-wavelength C-band. Human-made objects (e.g., large buildings, roads, and bridges) often cause problems in radar imaging, such as layover, shadowing, and multipath artifacts. Thus, the height of any urban pixel will represent an average urban surface rather than the ground, as discussed by Farr et al. (2007).

Lastly, we describe GTOPO30, a global DEM dataset derived from heterogeneous sources. GTOPO30 was developed over a 3-year period through a collaborative effort led by the staff at the USGS EROS Data Center (http://eros.usgs.gov/#/Find_Data/Products_and_Data_Available/GTOPO30). GTOPO30 is based on data derived from various sources of elevation information, including vector and raster datasets, such as Digital Terrain Elevation Data (DTED), Digital Chart of the World (DCW), USGS 1° DEMs, and others. For instance, DTED, a raster topographic database with a horizontal grid spacing of 3 arcseconds (approximately

90 m) produced by the National Imagery and Mapping Agency (NIMA) (formerly the Defense Mapping Agency and now the National Geospatial-Intelligence Agency), was used as the source for most of Eurasia and large parts of Africa and the Americas. The DCW was used as the primary source for filling gaps in the DTED coverage, whereas USGS 1° DEMs (with a horizontal grid spacing of 3 arcseconds) were used as the source data for the United States. Processing of the raster source data (e.g., DTED and USGS DEMs) involved generalizing the finer-resolution data to the 30-arcsecond horizontal grid spacing (about 1 km at the equator), typically by selecting a representative elevation value from 10 × 10 fine-resolution cells covering the coarse-resolution cell, using methods of generalization, such as the systematic subsampling for North and South America, the median-value method for Eurasia, and the breakline emphasis approach for Africa. The topographic information from the vector sources, such as the DCW, was converted into elevation grids through a vector-to-raster gridding procedure, incorporating contours, spot heights, stream lines, lake shorelines, and ocean coastlines. Further details can be found at http://webgis.wr.usgs.gov/globalgis/gtopo30/gtopo30.htm.

The absolute vertical accuracy of GTOPO30 varies by location according to the source data, given the variability of source data used in production. Generally, the areas derived from the raster source data have higher accuracy than those derived from the vector source data (http://webgis.wr.usgs.gov/globalgis/gtopo30/gtopo30.htm). The full-resolution, 3-arcsecond DTED and USGS DEMs have a vertical accuracy of ±30 m (linear error at 90% confidence level) (USGS 1993), which, if one assumes a Gaussian error distribution with a mean of zero, corresponds to an RMSE of 18 m. The areas of GTOPO30 derived from DTED and USGS DEMs retain that same level of accuracy because the cell of reduced resolution is assigned a value representative of the collocated full-resolution cells. The absolute vertical accuracy of the DCW (with primary contour interval of 305 m) is stated in its product specification as ±650 m (linear error at 90% confidence level) (Defense Mapping Agency 1990), although the RMSE of the DCW grid was evaluated at 97 m, using averaged DTED as reference, which, assuming Gaussian error distributions as above, can be expressed as ±160 m (linear error at 90% confidence level). The accuracy of GTOPO30 based on the other sources can only be estimated based on what is known about each source, and, for instance, it is often assumed that 90% of elevations should be within one-half of the contour interval, according to commonly used accuracy standards. It should be noted that the accuracy levels described above are estimates and that accuracies for specific locations can vary from the estimated values (for details, see http://webgis.wr.usgs.gov/globalgis/gtopo30/gtopo30.htm).

There is a distinction between grid spacing and resolution in GTOPO30. Whether the topographic features in the source data can be resolved by this 30-arcsecond grid spacing is determined by the source's level of detail, implying that not all topographic features expected to be resolved at that spacing will be represented (http://webgis.wr.usgs.gov/globalgis/gtopo30/gtopo30.htm). This 30-arcsecond grid spacing is appropriate for the areas derived from DTED, USGS DEMs, and grids derived from DCW hypsography. However, coverage of DCW contours is not complete, and there are areas for which elevations were interpolated based only on very sparse

DCW point data or distant contours. Therefore, GTOPO30 horizontal resolution (i.e., measurement scale) should be conceived as being spatially variable as indicated previously for vertical accuracy (http://webgis.wr.usgs.gov/globalgis/gtopo30/gtopo30.htm). Clearly, as with SRTM DEMs, GTOPO30 DEMs have their own characteristic error structures, effective horizontal resolution, and meanings of elevation, which often vary spatially. Although a variety of convolution functions was involved in SRTM data production (i.e., in the steps of interferometric processing, smoothing, coarsening, and void filling), the convolution involved in GTOPO30 DEMs can be very complicated.

In summary, detailed description of terrain datasets and the measurement processes involved are meant to provide clues to the measurement scales of the elevation data, their semantics, and their accuracies. This will provide clues as to how a measurement model, such as $Z = KNL * S + \varepsilon$, where Z, KNL, S, and ε stand for the elevation measurement, the convolution kernel, the true elevation value, and the additive error term, respectively, may be prescribed for a given elevation dataset. It is easy to see the spatial variation in effective resolution of the DEM datasets. Such spatial variability in effective measurement scale in terrain elevation data will be modified when they are further geo-processed, as shown in the next section; this should be kept in mind when using such processed data and for research on scale-related issues.

9.2 TERRAIN DERIVATIVES

Many terrain properties, such as slope, specific catchment area, aspect, and plan and profile curvature, can be derived from DEM data for each and every location as a function of its neighbors. Some algorithms perform operations on local neighborhoods (i.e., 3 × 3 moving windows), and others on extended neighborhoods, such as calculation of upslope drainage areas and viewsheds (Sharpnack and Akin 1969; Skidmore 1989; Burrough and McDonnell 1998; Florinsky 1998; Wilson and Gallant 2000).

This section will describe the methods for deriving terrain properties from DEMs. We will see how the terrain derivatives behave in terms of scale, that is, how their scale is related to the neighborhood they are calculated from and how scale in terrain derivatives is related to scale in DEMs. Grid resolution of digital elevation data influences both the spatial pattern and the frequency distribution of derived topographic properties such as slope and, therefore, influences environmental models built on these terrain derivatives, as discussed by Gallant and Hutchinson (1996).

9.2.1 SLOPE AND ASPECT

Slope and aspect are related parameters for terrain surfaces, with the former defined as the angle β between the normal $\vec{N}$ (a vector normal to a grid cell surface) at a location x and the zenith, and the latter the azimuth α of $\vec{N}$ projected on the horizontal plane with respect to the north, as shown in Figure 9.1.

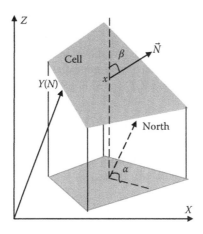

FIGURE 9.1 Slope and aspect at a location x.

Suppose a surface is represented mathematically as $Z = Z(X, Y)$. Gradient at location (X, Y):

$$\mathbf{grad}(X,Y) = Z_X \vec{i} + Z_Y \vec{j} \qquad (9.1)$$

where Z_X and Z_Y are partial derivatives of Z along the X and Y directions, respectively, and $\vec{i}$ and $\vec{j}$ are unit vectors along the X and Y directions, respectively. Thus, we have

$$\tan \beta = \sqrt{Z_X^2 + Z_Y^2} \qquad (9.2)$$

The direction of the gradient is the aspect

$$\tan \alpha = \frac{Z_Y}{Z_X} \qquad (9.3)$$

Aspect is measured clockwise, taking the north as zero azimuth. The sign of the coordinates needs to be taken into account to determine in which quadrant the vector **grad** (X, Y) lies (Watson 1966).

Based on triangulated irregular network (TIN) models, for a triangular element, we can obtain an equation, $Z(X, Y) = aX + bY + c$. From this triangular element equation, we can derive $Z_X = a$, $Z_Y = b$, which can be inserted into Equations 9.2 and 9.3 to calculate the slope and aspect, respectively.

As discussed above, for estimating terrain derivatives, we need to compute Z_X and Z_Y. DEMs are often represented as discretized versions of the earth's surface, whose true functions are rarely known in advance. Thus, calculation of Z_X and Z_Y must be based on numerical methods, surface fitting, or normal vectors. As will be shown below, to compute Z_X and Z_Y on gridded DEMs, we often perform the computation in a moving window of 3 × 3 cells and weighting is often inverse distance based (Zhou and Liu 2006; Zhang 2010).

First, consider numerical methods. As shown in Figure 9.2, the central grid cell is (IX, IY). If the local terrain surface is represented as $Z = Z(X, Y)$, then the first-order Taylor series expansion for Z at $(X, Y) = (IX, IY)$ is

$$Z(IX + k_{IX} \, pix, IY + k_{IY} \, pix) = Z(IX, IY) + k_{IX} pix Z_X + k_{IY} pix Z_Y \qquad (9.4)$$

where pix is the grid cell size (square, although rectangular cells with differing sizes in X and Y can be easily formulated), and $k_{IX}, k_{IY} \in \{-1, 0, 1\}$ signaling different ways for calculating finite differentials.

With different values for k_{IX} and k_{IY} and different weightings, there will be different formulae for slope and aspect. Taking $(IX, IY - 1)$, $(IX, IY + 1)$, $(IX - 1, IY)$, and $(IX + 1, IY)$ as the centers for expansion using Equation 9.4, and averaging the differentials, we get

$$\begin{cases} Z_X = \dfrac{z_{IX-1,IY+1} - z_{IX-1,IY-1} + z_{IX+1,IY+1} - z_{IX+1,IY-1}}{4\,pix} = \dfrac{z_3 - z_1 + z_9 - z_7}{4\,pix} \\[4mm] Z_Y = \dfrac{z_{IX+1,IY-1} - z_{IX-1,IY-1} + z_{IX+1,IY+1} - z_{IX-1,IY+1}}{4\,pix} = \dfrac{z_7 - z_1 + z_9 - z_3}{4\,pix} \end{cases} \qquad (9.5a)$$

$(IX + 1, IY - 1)$ 7	$(IX + 1, IY)$ 8	$(IX + 1, IY + 1)$ 9
$(IX, IY - 1)$ 4	(IX, IY) 5	$(IX, IY + 1)$ 6
$(IX - 1, IY - 1)$ 1	$(IX - 1, IY)$ 2	$(IX - 1, IY + 1)$ 3

FIGURE 9.2 Numbering the grid cells in a local 3 × 3 moving window.

Similar derivation was presented by Zhou and Liu (2006). Horn (1981) proposed an unequal weighting scheme for estimating differentials where neighbors 2, 4, 6, and 8 receive double weights:

$$
\left\{
\begin{aligned}
Z_X &= \frac{(z_{IX-1,IY+1} + 2z_{IX,IY+1} + z_{IX+1,IY+1}) - (z_{IX-1,IY-1} + 2z_{IX,IY-1} + z_{IX+1,IY-1})}{8\,pix} \\[4pt]
&= \frac{(z_3 + 2z_6 + z_9) - (z_1 + 2z_4 + z_7)}{8\,pix} \\[6pt]
Z_Y &= \frac{(z_{IX+1,IY-1} + 2z_{IX+1,IY} + z_{IX+1,IY+1}) - (z_{IX-1,IY-1} + 2z_{IX-1,IY} + z_{IX-1,IY+1})}{8\,pix} \\[4pt]
&= \frac{(z_7 + 2z_8 + z_9) - (z_1 + 2z_2 + z_3)}{8\,pix}
\end{aligned}
\right.
\tag{9.5b}
$$

Secondly, consider local surface fitting for estimating terrain derivatives. We may specify three different types of models for surface fitting, which are linear, quadratic, and partial quadratic (Zhou and Liu 2006):

$$Z = aX + bY + c \tag{9.6a}$$

$$Z = aX^2 + bY^2 + cXY + dX + eY + f \tag{9.6b}$$

$$Z = aX^2Y^2 + bX^2Y + cXY^2 + dX^2 + eY^2 + fXY + gX + hY + i \tag{9.6c}$$

For linear and quadratic models, there are more equations than unknowns, so we apply least squares. Zevenbergen and Thorne (1987) proposed the partial quadratic model (Equation 9.6c), in which there are nine unknowns, the same number as for the equations. The resultant curved surface will pass through all grid cells (centers). But in the 3×3 local coordinate system centered at the grid cell under study, the formulae are the same for the quadratic and partial quadratic cases.

Finally, there are normal-vector-based methods for computing slope and aspect (Corripio 2003). If we can derive a normal vector $\vec{N} = \mathbf{n}(n_X, n_Y, n_Z)$, we can calculate slope and aspect as

$$\tan \beta = \frac{\sqrt{n_X^2 + n_Y^2}}{n_Z} \tag{9.7a}$$

$$\tan \alpha = \frac{n_Y}{n_X} \tag{9.7b}$$

Ritter (1987) derived formulae for computing slope and aspect based on the normal vector for use with gridded DEMs, where the normal vector $\mathbf{n}$ is the vector product of west–east $\mathbf{n}_{\text{WE}} = (2\,pix, 0, z_{IX,IY+1} - z_{IX,IY-1})$ and south–north $\mathbf{n}_{\text{SN}} = (0, 2\,pix,$ $z_{IX+1,IY} - z_{IX-1,IY})$:

$$\mathbf{n} = \mathbf{n}_{\text{EW}} \times \mathbf{n}_{\text{SN}}$$

$$= \begin{vmatrix} \vec{i} & \vec{j} & \vec{k} \\ 2\,pix & 0 & z_{IX,IY+1} - z_{IX,IY-1} \\ 0 & 2\,pix & z_{IX+1,IY} - z_{IX-1,IY} \end{vmatrix} \tag{9.8}$$

$$= [-2\,pix(z_{IX,IY+1} - z_{IX,IY-1}), -2\,pix(z_{IX+1,IY} - z_{IX-1,IY}), 4\,pix^2]$$

For TINs, consider a triangle that comprises three vertices, $A(X_1, Y_1, Z_1)$, $B(X_2, Y_2, Z_2)$, and $C(X_3, Y_3, Z_3)$. $\mathbf{n}$ will be the vector product of vector $AB(X_2 - X_1, Y_2 - Y_1, Z_2 - Z_1)$ and $AC(X_3 - X_1, Y_3 - Y_1, Z_3 - Z_1)$. Thus, we have

$$\begin{pmatrix} n_X \\ n_Y \\ n_Z \end{pmatrix} = \begin{pmatrix} (Y_2 - Y_1)(Z_3 - Z_1) - (Y_3 - Y_1)(Z_2 - Z_1) \\ (Z_2 - Z_1)(X_3 - X_1) - (Z_3 - Z_1)(X_2 - X_1) \\ (X_2 - X_1)(Y_3 - Y_1) - (X_3 - X_1)(Y_2 - Y_1) \end{pmatrix} \tag{9.9}$$

as described by Zhou and Liu (2006).

9.2.2 CURVATURE

Computation of curvatures is more complicated than that of slope or aspect because, in general, the surface has different curvatures in different directions. Moreover, second-order and combined differentials are involved in the calculation of curvature. For applications in the geosciences, two types of curvature are important, as shown below. The curvature in the gradient direction (profile curvature) is important because it reflects the change in slope angle and, thus, controls the change in velocity of mass flowing down along the slope curve. The curvature in a direction perpendicular to the gradient (plan curvature) reflects the change in aspect angle and influences the divergence (or convergence) of water flow, and is usually measured in the horizontal plane as the curvature of contours (Zevenbergen and Thome 1987; Mitasova and Hofierka 1993; Florinsky 1998).

On a terrain surface $Z = Z(X, Y)$, for a given direction θ having an angle φ with the X-axis, the first and second directional derivatives are

$$\frac{\partial Z}{\partial \theta} = Z_X \cos \varphi + Z_Y \sin \varphi \tag{9.10a}$$

$$\frac{\partial^2 Z}{\partial \theta^2} = \frac{\partial^2 Z}{\partial X^2} \cos^2 \varphi + 2 \frac{\partial^2 Z}{\partial X \partial Y} \cos \varphi \sin \varphi + \frac{\partial^2 Z}{\partial Y^2} \sin^2 \varphi \tag{9.10b}$$

$$= Z_{XX} \cos^2 \varphi + 2 Z_{XY} \cos \varphi \sin \varphi + Z_{YY} \sin^2 \varphi$$

where $\cos \varphi$ and $\sin \varphi$ are measured as

$$\begin{cases} \cos \varphi = \dfrac{\dfrac{\partial Z}{\partial X}}{\sqrt{\left(\dfrac{\partial Z}{\partial X}\right)^2 + \left(\dfrac{\partial Z}{\partial Y}\right)^2}} = \dfrac{Z_X}{\sqrt{Z_X^2 + Z_Y^2}} \\[3em] \sin \varphi = \dfrac{\dfrac{\partial Z}{\partial Y}}{\sqrt{\left(\dfrac{\partial Z}{\partial X}\right)^2 + \left(\dfrac{\partial Z}{\partial Y}\right)^2}} = \dfrac{Z_Y}{\sqrt{Z_X^2 + Z_Y^2}} \end{cases} \tag{9.11}$$

Thus, for a point x, the curvature along the direction φ is

$$Curv = \frac{\dfrac{\partial^2 Z}{\partial \theta^2}}{\left[1 + \left(\dfrac{\partial Z}{\partial \theta}\right)^2\right]^{\frac{3}{2}}} \tag{9.12}$$

For profile curvature, let $\varphi = \alpha$ (α being the aspect at a point); we have

$$Curv_p = \frac{Z_X^2 Z_{XX} + 2 Z_X Z_Y Z_{XY} + Z_Y^2 Z_{YY}}{\left(Z_X^2 + Z_Y^2\right)\left(1 + Z_X^2 + Z_Y^2\right)^{\frac{3}{2}}} \tag{9.13}$$

as described by Zevenbergen and Thome (1987) and Mitasova and Hofierka (1993).

Plan curvature is defined as the curvature of the intersection between the contour plane and the terrain surface. Thus, a contour can be expressed as $Z = Z(X, Y) = cnst$ (*cnst* being a constant). Along the contour, we have

$$\frac{dY}{dX} = -\frac{Z_X}{Z_Y} \tag{9.14a}$$

$$\frac{d^2 Y}{dX^2} = -\frac{Z_{XX} Z_Y^2 - 2 Z_{XY} Z_X Z_Y + Z_{YY} Z_X^2}{Z_Y^3} \tag{9.14b}$$

Hence, inserting Equations 9.14a and 9.14b into Equation 9.12 gives rise to

$$Curv_c = \frac{Z_{XX}Z_Y^2 - 2Z_{XY}Z_XZ_Y + Z_{YY}Z_X^2}{\left(Z_X^2 + Z_Y^2\right)^{\frac{3}{2}}} \tag{9.15}$$

Positive and negative values of $Curv_p$ and $Curv_c$ can be combined to define the basic geometric relief forms and hence different types of flow (Mitasova and Hofierka 1993).

As shown in Equations 9.13 and 9.15, the calculation of curvature involves first-order, second-order, and combined differentials. We can usually estimate the differentials by using differencing or local surface fitting.

Consider differencing first. For a line with three equidistant vertices, $X - pix$, X, and $X + pix$ (pix being the interval), the first- and second-order differentials for $Z(X)$ at the central point X are

$$\begin{cases} Z'(X) = \dfrac{Z(X + pix) - Z(X - pix)}{2pix} \\ Z''(X) = \dfrac{Z(X + pix) - 2Z(X) + Z(X - pix)}{pix^2} \end{cases} \tag{9.16}$$

For a 3 × 3 local window, as shown in Figure 9.2, we have

$$\begin{cases} Z_X = \dfrac{z_6 - z_4}{2pix} \\ Z_Y = \dfrac{z_8 - z_2}{2pix} \\ Z_{XX} = \dfrac{z_4 + z_6 - 2z_5}{pix^2} \\ Z_{YY} = \dfrac{z_2 + z_8 - 2z_5}{pix^2} \\ Z_{XY} = Z_{YX} = \dfrac{z_9 + z_1 - z_7 - z_3}{4pix^2} \end{cases} \tag{9.17}$$

where $z_1 \sim z_9$ represent elevation values at grid nodes as numbered in Figure 9.2 (Moore et al. 1993; Zhou and Liu 2006).

Assume the surface fitting function is

$$Z(X,Y) = \sum_{k=1}^{K} \sum_{k'=1}^{K'} a_{kk'} X^k Y^{k'} \tag{9.18}$$

With all spot height data (i.e., (X_l, Y_l, Z_l) ($l = 1,2, \ldots, L$)), within the local moving window, we can calculate the partial differential and, thus, the desired curvature. For a 3×3 local window, L is 9. Thus, the order of the polynomials can be no more than 4. For surface-fitting-based estimation of differentials, we next review briefly the quadratic method (Evans 1980), the constrained quadratic method (Shary 1995), and the partial quadratic method (Zevenbergen and Thorne 1987).

First, we have the quadratic method (Evans 1980), $Z(X, Y) = aX^2 + bY^2 + cXY + dX + eY + f$, as shown in Equation 9.6b. The six unknown coefficients are solved through least squares:

$$\begin{cases} a = \dfrac{z_1 + z_3 + z_4 + z_6 + z_7 + z_9 - 2(z_2 + z_5 + z_8)}{6 pix^2} \\[2mm] b = \dfrac{z_1 + z_2 + z_3 + z_7 + z_8 + z_9 - 2(z_4 + z_5 + z_6)}{6 pix^2} \\[2mm] c = \dfrac{z_9 + z_1 - z_7 - z_3}{4 pix^2} \\[2mm] d = \dfrac{z_3 + z_6 + z_9 - z_1 - z_4 - z_7}{6 pix} \\[2mm] e = \dfrac{z_7 + z_8 + z_9 - z_1 - z_2 - z_3}{6 pix} \\[2mm] f = \dfrac{2(z_2 + z_4 + z_6 + z_8) - (z_1 + z_3 + z_7 + z_9) + 5z_5}{9} \end{cases} \tag{9.19}$$

The derivatives for $Z(X, Y) = aX^2 + bY^2 + cXY + dX + eY + f$ are

$$\begin{cases} Z_X = 2aX + cY + d \\ Z_Y = 2bY + cX + e \\ Z_{XX} = 2a \\ Z_{YY} = 2b \\ Z_{XY} = Z_{YX} = c \end{cases} \tag{9.20}$$

Substituting the central cell's coordinates into Equation 9.20, we have

$$
\begin{cases}
Z_X = \dfrac{z_3 + z_6 + z_9 - z_1 - z_4 - z_7}{6\,pix} \\[2ex]
Z_Y = \dfrac{z_7 + z_8 + z_9 - z_1 - z_2 - z_3}{6\,pix} \\[2ex]
Z_{XX} = \dfrac{z_1 + z_3 + z_4 + z_6 + z_7 + z_9 - 2(z_2 + z_5 + z_8)}{3\,pix^2} \\[2ex]
Z_{YY} = \dfrac{z_1 + z_2 + z_3 + z_7 + z_8 + z_9 - 2(z_4 + z_5 + z_6)}{3\,pix^2} \\[2ex]
Z_{XY} = Z_{YX} = \dfrac{z_9 + z_1 - z_7 - z_3}{4\,pix^2}
\end{cases}
\tag{9.21}
$$

as reviewed by Zhou and Liu (2006).

The constrained quadratic (Shary 1995) imposes the condition that the fitted surface passes through the central cell, so, $Z = z_5$. The resulting formulae are (Zhou and Liu 2006; Zhang 2010)

$$
\begin{cases}
Z_{XX} = \dfrac{z_1 + z_3 + z_7 + z_9 + 3(z_4 + z_6) - 2(z_2 + 3z_5 + z_8)}{5\,pix^2} \\[2ex]
Z_{YY} = \dfrac{z_1 + z_3 + z_7 + z_9 + 3(z_2 + z_8) - 2(z_4 + 3z_5 + z_6)}{5\,pix^2} \\[2ex]
Z_{XY} = Z_{YX} = \dfrac{z_9 + z_1 - z_7 - z_3}{4\,pix^2} \\[2ex]
Z_X = \dfrac{z_3 + z_6 + z_9 - z_1 - z_4 - z_7}{6\,pix} \\[2ex]
Z_Y = \dfrac{z_7 + z_8 + z_9 - z_1 - z_2 - z_3}{6\,pix}
\end{cases}
\tag{9.22}
$$

Evan's quadratic equation need not pass through the nine elevations of the 3×3 altitude submatrix. The first modification to Evan's equation by Zevenbergen and Thorne (1987) chooses a surface that passes through the nine points, resulting in the following partial quadratic equation, $Z(X, Y) = aX^2Y^2 + bX^2Y + cXY^2 + dX^2 + eY^2 + fXY + gX + hY + i$, as shown in Equation 9.6c. The nine parameters, a–i, can be determined from the nine elevations of the 3×3 submatrix by Lagrange polynomials (Zhou and Liu 2006).

The relationships between the nine parameters, a–i, and the nine submatrix elevations are

$$
\begin{cases}
a = \dfrac{\dfrac{1}{4}(z_1 + z_3 + z_7 + z_9) - \dfrac{1}{2}(z_2 + z_4 + z_6 + z_8) + z_5}{pix^4} \\[4mm]
b = \dfrac{\dfrac{1}{4}(z_7 + z_9 - z_1 - z_3) - \dfrac{1}{2}(z_8 - z_2)}{pix^3} \\[4mm]
c = \dfrac{\dfrac{1}{4}(z_3 + z_9 - z_1 - z_7) + \dfrac{1}{2}(z_4 - z_6)}{pix^3} \\[4mm]
d = \dfrac{\dfrac{1}{2}(z_2 + z_8) - z_5}{pix^2} \\[4mm]
e = \dfrac{\dfrac{1}{2}(z_4 + z_6) - z_5}{pix^2} \\[4mm]
f = \dfrac{-z_7 + z_9 + z_1 - z_3}{4\,pix^2} \\[4mm]
g = \dfrac{z_6 - z_4}{2\,pix} \\[4mm]
h = \dfrac{z_8 - z_2}{2\,pix} \\[4mm]
i = z_5
\end{cases} \tag{9.23}
$$

Thus, we calculate partial differentials at z_5 as

$$
\begin{cases}
Z_X = \dfrac{z_6 - z_4}{2\,pix} \\[4mm]
Z_Y = \dfrac{z_8 - z_2}{2\,pix} \\[4mm]
Z_{XX} = \dfrac{z_4 + z_6 - 2z_5}{2\,pix^2} \\[4mm]
Z_{YY} = \dfrac{z_2 + z_8 - 2z_5}{2\,pix^2} \\[4mm]
Z_{XY} = Z_{YX} = \dfrac{z_9 + z_1 - z_7 - z_3}{4\,pix^2}
\end{cases} \tag{9.24}
$$

(Zhou and Liu 2006; Zhang 2010).

9.2.3 SECONDARY PROPERTIES

Slope, aspect, plan, and profile curvature belong to the so-called primary topographic properties, which are those computed directly from DEMs. Other primary properties include flow-path length and upslope contributing area. Most of these topographic properties can be computed directly with a second-order finite difference scheme or by fitting a bivariate interpolation function $Z = Z(X, Y)$ to the DEM and then calculating the derivatives of the function, as reviewed by Wilson and Gallant (2000). Flow-path lengths and upslope contributing areas are calculated by specifying one or more rules to determine drainage directions and the connectivity of individual elements (Band 1986; Jenson and Domingue 1988; Wilson and Gallant 2000).

Secondary properties involve combinations of primary properties. The secondary properties help to describe pattern as a function of process. For example, these properties may affect the distribution and abundance of soil water, and the distribution and abundance of flora and fauna, as reviewed by Wilson and Gallant (2000). Thus, their derivation and analysis contribute to process-pattern modeling (Beven and Kirkby 1979; Moore et al. 1991). Below, we discuss some compound topographic indices.

A topographic wetness index Wet_T describes the spatial distribution and extent of zones of saturation (i.e., variable source areas) for runoff generation as a function of upslope contributing area (A_s), soil transmissivity (*Trnsms*), and slope gradient (β). It is calculated as

$$Wet_T = \ln\left(\frac{A_s}{Trnsms \tan\beta}\right) \tag{9.25a}$$

Assuming steady-state conditions and uniform soil properties (i.e., transmissivity is constant throughout the catchment and equal to unity), we can rewrite Equation 9.25a as

$$Wet = \ln\left(\frac{A_s}{\tan\beta}\right) \tag{9.25b}$$

as described by Beven and Kirkby (1979), Moore et al. (1991, 1993), and Wilson and Gallant (2000).

It can be predicted from Equations 9.25a and 9.25b that points lower in the catchment, points along drainage paths, and points near the outlets of the main channels are the wettest points in the catchment (Wilson and Gallant 2000). This can be seen from Equations 9.25a and 9.25b where large values of wetness index corresponding to zones of saturation occur where A_s is large (typically in converging segments of landscapes), β is small (at the base of concave slopes where slope gradient is reduced), and *Trnsms* is small (on shallow soils).

The above indices are used in TOPMODEL (Beven and Kirkby 1979) and have been used to characterize hydrologic similarity (Sivapalan et al. 1987; Lyon and

Troch 2010). Quinn et al. (1995) summarized various problems and described how steady-state topographic wetness indices can be calculated and used effectively in the TOPMODEL hydrologic modeling framework, as reviewed by Wilson and Gallant (2000).

As argued by Wilson and Gallant (2000), the topographic wetness index is based on the assumption that the soil hydraulic conductivity decreases exponentially with depth so that subsurface flow is confined to a shallow layer. If this is not the case, the steady-state and quasi-dynamic topographic wetness indices will be poor predictors of the spatial distribution of soil water.

As reviewed by Wilson and Gallant (2000), topography may exert a large impact on the amount of solar energy incident at a location on the earth's surface and can cause substantial differences in solar radiation and thereby affect such biophysical processes as air and soil heating, evapotranspiration, and primary production. Topographic solar radiation modeling and applications are discussed by other authors including Dubayah and Rich (1995) and Corripio (2003). Ruiz-Arias et al. (2009) tested solar radiation models based on DEM data in a Mediterranean environment characterized by a complex topography. The models' estimates were evaluated against 40 days of radiometric data collected at 14 stations. Analyzed sky conditions ranged from completely overcast conditions to clear skies. The role of the spatial resolution of the DEM was also evaluated, with results suggesting that the reliability of the estimates for areas of complex topography under clear-sky conditions increases using a finer spatial resolution.

It has also been shown that the distributions of solar radiation and vegetation are highly correlated (Kumar and Skidmore 2000). There remain challenges in building accurate, physically based topographic solar radiation models, as discussed in Dubayah and Rich (1995) and Scheifinger and Kromp-Kolb (2000).

9.3 MODELS OF SCALE IN TOPOGRAPHY

The variogram method is a well-known technique for determining the spatial structure of elevation data (i.e., the statistical variation in elevation between samples as some function of the distance between them). Chapter 4 described variograms as models of scale in spatially distributed properties, such as elevation. In this section, we will describe alternative approaches to modeling spatial scale in topography after a brief description of local variogram models.

Local variogram models are suited to characterizing scales of variation that vary spatially (Chen and Li 2012). There are several possible approaches for implementing a nonstationary variogram, which include locally adaptive kriging, spatial deformation of the data, and segmentation, as reviewed by Atkinson and Lloyd (2007). Locally adaptive kriging involves predicting and modeling the local experimental variogram and using the coefficients of the locally fitted model in (the local) kriging. This approach is difficult to implement because completely automatic fitting of models to variograms can be problematic (Webster and Oliver 1990). Surface deformation aims to distort a surface such that a stationary variogram results from data in transformed space (Sampson et al. 2001). Segmentation is the approach employed by Atkinson and Lloyd (2007). It involves dividing the region of interest into smaller segments within which the variogram can be considered stationary, thus allowing for local application of geostatistical optimal sampling design in their study.

Local variance (*LV*), which is calculated as the mean value of standard deviation computed in a small neighborhood (say a 3 × 3 moving window) (Woodcock and Strahler 1987; Drăguţ and Strasser 2011), measures the spatial variability of values in their neighborhood. The procedures for plotting the mean *LV* are as follows: (1) upscaling the variable under consideration to multiple "scale levels" (i.e., spatial resolutions), (2) calculating *LV* as the mean standard deviation (*SD*) within a 3 × 3 moving window for each scale level, and (3) plotting values so obtained against the scale level. The peaks in the *LV* plot are interpreted as markers of predominant scales of spatial variation. Here, scale refers to grain size (e.g., moving-window size or mean object size), which specifies the neighborhood size for calculating *LV* (Woodcock and Strahler 1987; Drăguţ and Strasser 2011).

LV can be pixel based or object based, with resampling and segmentation applied, respectively. Resampling changes the input pixel size to a user-specified pixel size. Object-based image analysis is based on segments (image objects) derived from image segmentation (i.e., image regions of minimum heterogeneity based on several user-defined parameters, such as scale parameters defining the threshold of the maximum increase in heterogeneity when two objects are merged). The value assigned to an image object is the mean of the aggregated pixel values. In comparison, segmentation produces more realistic objects than resampling (Drăguţ and Strasser 2011).

Drăguţ and Strasser (2011) applied the *LV* method for multiscale analysis in geomorphometry. The proposed method was applied to LiDAR DEMs in two test areas of different roughness: low relief and mountainous, respectively. For each test area, slope gradient, plan, and profile curvatures were produced at constant increments in scale levels with either resampling or image segmentation. Slope gradient, and plan and profile curvatures, derived from the initial 1-m spatial resolution LiDAR DEM in a standard 3 × 3 window, were used as input for multiresolution segmentation. The value assigned to an object is the mean of the aggregated pixels making up that object. They selected 200 as the upper threshold, because at this level the mean object size was comparable to the size of the maximum moving window in pixel-based scaling. Based on previous experiences in terrain segmentation, the weights for value and shape homogeneity were set to 0.9 and 0.1, respectively. At each level, *LV* was calculated as the mean *SD* of objects. The *LV* method was found to be more effective in delineating multiscale pattern elements on scale levels generated through segmentation as compared to upscaling through resampling (Drăguţ and Strasser 2011).

By making the assumption that terrain surfaces have statistical properties similar to those of fractional Brownian surfaces, it is possible to obtain fractal dimensions of terrain surfaces directly from variograms (Mandelbrot 1967; Goodchild 1982). The fractal dimension $D_{fractal}$ is directly related to the slope of the best-fitting line produced when the log of the distance between samples is regressed against the log of the mean-squared difference in the elevations for that distance. In addition to the slope of the variogram (used to determine the fractal dimension), the intercept on the ordinate, which hereafter is referred to as gamma, is also important, as the gamma value represents the expected difference in elevation for points a unit distance apart. It is anticipated that the combination of fractal dimension and gamma value will capture the essential characteristics of the land surfaces (Klinkenberg and Goodchild 1992). Thus, two terrain surfaces may have similar forms (i.e., values of $D_{fractal}$) but different gamma, indicating different roughness.

Fractal dimension can also be determined by the dividers method (Goodchild 1982). It works by walking a pair of dividers along a linear feature to determine its length (Maling 1968). As reviewed by Klinkenberg and Goodchild (1992), the fractal dimension is obtained from the slope of the linear regression of the log of the dividers' spacing against the log of the measured length of the line. The method varies with respect to how the remainder is treated: almost all "walks" along a line result in a noninteger number of dividers being required to cover the line completely. The most common approach is to add the remainder as a proportion of a divider (e.g., Shelberg et al. 1982). For elevation data, the divider method works by using contour lines converted from the gridded DEM. The string of coordinates for each contour line is then used in the subsequent analyses. It is important to ensure that an adequate number of data points are available for use in determining the slope, which is used to determine the fractal dimension of the contour line. The dimension of the surface can be obtained by adding one to the value obtained for the dimension of the contour lines, as described by Klinkenberg and Goodchild (1992).

In addition to the variogram and dividers methods for estimating fractal dimension, there is the cell-counting method for determining the fractal dimensions of surfaces as well as of linear features (Goodchild 1980a). The method works as follows: (1) The surface is sliced by a horizontal plane at a given elevation. (2) Those cells with elevations that are above or on the plane are coded as black; those below are coded as white. (3) A count is made of the number of adjacencies (i.e., where a black cell occurs next to a white cell). (4) The cell size is increased (by replacing four cells with one cell that is assigned the average elevation of the four original cells) and the process repeated until the cell size is equal to the size of the array. The surface fractal dimension determined using the cell-counting technique can be derived from the slope of the best-fitting line from the plot showing the log of the average number of boundary cells plotted against the log of the cell size (Klinkenberg and Goodchild 1992).

There are also scale descriptors based on the concept of *representative elementary area* (REA). Wood et al. (1988) modeled the hydrologic response of a catchment using a modified version of TOPMODEL (Quinn et al. 1995), which is capable of modeling both infiltration excess and saturation excess runoff and incorporating the spatial variability of soils, topography, and rainfall. They confirmed the existence of REAs. REA indicates a spatial scale over which the process representations can remain simple. Thus, REA is defined as the minimum area of a sample from which a given parameter becomes independent of the size of the sample. The size of an REA ranges from a minimum bound, which is the transition from the microscale to the macroscale level, to a maximum bound, which is the transition from a homogenous to a heterogeneous state. Blöschl et al. (1995) further investigated the REA concept and reassessed its utility for distributed parameter rainfall-runoff modeling.

If a terrain includes heterogeneous patches, then a landscape property can be characterized by different REAs within adjacent patches. Based on the concept of REA, Florinsky and Kuryakova (2000) defined the adequate interval of areas, which refers to an interval of observation areas wherein terrain property values vary smoothly or have a constant value. A continuous surface with a limiting frequency $1/s$ can be uniquely determined by its values with sampling step $s/2$, where s is the smallest planimetric size of the elements of the surface. Therefore, $s/2$ is an adequate area to represent a landscape property. Florinsky and Kuryakova (2000) developed an experimental statistical method to determine an adequate w for DEMs applied to landscape studies. The method includes the following steps: (1) derivation of a DEM set using a series of w_i, (2) a correlation analysis of data on a landscape property and a topographic variable estimated with various w_i, (3) plotting of correlation coefficients obtained vs. w_i, and (4) determination of smoothed plot portions indicating intervals of an adequate w.

Spectral analysis provides information on relative amounts of variation at different wavelengths or spatial frequencies (Pike and Rozema 1975), with wavelength approximately equivalent to spatial scale, as mentioned in Chapter 1. The analysis by Pike and Rozema (1975) was limited to one-dimensional topographic profiles and may be extended to two-dimensional terrain data. As reviewed by Perron et al. (2008), the discrete Fourier transform (DFT) of a two-dimensional dataset $Z(X, Y)$ consisting of N_X by N_Y measurements spaced at even intervals ΔX and ΔY can be written as

$$\mathbb{F}_Z(k_X, k_Y) = \sum_{l_X=0}^{N_x-1} \sum_{l_Y=0}^{N_y-1} Z(l_X \Delta Y, l_Y \Delta Y) e^{-12\pi(k_X l_X/N_X + k_Y l_Y/N_Y)} \tag{9.26}$$

where k_X and k_Y are the wavenumbers in the X and Y directions, respectively; l_X and l_Y are Zs location indices ($X = l_X \Delta X$, $Y = l_Y \Delta Y$); and $I^2 = -1$. The complex DFT expresses the amplitude and phase of the sinusoidal components of Z. The DFT is an $N_X \times N_Y$ array, and the wavenumbers are the indices of the array. The output of most DFT algorithms must be rearranged to place the zero wavenumber element near the center of the array. With even N_X and N_Y, the output array can be divided into four equal quadrants. We can then exchange the nonadjacent quadrants so that the zero wavenumber element will be positioned at ($N_X + 1$, $N_Y - 1$) in the new array. With wavenumbers referenced to this location, an element at (k_X, k_Y) in the DFT array corresponds to the two orthogonal frequency components: $f_X = \dfrac{k_X}{N_X \Delta X}$ and $f_Y = \dfrac{k_Y}{N_Y \Delta Y}$. If X and Y have units of length, as in the case of topographic data, f_X and f_Y have units of cycles per unit length. The element at $\mathbb{F}_Z(k_X, k_Y)$ describes a wave with a wavelength

$$\lambda = 1 \Big/ \sqrt{f_X^2 + f_Y^2} \tag{9.27a}$$

and an orientation ζ, measured counterclockwise from the positive X direction and given by

$$\tan \zeta = k_Y \Delta Y / (k_X \Delta X) \tag{9.27b}$$

The power spectrum provides a measure of how the variance of Z varies with frequency. As described by Perron et al. (2008), one common way of estimating the power spectrum is the DFT periodogram *Prdgram*:

$$Prdgram_Z(k_X,k_Y) = \frac{1}{N_X^2 N_Y^2}\left|\mathbb{F}_Z(k_X,k_Y)\right|^2 \tag{9.28}$$

Prdgram has units of amplitude squared. It is linearly related to the power spectral density (PSD), which has units of amplitude squared per unit X frequency per unit Y frequency, or amplitude squared per frequency squared:

$$PSD_Z(k_X, k_Y) = Prdgram_Z(k_X, k_Y)N_X N_Y \Delta X \Delta Y \tag{9.29}$$

Perron et al. (2008) investigated quantitatively the existence of characteristic landscape scales by analyzing two-dimensional Fourier power spectra derived from fine-resolution topographic maps of two landscapes in California. In both cases, they found that spectral power declines sharply above a frequency that corresponds roughly to hillslope length, implying that the landscape is relatively smooth at finer scales. The spectra also show that both landscapes contain quasi-periodic ridge-and-valley structures, and they derived a robust measure of the ridge–valley wavelength. In addition to the identification of characteristic spatial scales, spectral analysis may be applied for filtering in geomorphology.

Application of the discrete Fourier transform to topographic data is, however, problematic. First, the data are usually spatially varying, implying the need for a nonstationary model. Second, topographic features such as ridges and valleys are not sinusoids but complex in shape (i.e., they must be described by a range of frequencies). Third, a single feature in the landscape might be represented in the Fourier transform by a great number of sinusoidal components, and, conversely, a single component in the Fourier transform contains contributions from many surface features (Gallant and Hutchinson 1996).

Wavelet analysis uses a single basis function that is localized in both position and frequency and can be translated and dilated to cover the entire position-frequency plane, as reviewed by Lark and Webster (1999). This is in contrast with Fourier analysis, which uses many basis functions, each at a different frequency. Wavelet analysis could produce a power spectrum at every point in the dataset, which overcomes the global model restriction of the Fourier transform (Gallant and Hutchinson 1996).

Consider one-dimensional data $z(x_i)$ $(i = 1, 2, ..., n)$, where x_i is a measure of position along the transect. As described in Chapter 8, the detail signal of Z at resolution 2^j can be computed as an inner product: $D_{2^j}Z = \langle Z(x), \psi_{2^j}(x - 2^{-j}n)\rangle$ (Mallat 1989). A general form for the wavelet can be written as

$$\psi_{rsl,sft}(x) = rsl^{-1/2}\psi\left[\frac{x - sft \times rsl}{rsl}\right] \tag{9.30}$$

where *rsl* stands for resolution and *sft* for translation (taking values as x_i, $i = 1, 2, ..., n$). The wavelet transform $D_{rsl,sft}Z$ is calculated as

$$D_{rsl,sft}Z = \sum_{j=1}^{n} z(x_j) \psi_{rsl,sft}(x_j) \tag{9.31}$$

where *rsl* is the scale (resolution) being tested.

Wavelet variance, for scale *rsl*, is the square of the wavelet coefficients averaged over position:

$$\text{var}_Z(rsl) = \frac{1}{n} \sum_{i=1}^{n} \left[D_{rsl,x_i}Z \right]^2 \tag{9.32}$$

which is referred to as the scalogram (Ollier et al. 2003). It can be used to determine the scales of pattern in the data. Wavelet variance responds to the scale of the pattern, which is the average of the sizes of the patches and of the gaps, as reviewed by Dale and Mah (1998).

We can also compute a position variance by averaging the wavelet coefficients over all scales, from 1 to *RSL* (*RSL* represents the number of resolution levels being considered or the resolution at the coarsest scale), for any particular position x_i along a transect:

$$\text{var}_Z(x_i) = \frac{1}{RSL} \sum_{res=1}^{RSL} D_{rsl_{res},x_i}Z \tag{9.33}$$

The position variance can be used to identify density peaks or density transitions in the data, even with random errors added. The wavelet method allows great flexibility in analysis, because the data can be analyzed using different wavelet forms (Dale and Mah 1998).

As discussed by Dale and Mah (1998), when wavelet variance is plotted as a function of scale, the peak variance height is determined by pattern intensity and does not increase with scale and, depending on the wavelet chosen, the position of the variance peak matches the scale exactly. Moreover, by using several different wavelet forms, different characteristics of the pattern can be investigated; the method is able to portray trends in the data, when the pattern is nonstationary; and the wavelet position variance can be used to identify both patches and gaps in data contaminated with random error.

Various indices have been proposed to measure terrain *ruggedness*. A widely recognized method for quantifying ruggedness among biologists is the land-surface ruggedness index developed by Beasom et al. (1983). This index is based on the assumption that ruggedness is a function of total length of topographic contour lines in a given area. Terrain ruggedness index (TRI) (Riley et al. 1999) provides a quantitative measure of topographic heterogeneity. TRI values for individual grid cells of a DEM can be computed by taking the square root of the sum of squared differences in elevation between a center grid cell and its eight neighbor grid cells. Riley et al. (1999) used an equal-area classification method to group continuous ranges of TRI values into seven classes of unequal range: level, nearly level, slightly rugged, intermediately rugged, moderately rugged, highly rugged, and extremely rugged. There is also a terrain ruggedness index defined as the mean difference (absolute value) between a central pixel and its surrounding cells (Wilson et al. 2007). Wilson et al. (2007) applied TRI for seabed habitat mapping in which the TRI value is a measure of the local variation in seabed terrain about a central pixel. Since variation in the seabed terrain is associated with changes in habitat and colonization, such TRI values may be particularly relevant for deep sea habitat mapping studies where opportunities to survey the benthic fauna directly remain limited.

Unfortunately, neither of the above indices measures directly the variability in topographic aspect and gradient. Thus, these indices may not be able to distinguish steep, even terrain (high slope and low ruggedness) from steep terrain that is uneven and broken (high slope and high ruggedness). The *vector ruggedness measure* (VRM) measures terrain ruggedness as the variation in the three-dimensional orientation of grid cells within a neighborhood (Sappington et al. 2007). VRM was adapted from a method first proposed by Hobson (1972). By measuring vector dispersion in three dimensions, VRM combines variation in slope and aspect into a single measure and can thus give a more informative picture of heterogeneity of terrain than indices based only on slope or elevation, such as TRI. This method effectively captures variability in slope and aspect into a single measure.

Computation of VRM is completed by (1) decomposition of unit vectors normal (orthogonal) to each grid cell into their X, Y, and Z components using standard trigonometric operators and the slope and aspect of the cell, and (2) calculation of a resultant vector over a 3×3 neighborhood centered on each cell, using a moving-window routine (Sappington et al. 2007). The magnitude of the resultant vector in standardized form (vector strength divided by the number of cells in the neighborhood) is a measure of the ruggedness of the landscape for the selected scale (Hobson 1972). Subtracting this magnitude in standardized form from 1 results in a dimensionless ruggedness number that ranges from 0 (flat) to 1 (most rugged) (for details, see Sappington et al. 2007).

9.4 METHODS FOR SCALING TERRAIN VARIABLES

9.4.1 UPSCALING

Topographic information derived from DEMs is frequently utilized for spatial analysis and modeling in various subject areas including biogeography and water conservancy. As shown in the first section, global DEMs with coarser resolution are often of mixed quality as they were acquired from heterogeneous sources (Rabus et al. 2003; Wechsler 2007). Recently, fine-resolution DEMs, such as SRTM, became available for many large areas. However, spatial analysis over large areas with DEM data at fine resolution is computationally expensive. Thus, fine-resolution DEMs often need to be aggregated to coarser resolutions to streamline geo-computation with terrain data. In other words, upscaling or aggregation is necessary for performing spatial analysis and input to simulations at regional and global scales (Qi and Hesketh 2005; Le Coz et al. 2009; Milzow and Kinzelbach 2010).

As reviewed by Le Coz et al. (2009), DEM aggregation significantly affects representations, analyses, and modeling of topography and related properties. For example, at various resolution levels, DEM aggregation tends to decrease slopes, but increase drainage areas per contour unit (Zhang and Montgomery 1994). The drainage network structure and the basin extent are also influenced by changes in DEM resolution (Garbrecht and Martz 1994). Model efficiency generally decreases with DEM aggregation in hydrological simulations (Le Coz et al. 2009). Therefore, upscaling should be approached with due consideration to the scale dependence (also known as grid dependence) of the specific properties being analyzed. As explained by Milzow and Kinzelbach (2010) in the context of MODFLOW-based hydrological modeling in the Okavango Wetlands, the variability and spatial correlation of elevation data should be preserved in upscaling, at least to some extent.

Various methods can be explored for upscaling terrain data. Image resampling methods, such as nearest neighbor, bilinear interpolation, or cubic convolution (Lehmann et al. 1999), may be applied. As discussed by Parker et al. (1983) in the context of medical imaging, the nearest-neighbor method (i.e., central pixel resampling) tends to shift the image up to one-half a pixel, whereas the linear and cubic B-spline interpolation method tends to smooth the image. At the expense of some increased computation, image quality can be improved by resampling using the so-called fine-resolution cubic spline function as compared to the nearest-neighbor, linear, or cubic B-spline functions. However, for slope calculation, the nearest-neighbor method was found to be the most accurate algorithm (Takagi 1998). Area-weighted averaging can also be used for upscaling, as it is believed to be more informative when aggregating fine-resolution data to produce coarse-resolution data.

Consider *statistical methods* for upscaling, which were reviewed in general terms in Chapter 7. One of the most commonly used methods for DEM aggregation is to take the mean value over a block (at coarser resolution) of finer-resolution grid cells. Aggregation can also be performed with other statistical methods (e.g., median, mode, maximum, or minimum values). Each method extracts a value over a moving window (say, 3 × 3) in the baseline DEM. The maximum and minimum methods take, respectively, the highest and the lowest elevation values in the window. These two methods are best suited to preserve valleys (minima) and crests (maxima) and, therefore, the drainage network and basin boundaries (Le Coz et al. 2009). The mean, median, and mode methods compute a value that characterizes the middle of the statistical distribution over the window. These methods may be deemed more appropriate for remotely sensed data, because a pixel value is assumed to be the integrated value over the corresponding area on the ground (Bian and Butler 1999). In particular, the median value is the most suitable aggregation method for fine-resolution modeling in urban areas, as reviewed by Le Coz et al. (2009; Ulrich et al. 2002).

Le Coz et al. (2009) examined appropriate methods to aggregate SRTM DEMs within the framework of modeling the water balance of the Lake Chad basin (LCB) (2.5 Mkm²). The Terrestrial Hydrology Model with Biogeochemistry (THMB), developed by Coe (2000), was employed, which simulates seasonal river discharges and changes in surface-water level, area, and volume. The processed 3-arcsecond SRTM DEM (subsequently called baseline DEM) was aggregated from 3 arcseconds (~90 m) to 5 arcminutes (~10 km) so as to be used as input to THMB, which currently operates on a 5 × 5-arcminute spatial resolution for computational efficiency and because of the unavailability of fine-resolution runoff and climatic data in LCB. Six aggregation methods (the mean, the median, the mode, the nearest neighbor, the maximum, and the minimum) were selected for aggregating the SRTM DEM from 3 arcseconds to 5 arcminutes, They gave rise to contradictory results depending on the entity selected: for the drainage network connectivity, only the mean and median methods produced consistent results; for the Lake Chad bottom topography, nearest neighbor, median, and mean were the most suitable; for the representation of depressions governing floodplain dynamics, nearest neighbor and maximum were the most useful. Therefore, the selection of an aggregation procedure is a crucial step that needs to be carried out in accordance with the intended use of the DEM.

Below, we review orography-based methods for upscaling, as described by Bindlish and Barros (1996). Mean orography can be obtained by using a low-pass filter that suppresses peaks and valleys. The low-pass filtering method preserves the overall volume of topographic features, but with possible areal spreading. With low-pass filtering, the high-frequency roughness of the terrain is smoothed, resulting in underestimation of the peaks (Bindlish and Barros 1996).

Envelope orography is created by adding a multiple of the local standard deviation σ of the fine-resolution DEM to the mean orography

$$\sigma = [(z_{max} - \bar{z})(z_{min} - \bar{z})]^{1/2} \tag{9.34}$$

where $\bar{z}$ is the mean terrain height at the coarser resolution, and z_{max} and z_{min} are, respectively, the maximum and minimum terrain height within the area corresponding to z. Although envelope orography captures topographic peaks more accurately than the corresponding mean orography, low-elevation plains and valleys are not well resolved, and the total volume of orographic features is not preserved (Bindlish and Barros 1996).

By the Cressman-type objective analysis (Bindlish and Barros 1996), the terrain height $z_c(x)$ on a coarse-resolution grid is obtained by a weighted average of the heights $z(x)$ at fine resolution:

$$z_c(x) = \sum_{l=1}^{np} wt_l z(x_l) \bigg/ \sum_{l=1}^{np} wt_l \tag{9.35}$$

where wt_l is a weighting function that depends on the radius of influence RDS and on the distance rds from the point x where the elevation is being computed to fine-resolution data point x_l (i.e., $rds = |x_l - x|$), and np represents the total number of points within a radius rds smaller than or equal to the user-specified radius of influence RDS. The weight wt_l is calculated as

$$wt_l = (RDS^2 - rds^2)/(RDS^2 + rds^2) \qquad (9.36)$$

where $rds \leq RDS$, as described by Bindlish and Barros (1996). One limitation of this method is the need to specify the proper radius of influence on a case-by-case basis and for distinct aggregation levels, in order to obtain sufficiently accurate results.

Bindlish and Barros (1996) also proposed a modified fractal interpolation scheme to produce consistent digital terrain models at different spatial resolutions. Comparisons with alternative approaches such as sampling, the mean and envelope orography methods, and the Cressman-type objective analysis confirmed the superiority of the modified fractal approach, as it preserves well both the spatial structure in elevation and the orographic gradients and performed consistently more accurately than others, especially for high-level aggregation.

9.4.2 DOWNSCALING

When calculating slope from regional- to global-scale DEMs, it is important to recognize the fact that slopes derived from DEMs vary with the spatial resolution and that slopes estimated from coarse-resolution data are often underestimates of the true slope, as mentioned previously. However, downscaling can restore slopes at finer resolution, to some extent, as shown below.

Klinkenberg and Goodchild (1992) demonstrated that topography generally exhibits fractal characteristics, although not perfectly fractal. This fact implies that the linkage between the topography observed and the scale used for the observation should be self-similar for some landscapes. Given that slope is the derivative of the local topography, its measurement should therefore also be a function of scale. Fractal techniques are therefore potentially useful in providing more accurate estimates of local slope using only coarse-resolution data.

We can calculate the fractal dimension in a region when the log of distance between samples is regressed against the log of the mean-squared difference in the elevations for that distance, as shown in the preceding subsection. Through variogram modeling, the relationship between the elevations of two points and their distance can be converted to the following formula (Klinkenberg and Goodchild 1992):

$$(Z_x - Z_{x'})^2 = k \times distance^{4-2D_{fractal}} \qquad (9.37a)$$

So that

$$slope = (Z_x - Z_{x'})\big/distance = \pm\sqrt{k}\,distance^{1-D_{fractal}} \qquad (9.37b)$$

where Z_x and $Z_{x'}$ are the elevations at points x and x', *distance* is the distance between x and x', k is a constant, and $D_{fractal}$ is the fractal dimension. The value $(Z_x - Z_{x'})/$ *distance* is actually the surface slope. Thus, the percentage slope is related to its corresponding scale (grid size) *distance* by Equation 9.37b (for further details, see Zhang et al. 1999).

The above relationship implies that if topography is fractal, then slope will also be a function of the scale of measurement represented by *distance* in Equation 9.37b. Thus, the average slope values in a DEM can be estimated at different resolutions when using the slope scaling Equation 9.37b. It is reasonable to suppose that different kinds of terrain might have characteristic topography that can be expressed in terms of different values of k and $D_{fractal}$. Bindlish and Barros (1996) proposed a modified fractal interpolation scheme to produce consistent digital terrain models at different spatial resolutions, which is applicable for both upscaling and downscaling.

As shown in the previous subsection, upscaling is necessary for applying hydrological models developed at the plot scale or the catchment scale at coarser scales. Upscaling leads to loss of spatial heterogeneity due to reduction of spatial scale (i.e., coarsening of spatial resolution) and can have great impacts on model parameters and, thus, the results (Zhang et al. 2002). The spatial heterogeneity may itself be an important factor in spatial analysis and modeling. Thus, it is important to consider reproducing such fine-scale heterogeneity from coarse-scale data. Zhang et al. (2002) implemented a soil-erosion model developed from the analysis of the results of plot-scale erosion experiments (1 to 30 m²). The model employed involves slope and vegetation cover, for which the model exhibits strong nonlinear scaling. The approach developed by them uses the frequency distribution of a variable, say, vegetation cover, within a large pixel to describe the variable's spatial variation at a finer resolution. Their approach to downscaling is special for three reasons. First, model parameters, rather than model results, are downscaled to the scale at which the model is known to be applicable. Second, the aim of the downscaling is not to provide more detail over limited areas, but to account for the causes of nonlinear scaling of the model in question for global datasets. Third, the methods consider scale explicitly as a parameter in the downscaling equations, as described by Zhang et al. (2002).

We should consider the issue of resolution *vs.* accuracy for downscaling carefully. To define an appropriate grid size for simulations using topographically driven models, we should base the selection of resolution on the accuracy and density of the original elevation data from which the DEM was constructed and the terrain processes under consideration (Zhang and Montgomery 1994). Firstly, it is necessary to consider the spot elevation data used to create a DEM. The spacing of the original data used to construct a DEM effectively limits the resolution of the DEM. Decreasing the grid size beyond the resolution of the original survey data does not provide much additional information nor increase the accuracy of the DEM for surface presentation and it potentially introduces interpolation errors, as explained by Zhang and Montgomery (1994). In addition to the average spacing of the data used to derive a DEM, the length scale of the primary landscape features of interest provides a natural guide to an appropriate grid size. For example, a grid size smaller than the hillslope length is necessary to simulate adequately processes controlled by landform, as it would be unreasonable to use a 30-m or a 90-m grid spacing to model hillslope or runoff generation processes in moderate to steep gradient topography without some calibration of the process model. On the other hand, a coarse (e.g., 90 m) grid spacing may be most appropriate for modeling orogenic processes over large areas and long time scales (Zhang and Montgomery 1994). Zhang and Montgomery

(1994) generated a series of simulated landscapes by gridding elevation data at 2, 4, 10, 30, and 90 m of scales. Frequency distributions of slope (tan β), drainage area per unit contour length (a), and the topographic index (a/tan β) were calculated for each grid size model. Their analyses suggested that, for many landscapes, a grid size of 10 m would suffice for simulating geomorphic and hydrological processes.

9.5 DISCUSSION

As shown in earlier sections of this chapter, topography manifests varied scale dependencies. But there are different scale effects with different topographic properties. Furthermore, topographic properties are often more effectively characterized at multiple scales. Schmidt and Andrew (2005) performed local interpolation of landform surfaces at different window sizes to detect landform characteristics at different scales. The terrain surface was approximated by polynomial models. Values for gradient and curvature for each point at different window sizes can be derived. Additionally, the mean square error of the least-squares fit provides information about the changing surface characteristics at different scales (window sizes). These values can be plotted against the window size (scale) representing the local scale behavior of each point in a DEM. Dominant scale ranges of the surface for each point were defined by examining breakpoints in the scale plots. These breakpoints define fundamental changes in the characteristics of the surface with scale (e.g., from convex to concave) and can be used to identify scale ranges (for details, see Schmidt and Andrew 2005).

Methods for both upscaling and downscaling in terrain data and analysis have been reviewed. Clearly, statistical approaches dominate the list, whereas links between geostatistics and fractals have also been exploited for scaling. Physically based methods for scaling, such as those used in remote sensing, may be usefully explored for scaling terrain data (Merlin et al. 2009). Further developments in scaling lie in the combined use of statistical and physical models (Luoto and Hjort 2008).

Downscaling is relatively more challenging than upscaling. Thus, collateral information may be valuable for downscaling. Saito and Grasselli (2010) proposed a geostatistical method that accounts for roughness characteristics when downscaling fracture-surface topography. In the proposed approach, the small-scale fracture surface roughness is described using a local roughness pattern that indicates the relative height of a location compared to its surrounding locations, whereas the large-scale roughness is considered using the surface semivariogram. Simulated annealing was performed, accounting for both minimization of the local error variance and the reproduction of the local roughness characteristics. The fracture surface topography downscaling process resulted in greater accuracy compared to ordinary kriging and sequential Gaussian simulation.

Although scales in discrete objects have not been discussed in depth, we can hint at some of their scale-related issues. Discrete objects, such as channels and ridges (Douglas 1986), may be derived from DEMs. These objects need to be interpreted differently from objects that are mere products of terrain abstractions and representations. For example, flow paths are different from contour lines, with the latter being artifacts, abstracted for surface representations. Thus, although fractals may be applied to these different kinds of objects extracted from terrain surfaces, their underlying semantics must be understood for proper interpretation and application.

Fine-resolution river networks have a level of detail appropriate for analysis at the watershed scale, but are too detailed for global hydrologic studies. River-network upscaling consists of coarsening the resolution of fine-resolution networks to make them compatible with regional- and global-scale models, such as global climate models. A river network upscaling algorithm, which processes fine-resolution DEMs to determine the flow directions that best describe the flow patterns in a coarser user-defined scale, was presented by Olivera et al. (2002). Their algorithm was called the double maximum method (DMM). The resulting river networks of the DMM reproduce well the river network pattern and the watershed boundaries of observed fine-resolution data. Compared to other coarse-resolution river networks, the DMM produces similar or more accurate results. More recently, Wu et al. (2011) developed a hierarchical dominant river tracing (DRT) algorithm for automated extraction and spatial upscaling of basin flow directions and river networks using fine-scale hydrography inputs. Unlike upscaling methods, the DRT algorithm utilizes information on global and local drainage patterns from baseline fine-scale hydrography to determine upscaled flow directions and other critical variables including upscaled basin area, basin shape, and river lengths. Information about the original baseline hierarchical drainage structure is preserved by tracing each entire flow path from headwater to river mouth at fine scale while prioritizing successively higher-order basins and rivers for tracing.

In addition to primary (e.g., slope and aspect) and secondary (e.g., wetness index) topographic properties and terrain features (e.g., channels, ridges, and river networks), landform maps (Hammond 1954) can also be derived from DEMs. Landform classifications are often based on combinations of terrain variables (Iwahashi and Pike 2007). As shown by Pike (1988), topography associated with different landslide processes can be discriminated numerically by a geometric signature (i.e., selection of terrain derivatives, such as elevation and elevation variance spectrum) tailored to the problem. The signature approach to solving geomorphic problems with DEM data should be adaptable to geologic hazard mapping and, ultimately, to general geomorphic analysis including such applications as numerical mapping of physiographic divisions and correlating landform with remotely sensed environmental data. Drăguţ and Eisank (2012) described an object-based approach to classifying topography at broader scales into landform types, with the proposed approach implemented as a customized process for the eCognition software (http://www.ecognition.com/products). With this object-based approach, classification is performed at three scale levels that represent domains of complexity, and, for each domain, scales in the data are detected with the help of local variance and segmentation is performed at these appropriate scales. Objects resulting from segmentation are partitioned into subdomains based on thresholds given by the mean values and standard deviation of elevation, respectively. Statistical evaluation indicates that most of the classes satisfy the regionalization requirements of maximizing internal homogeneity while minimizing external homogeneity, with most objects having boundaries matching natural discontinuities at the regional level. Landform classification driven by numerical analysis of DEMs and their derivatives falls in the realm of area-class mapping, the focus for the next chapter, for which issues of scale are even more challenging.

10 Scale in Area-Class Mapping

As indicated at the beginning of Chapter 9, scale issues in terrain analysis and area-class mapping are important topics for geospatial research and applications, for at least two good reasons. One is that terrain analysis and area-class mapping are key components in many spatial problem-solving scenarios, in particular those concerning scale and scaling in spatial analysis and modeling. The other is the impressive accumulation of research achievements in terrain analysis and area-class mapping, which can shed light on further research on related subjects. Therefore, this chapter seeks to answer questions about scale and scaling in area-class mapping, continuing the previous discussion of terrain analysis.

Information about land cover and related surface processes is becoming increasingly important for environmental modeling, landscape analysis, and global change research (Skole and Tucker 1993; Goodchild 1994; Gallant et al. 2004; Lawrence and Chase 2010). The sophistication of remote sensing and geographic information systems (GIS) has contributed to the proliferation of information products (DeFries and Townshend 1994; Lambin and Strahler 1994; Lunetta and Elvidge 1998; Coppin et al. 2004; Roy et al. 2008; Drummond et al. 2012). For instance, the European Commission TRopical Ecosystem Environment observation by Satellite (TREES) project has produced a map of the forest cover of the humid tropics based on satellite sensor data at 1 km spatial resolution with consistency in data acquisition and interpretation. There is general agreement between this map product and other inventories, such as the Food and Agriculture Organization's (FAO's) Forest Resources Assessment 90 and the National Aeronautics and Space Administration's (NASA's) Landsat Pathfinder datasets, using different approaches ranging from compilation of existing data and statistical sampling to exhaustive survey with satellite sensor data (Mayaux et al. 1998). Hansen et al. (2005) tested multiple single-date images, 40-day composites, and multitemporal annual metrics from the MODIS sensor for mapping percent tree cover at global, continental, and regional/local scales, using global training data and a SAFARI 2000 validation database from western Zambia. The Forest Inventory and Analysis (FIA) National Program (http://fia.fs.fed.us/) of the US Department of Agriculture's (USDA's) Forest Service conducts detailed surveys of forests across all land ownerships, based on remote sensing for forest and nonforest land classification and making ancillary measurements, such as fragmentation, urbanization, and distance variables (Nelson et al. 2009).

As reviewed by Goodchild et al. (2009), land cover maps and those depicting distributions of categorical fields are called area-class maps or categorical maps (Mark and Csillag 1989). They are usually represented in vector format as polygons associated with categorical description or as contiguous blocks of colored raster cells, perhaps to cater for visual interpretation and computer-based processing, respectively. The advent of geospatial techniques such as remote sensing and GIS has enhanced cost-effectiveness in area-class mapping, as shown above. More importantly, as discussed by Goodchild et al. (2009), technological developments have prompted renewed interest in some of the fundamental issues associated with area-class mapping, to which a key hurdle concerns how area classes are defined. As different classifications are performed using different criteria and to serve different purposes, it is rare for classifications derived from different systems, as different generalizations, to be comparable (Robinove 1981). It is important to develop classifications using simple, observable, unambiguous covariates that are preferably remotely sensible and field measurable, with the defined classes directly translatable to the parameters sought (Robinove 1981; Goodchild et al. 2009).

Historical developments in soil science and eco-geographic regionalization may be looked to for insights. Different soils originate from different external environmental conditions, primarily of vegetation and climate, according to the Russian tradition, whereas in the USA relationships between soils and climate were emphasized, resulting in the concept of climate zones (Jenny 1941), as also discussed by Goodchild et al. (2009). The physiographic regions of the world are area-class definitions of the earth's landforms based upon Fenneman's (1928) three-tiered approach: divisions, provinces, and sections. Ecoregions of the United States, as large areas of similar climate, were mapped to depict ecosystems of regional extent according to the Crowley (1967) classification (Bailey 1983), with climate and vegetation as indicators of the extent of each unit. Four ecological levels are shown on the map: domain, division, province, and section. Domains are based on observable differences having developed largely because of prevailing climatic conditions, which, on the basis of further climatic criteria, are broken down into divisions, which, on the basis of the climax plant formation, are then subdivided into provinces, which are subdivided into sections differing in the floristic composition of the climax plant formation. Highland provinces and sections are distinguished where the climatic regime differs substantially from those in adjacent lowlands, so as to cause a complex vertical climate-vegetation-soil zonation because of altitude (Bailey 1983).

Goodchild and Dubuc (1987) proposed a phase-space-based logic for categorical mapping, by which land-cover types may be differentiated into zones of unique combinations of controlling factors. Classification derived from the phase space can be interpreted more objectively than conventionally derived equivalents that are largely subject to subjectivity in class clustering and labeling, leading to analysis that can be performed quantitatively by taking advantage of class-forming covariates embedded in the phase space. Along a similar line, Busby (2002) reported a test using a tool known as BIOCLIM that uses climate data to estimate distributions of species, such as temperate rainforest trees and bats. BIOCLIM works on the basis of climate surfaces, which are used to produce site-specific estimates of monthly temperature and precipitation for places where animal or plant species have been recorded. The climate estimates are then aggregated into a climate profile, using parameters indicative of annual mean, and seasonal and monthly extreme values. Prediction of species occurrences is based on the similarity of climates at grid nodes being mapped to the climate profile.

After class definition, there is the issue of scale for area classes, which is an important dimension in the logic framework for area-class mapping. Area classes are definable because they occupy measurable areal extents (Goodchild et al. 2007). Also, semantically identical classes can be defined and mapped at multiple spatial scales. For example, a variety of satellite sensor images are now used routinely (e.g., Ikonos, Quickbird, SPOT, and Landsat), whose pixel sizes (indicative of spatial resolution) range from less than 1 m for Quickbird to 30 m for Landsat. The multiplicity of data sources and scales has implications for area-class mapping: the spatial heterogeneity of the underlying scene being imaged, the differences in sensor mechanisms, local variation in the sun-surface-sensor geometry, and the scale dependence of image classification all affect the accuracy and consistency in analysis and applications (Lu and Weng 2007; Gardner et al. 2008). Furthermore, incompatibilities in classification should be considered as limiting factors for analysis and accuracy assessment (Congalton 1991).

To deal with the scale multiplicity inherent to class definitions, we first need to make sure that all maps have equivalent or transferable class naming. For this, extensive protocols exist to ensure that this will be possible, including the Anderson et al. (1976) classification, the National Land Cover Database (Vogelmann et al. 2001), and the National Vegetation Classification System (Grossman et al. 1994), as argued by Gardner et al. (2008). Nevertheless, semantic scale remains an important topic for further research.

Secondly, it is important to identify and describe scales in area-class maps. Wu et al. (2000) reviewed two approaches to multiscale analysis of landscape heterogeneity: the direct and indirect approaches. The former includes variogram and scale variance methods, whereas the latter includes numerous landscape metrics. Variogram models reviewed in Chapter 4 can be extended to describe scale in categorical data. Scale variance introduced in Chapter 9 is potentially a powerful method to detect and describe scales, in particular those of multiple structures. Landscape metrics are important as they respond to changing grain size (i.e., change of measurement scale) rather differently, and these changes are indicative of the multiple-scale structures in landscape patterns (Wu et al. 2000). According to Wu et al. (2000), some metrics (e.g., the number of patches, patch density, total edge, edge density, mean patch size, patch size coefficient of variation) exhibit consistent, predictable patterns over a wide range of grain sizes, whereas others (e.g., patch diversity, contagion, landscape fractal dimension) have more complex responses. The direct and indirect approaches mentioned above are complementary and should be explored in combination. For this, the phase-space model is valuable for exploring the relationships between geostatistical models and landscape pattern indices, bridging the gap between continuous covariates and discrete classes in regard to scale modeling in area-class maps.

Thirdly, as shown in the context of interval/ratio fields in Chapter 9, both upscaling and downscaling are important, and relevant methods should be devised for area-class maps whose scaling needs to cater for different semantics. The aggregation of fine-resolution data has been employed widely with several methods, such as nearest-neighbor resampling and majority-rule aggregation (Dodgson 1997; He et al. 2002). Numerous studies have assessed spatial aggregation errors (e.g., Turner et al. 1989; Costanza and Maxwell 1994; Moody and Woodcock 1995, 1996), and model-based corrections have been suggested (e.g., Walsh and Burk 1993; Mayaux and Lambin 1995; Moody 1998). There have also been research efforts investigating methods for downscaling categorical data (Riitters 2005). However, the issues associated with both upscaling and downscaling area-class maps have not been systematically explored.

This chapter will first describe discriminant space, a renaming and extension of phase space (Goodchild and Wang 1989; Goodchild et al. 2009), as a conceptual framework for area-class mapping. Methods for realizations of discriminant spaces will then be discussed, ranging from the feature space concept in remote sensing to heuristic and quantitative methods for environmental-species correlation and modeling. This is followed by a description of pattern classification methods, such as nearest neighbors and generalized linear modeling, which is not meant to be comprehensive but constructive for the mapping from class-defining discriminant variables to area classes.

For modeling scales, the rich set of landscape spatial pattern indices (e.g., Farina 2006) will be reviewed, for which not only calculation, but also interrelationships are explained. It is important to view discriminant space-based scale descriptors as integrals of individual classes' scales. Thus, landscape pattern indices can be seen as a manifestation of the scales of the underlying controlling factors.

For upscaling area-class maps, we will describe alternative methods, including majority rule, averaging, and random selection. Effects of upscaling of area-class data should be analyzed as criteria for transfer of categorical information from fine resolution to coarse resolution. Methods for downscaling range from estimating areal extents of individual classes (at fine resolution) based on coarse-resolution data (either discrete classification or class proportions) to predicting fine-resolution classes at pixel level from coarse-resolution data and covariates. There seems to be merit in exploring discriminant-space models and geostatistical approaches for both upscaling and downscaling of area-class information.

In the concluding section, we will address three major issues. The first concerns semantic scale (or thematic detail), as it is important to have a classification scheme that matches well with the spatial scale of categorical information. The second issue is about H-resolution vs. L-resolution data, and implications for fields vs. objects in scale modeling and scaling. Lastly, we discuss the relationships between scale and uncertainty in categorical information.

10.1 AREA-CLASS MAPPING

10.1.1 Discriminant-Space Model

The preceding reassessment of conceptual issues in categorical mapping leads to the recognition of a paradigmatic shift from mere class labeling to understanding the mechanism by which classes are formed. Therefore, a conceptual framework is proposed for the mapping from class-shaping covariates to categorical information, which would allow for decomposition and reassembling of algorithms and reasoning underlying the transformation of raw data into meaningful information, as discussed by Goodchild et al. (2009).

Consider a set of properties $\{Z_1, Z_2, ..., Z_b\}$ (*b* being a positive integer) measured on an interval or ratio scale. For example, the properties might be those relevant to the occurrences of various plant species, such as mean annual temperature, mean annual precipitation, elevation, slope, and soil pH. Under Tobler's first law of geography (TFL), geographic properties will exhibit spatial dependence, with proximate locations being more similar than those further apart. Now, consider a *b*-dimensional space with dimensions defined by these properties, and assume that the space is partitioned into *K* zones, each corresponding to one class type, and defined by the range of conditions characteristic of that type. We might term this a *phase space* (Goodchild and Wang 1989) by analogy to the space defined by temperature and pressure that characterizes the gaseous, liquid, and solid phases of a chemical compound. Any point (or location) in this space will map to one of the *K* types, depending on the zone in which the point falls. Thus, any point in geographic space, characterized by a set of values on the *b* properties, maps to a point in the *b*-dimensional phase space, which, in turn, maps to one of the *K* types in geographic space (Goodchild et al. 2007).

For a location *x*, the class at that location is $C(x)$, and $Z_1(x)$ through to $Z_b(x)$ are the properties at *x*. The phase-space model was renamed the *discriminant-space* model, as it is in the phase space that individual locations are partitioned into classes by a process of discriminant analysis (Goodchild et al. 2009). The aforementioned discriminant covariates at *x* constitute a vector field $Z(x)$ of dimension *b*, that is, $Z(x) = [Z_1(x),...,Z_b(x)]^T$. The discriminant model provides a function η linking the **Z** values at a location *x* to the area class at *x*, so that any point in the **Z** space maps to a point in the geographic space (Goodchild et al. 2007). Thus, the mapping from measurement to class labels can be expressed as

$$\hat{C}(x) = \eta(\mathbf{Z}(x)) = \arg\max_{k=1,...,K} F_k(\mathbf{Z}(x)) \tag{10.1}$$

where F_k calculates measures of proximity to indicate categorical similarity to class *k*, with the predicted class $\hat{C}(x)$ having the maximum proximity (Goodchild et al. 2009). We will describe some methods for class probability estimation and classification in the next subsection.

In this model, $C(x)$ can be regarded as a nominal-scaled field or area-class map (Mark and Csillag 1989), or, alternatively, connected areas of the same value can be regarded as *bona fide* geo-objects (Goodchild et al. 2007). TFL ensures that these connected areas are of measurable size, which they would not be if the input geofields lacked positive spatial autocorrelation, as discussed by Goodchild et al. (2007). Moreover, because of TFL, only certain adjacencies can occur between types: only types that are adjacent in discriminant space can be adjacent in geographic space. When $b = 1$, the boundaries of the types will be isolines of the input variable, and hence the map of $C(x)$ will have a characteristic appearance with no nodes in the boundary network (Goodchild et al. 2009). The kind of model just described is widely used in biogeography (Holdridge 1947) and is also the model used in remote sensing to classify multispectral data, as discussed by Goodchild et al. (2007). This provides a general way of understanding how homogeneous patches and area-class maps arise in reality and of linking concepts of geo-objects to concepts of geo-fields (Goodchild et al. 2007), as also described in Chapter 2.

The fulfillment of the discriminant-space-based classification relies, however, on the identification and measurement of the discriminant variables Z, followed by Z-based class mapping. First, class-discerning Z space needs to be extracted from data space properly in the sense that class defining and mapping are oriented to the information classes desired rather than simply to data classes. For this, we describe approaches to Z space construction in remote sensing and predictive modeling below.

In pattern recognition, a *feature space* is an abstract space where each pattern sample is represented as a point in b-dimensional space, b being the number of features used to describe the patterns. Feature space is analogous to discriminant space. Similar samples are grouped together, allowing for pattern finding. There is a large literature on pattern recognition based on remote measurements for extraction of categorical information. Features that enhance classification accuracy need to be identified from raw datasets. For example, the normalized difference vegetation index and leaf area index are commonly used metrics, based on known spectral interactions in green vegetation canopies. They provide some limited data-volume compression and are more readily associated with scene (vegetation) properties than are the raw spectral data, as discussed by Crist and Cicone (1984).

Although techniques such as principal components analysis may be employed for feature extraction, the Tasseled Cap transformation of Landsat MSS data, developed by Kauth and Thomas (1976), better furnishes data-to-information transformation as it provides a direct link between features and physical scene characteristics, as explained below. The MSS Tasseled Cap transformation orients the data plane such that two features related directly to physical scene characteristics are extracted: (1) brightness, a weighted sum of all the bands, defined in the direction of principal variation in soil reflectance; and (2) greenness, a contrast between the near-infrared bands and the visible bands. The substantial scattering of infrared radiation resulting from the cellular structure of green vegetation and the absorption of visible radiation by plant pigments (e.g., chlorophyll) combine to produce large greenness values for surfaces with high densities of green vegetation, whereas the flatter reflectance curves of soils are expressed in small greenness values. Since the MSS Tasseled Cap transformation is an invariant transformation, features are consistent between scenes, and can be so interpreted once the data have been normalized for external effects such as haze level and viewing and illumination geometry (Kauth and Thomas 1976). The Tasseled Cap transformation was extended to Landsat TM data (Crist and Cicone 1984), with two dimensions equivalent to MSS Tasseled Cap transformed brightness and greenness features and a third one primarily related to soil characteristics. As with the MSS Tasseled Cap transformation, the TM Tasseled Cap transformation is intended to be an invariant transformation. The direct link between features and physical scene characteristics enhances both the interpretation of observed spectral variation and the prediction of the spectral effects of particular changes in scene characteristics (Crist and Cicone 1984).

Although remote sensing is mainly oriented for mapping actual distributions of area classes, such as vegetation, predictive modeling approaches are directed toward mapping potential or long-term distributions controlled by biophysical processes. Further to the developments in eco-regionalization (Bailey 1983), quantitative methods have also been explored. Direct gradient analysis relates species presence or abundance to environmental variables on the basis of species and environment data from the same set of sample plots (Gauch 1982), as discussed by Ter Braak (1987). The simplest methods of direct gradient analysis involve plotting each species' abundance values against values of an environmental variable or drawing isopleths for each species in a space of two environmental variables (Whittaker 1967; Ter Braak 1987). With these simple methods, one can easily visualize the relation between many species and one or two environmental variables.

As it is difficult to put results for several species together to obtain an overall graphical summary of species–environment relationships, a simple method is needed to analyze and visualize the relationships between many species and many environmental variables. Canonical correspondence analysis (CCA) is designed to help with this need (Ter Braak 1987). CCA is an eigenvector ordination technique that also produces a multivariate direct gradient analysis (Ter Braak 1986), which aims to visualize a pattern of community variation, as in standard ordination, and the main features of species distributions along the environmental variables, as described by Ter Braak (1987). A physiographic-space-based method based on CCA was described by Chokmani and Ouarda (2004).

Ouarda et al. (2001) provided a useful description of the mathematical background of CCA. The multivariate approach of CCA is most commonly used in the context where we have two sets of random variables $\mathbf{Z} = \{Z_1, Z_2, ..., Z_n\}$ and $\mathbf{Z}' = \{Z_1', Z_2', ..., Z_r'\}$, $n \geq r$. For instance, the set $\mathbf{Z}$ could contain biophysical variables and the set $\mathbf{Z}'$ could represent vegetation variables. We can first compute the correlation matrix for the variables of sets $\mathbf{Z}$ and $\mathbf{Z}'$. This leads to a numerical description of the interaction between any two variables. However, to obtain a clear view of the correlation structure, we have to identify the sources of interaction among these variables. CCA allows one to identify the dominant linear modes of covariability between the sets $\mathbf{Z}$ and $\mathbf{Z}'$ (Ouarda et al. 2001), as described below.

Consider any linear combination G and G' of the variables $\mathbf{Z} = \{Z_1, ..., Z_n\}$ and $\mathbf{Z}' = (Z_1', ..., Z_r')$, respectively,

$$G = \mathbf{a}^\mathrm{T}\mathbf{Z} \tag{10.2a}$$

$$G' = \mathbf{b}^\mathrm{T}\mathbf{Z}' \tag{10.2b}$$

where vectors are denoted in bold, and $\mathbf{a}^\mathrm{T}$ and $\mathbf{b}^\mathrm{T}$ denote the transposes of vectors $\mathbf{a}$ and $\mathbf{b}$, respectively. Let $\mathbf{C}$ be the covariance matrix of the set of variables $(Z_1, ..., Z_n, Z_1', ..., Z_r')$

$$\mathbf{C} = \begin{pmatrix} \mathbf{C}_{ZZ} & \mathbf{C}_{ZZ'} \\ \mathbf{C}_{ZZ'}^\mathrm{T} & \mathbf{C}_{Z'Z'} \end{pmatrix} \tag{10.3}$$

The correlation between the random variables G and G' is given by

$$\mathrm{corr}(G,G') = \frac{\mathrm{cov}(G,G')}{\sqrt{\mathrm{var}(G)}\sqrt{\mathrm{var}(G')}} = \frac{\mathbf{a}^\mathrm{T}\mathbf{C}_{ZZ'}\mathbf{b}}{\sqrt{\mathbf{a}^\mathrm{T}\mathbf{C}_{ZZ}\mathbf{a}}\sqrt{\mathbf{b}^\mathrm{T}\mathbf{C}_{ZZ'}\mathbf{b}}} \tag{10.4}$$

CCA allows one to identify vectors $\mathbf{a}$ and $\mathbf{b}$ for which corr(G, G') is maximal (Ouarda et al. 2001). Furthermore, when identifying the maximal value for corr(G, G'), we need only to consider vectors $\mathbf{a}$ and $\mathbf{b}$ for which G and G' have unit variances, as shown below.

We use the Lagrange multipliers technique to maximize corr(G, G') subject to the constraints that Var(G) = Var(G') = 1, as described by Ouarda et al. (2001). Note that, with these constraints, we have corr(G, G') = $\mathbf{a}^{\mathrm{T}}\mathbf{C}_{ZZ'}\mathbf{b}$. The Lagrangian function L is defined as

$$L = \mathbf{a}^{\mathrm{T}}\mathbf{C}_{ZZ'}\mathbf{b} - \mu_1(\mathbf{a}^{\mathrm{T}}\mathbf{C}_{ZZ}\mathbf{a} - 1) - \mu_2(\mathbf{b}^{\mathrm{T}}\mathbf{C}_{ZZ'}\mathbf{b} - 1) \tag{10.5}$$

where μ_1 and μ_2 are the Lagrangian parameters. The partial derivatives of the Lagrangian function with respect to $\mathbf{a}$ and $\mathbf{b}$ are set to zero to obtain the maximum of L. We have

$$\mathbf{C}_{ZZ'}\mathbf{b} = 2\mu_1\mathbf{C}_{ZZ}\mathbf{a} \tag{10.6a}$$

$$\mathbf{C}_{ZZ'}^{\mathrm{T}}\mathbf{a} = 2\mu_2\mathbf{C}_{Z'Z'}\mathbf{b} \tag{10.6b}$$

By premultiplying Equation 10.6a by $\mathbf{a}'$ and Equation 10.6b by $\mathbf{b}'$, and given that $\mathbf{a}^{\mathrm{T}}\mathbf{C}_{ZZ}\mathbf{a} = 1$ and $\mathbf{b}^{\mathrm{T}}\mathbf{C}_{Z'Z'}\mathbf{b} = 1$ (i.e., Var(G) = Var(G') = 1), we have

$$2\mu_1 = \mathbf{a}^{\mathrm{T}}\mathbf{C}_{ZZ'}\mathbf{b} \tag{10.7a}$$

$$2\mu_2 = \mathbf{b}^{\mathrm{T}}\mathbf{C}_{ZZ'}^{\mathrm{T}}\mathbf{a} \tag{10.7b}$$

Since $\mathbf{a}^{\mathrm{T}}\mathbf{C}_{ZZ'}\mathbf{b}$ in Equation 10.7a is a scalar and given that its transpose is equal to the right-hand side of Equation 10.7b, we can set $\lambda = 2\mu_1 = 2\mu_2 = \mathbf{a}^{\mathrm{T}}\mathbf{C}_{ZZ'}\mathbf{b}$ (Ouarda et al. 2001).

Isolating $\mathbf{a}$ in Equation 10.6a and substituting it into Equation 10.6b, we have

$$\frac{1}{\lambda}\mathbf{C}_{ZZ'}^{\mathrm{T}}\mathbf{C}_{ZZ}^{-1}\mathbf{C}_{ZZ'}\mathbf{b} = \lambda\mathbf{C}_{Z'Z'}\mathbf{b} \tag{10.8a}$$

which can also be written as

$$\mathbf{C}_{Z'Z'}^{-1}\mathbf{C}_{ZZ'}^{\mathrm{T}}\mathbf{C}_{ZZ}^{-1}\mathbf{C}_{ZZ'}\mathbf{b} = \lambda^2\mathbf{b} \tag{10.8b}$$

In a similar way, starting with vector $\mathbf{b}$ we obtain

$$\mathbf{C}_{ZZ}^{-1}\mathbf{C}_{ZZ'}\mathbf{C}_{Z'Z'}^{-1}\mathbf{C}_{ZZ'}^{\mathrm{T}}\mathbf{a} = \lambda^2\mathbf{a} \tag{10.9}$$

Thus, the vectors $\mathbf{a}$ and $\mathbf{b}$ are eigenvectors of $\mathbf{C}_{ZZ}^{-1}\mathbf{C}_{ZZ'}\mathbf{C}_{Z'Z'}^{-1}\mathbf{C}_{ZZ'}^{\mathrm{T}}$ and $\mathbf{C}_{Z'Z'}^{-1}\mathbf{C}_{ZZ'}^{\mathrm{T}}\mathbf{C}_{ZZ}^{-1}\mathbf{C}_{ZZ'}$, respectively. We can deduce that, for the optimal vectors $\mathbf{a}$ and $\mathbf{b}$, we have $corr(G, G') = \lambda$. If rk is the rank of $C_{ZZ'}$, then solving Equations 10.8b and 10.9 leads to rk solution triplets $(\lambda_i, \mathbf{a}_i, \mathbf{b}_i)$, each giving rise to a triplet (λ_i, G_i, G_i), according to Equations 10.2a and 10.2b. CCA provides us with new variables, known as canonical variables, $G_1, G_2, ..., G_{rk}$ and $G_1', G_2', ..., G_{rk}'$ (to replace $Z_1, Z_2, ..., Z_n$ and $Z_1', Z_2', ..., Z_r'$, respectively), along with the correlation coefficients $\lambda_1, \lambda_2, ..., \lambda_{rk}$. Further detail is provided by Ouarda et al. (2001).

The ordination diagram of CCA displays sites, species, and environmental variables. They display variation in species composition over the sites. The arrow for an environmental variable points in the direction of maximum change of that environmental variable across the diagram, and its length is proportional to the rate of change in this direction (Ter Braak 1986). In other words, each arrow representing an environmental variable determines a direction or axis in the diagram; the species points can be projected onto this axis. The order of the projection points corresponds approximately to the ranking of the weighted averages of the species with respect to that environmental variable (for illustration, see Ter Braak 1986). Clearly, CCA is an important tool for establishing species–environment relationships (i.e., class-**Z** space relationships) that are central to predictive mapping of area-class distributions and dynamics.

10.1.2 METHODS FOR MAPPING AREA CLASSES

We will describe a few techniques for mapping area classes from discriminant variables. Some are nonparametric techniques, such as the k-nearest neighbor (k-NN) and kernel density estimation methods, whereas others are parametric, such as generalized linear models.

One of the simplest methods for area-class mapping is thresholding, by which a continuous variable Z is discretized into a number of classes with thresholds, like in indicator transforms. Parallelepiped classifiers are an extension of simple thresholding of a one-dimensional quantity into two dimensions, as discussed by Goodchild et al. (2009).

The k-NN algorithm (k is a positive integer) is commonly used and based on a simple principle: an object (of unknown class membership) is classified by a majority vote of its neighbors of known class labels (because they come from the training dataset) in feature space, with the object being assigned to the class most common among its k nearest neighbors (Cover and Hart 1967; Duda et al. 2001; Mills 2011). The choice of a suitable k depends upon the data and may be based on cross-validation. Euclidean distance is usually used as the distance metric for continuous variables, whereas Hamming distance can be used for discrete data. The nearest-neighbor algorithm is the special case for k-NN when $k = 1$ (so that we predict the class of an unknown object to be the class of the closest training sample in feature space). The k-NN algorithm suffers from the drawback that the classes with more frequency in the training set tend to dominate the predicted classes at unsampled locations. One way to overcome this problem is to weight the class membership values (of the k nearest neighbors) by the distance in feature space from the point being predicted to its k nearest neighbors.

k-NN is a special kernel density estimator with a uniform kernel. Kernel density estimation is a nonparametric way of estimating the probability density function of a random variable (Parzen 1962). Let $(z_1, z_2, ..., z_n)$ be a sample drawn from some univariate distribution with an unknown density kd. We are interested in estimating the shape of this function kd. Its kernel density estimator is

$$\hat{kd}_{bwd}(z) = \frac{1}{n} \sum_{i=1}^{n} KNL_{bdw}(z - z_i) = \frac{1}{n \times bdw} \sum_{i=1}^{n} KNL\left(\frac{z - z_i}{bdw}\right) \qquad (10.10)$$

where $KNL()$ is the kernel, a symmetric function that integrates to one, which can be uniform, triangular, normal, and others, and $bdw > 0$ is a smoothing parameter called the bandwidth. The bandwidth bdw of the kernel affects the resulting density estimate: using a bdw too small is likely to generate spurious data artifacts, whereas using a bdw too large tends to oversmooth the data. The optimal choice of bdw remains a major issue (e.g., Zhang and You 2011). Adaptive or variable-bandwidth kernel density estimation seems to be promising, but much work remains to be done (Terrell and Scott 1992; Duda et al. 2001).

Let $\mathbf{Z}(x)$ be a random vector of observations. For a K-class ($K \geq 2$) problem where linear class boundaries in the discriminant space are involved, linear discriminant functions that take the following form may be helpful in demarcating the classes:

$$F_k(\mathbf{Z}(x)) = \mathbf{W}^T\mathbf{Z}(x) + w_0, k = 1, ..., K - 1 \qquad (10.11)$$

where $\mathbf{W}$ is a vector of linear weights for $\mathbf{Z}(x)$ and w_0 is a constant. The weights $\mathbf{W}$ in Equation 10.11 are found by identifying the transform vector that maximizes the ratio of between-class scatter to within-class scatter (Duda et al. 2001), as also reviewed by Goodchild et al. (2009). Location x with measurement $\mathbf{Z}$ is assigned to class i if $F_i(\mathbf{Z}(x)) > F_j(\mathbf{Z}(x))$ for all $j \neq i$.

Maximum-likelihood classification is widely applied for pattern classification. It results from decision theory with a 0–1 loss function and use of Bayes's theorem (Duda et al. 2001). Suppose $\{v_i, i = 1, ..., K\}$ is the set of possible classes. The maximum-likelihood classification is written as

$$\hat{C}(x) = v_k \quad \text{if and only if} \quad p(v_k | \mathbf{Z}(x)) = \max_i p(v_i | \mathbf{Z}(x)) \qquad (10.12)$$

where $\hat{C}(x)$ is the predicted class label for location x and $p(v_i | \mathbf{Z}(x))$ is the probability of class v_i occurring conditional on observation $\mathbf{Z}(x)$. The evaluation of conditional probability $p(v_i | \mathbf{Z}(x))$ is based on Bayes's theorem, which is a general relationship between prior probability $p(v_i)$, probability density $p(\mathbf{Z}(x))$, posterior probability $p(v_i | \mathbf{Z}(x))$, and conditional density $p(\mathbf{Z}(x) | v_i)$ (in the same event space), as described by Goodchild et al. (2009) and Zhang et al. (2009).

The generalized linear model (GLM) generalizes linear regression to the cases of both continuous and categorical response variables via link functions and allows the variance of each measurement to be a function of its predicted value (Nelder and Wedderburn 1972). In a GLM, each outcome of the dependent variables, Z', is assumed to be generated from a distribution in the exponential family, such as the normal, binomial, and Poisson distributions. The mean, μ, of the distribution depends on the independent variables, $\mathbf{Z}$, through

$$E(Z') = \mu = g^{-1}(\mathbf{Z\beta}) \tag{10.13}$$

where $E(Z')$ is the expected value of Z'; $\mathbf{Z\beta}$ is the linear predictor, a linear combination of unknown parameters, $\boldsymbol{\beta}$; and g is the link function providing the relationship between $\mathbf{Z\beta}$ and μ. As the response data in categorical mapping, Z', are binary (0 or 1 in value), the distribution function is generally chosen to be the binomial distribution, and the interpretation of μ is then the probability, p, of Z' taking on the value of 1. For binomial distributions, the canonical logit link is commonly used:

$$\mathbf{Z\beta} = \ln\left(\frac{\mu}{1-\mu}\right) \quad \text{or} \quad g(p) = \ln\left(\frac{p}{1-p}\right) \tag{10.14}$$

where a linear model is shown explicitly relating the linear predictor, $\mathbf{Z\beta}$, with the logit of mean of Z', $\ln\left(\frac{\mu}{1-\mu}\right)$. GLMs of the form in Equation 10.14 are also known as logistic regression models. The mean function in Equation 10.13 can be solved by inversion of Equation 10.14 for μ (or p):

$$\mu = \frac{\exp(\mathbf{Z\beta})}{1+\exp(\mathbf{Z\beta})} = \frac{1}{1+\exp(-\mathbf{Z\beta})} \tag{10.15}$$

Multinomial logistic regression generalizes logistic regression to cases concerning more than two discrete outcomes. Suppose there are K classes $\{1, \ldots, K\}$ (which are dependent variables in regression analysis). One regression is run for each class $k \in \{1, 2, \ldots, K\}$ to predict the probability of Z' being in class k:

$$p(Z' = k) = \frac{\exp(\mathbf{Z\beta}_k)}{1+\displaystyle\sum_{j=1}^{K}\exp(\mathbf{Z\beta}_j)} \tag{10.16}$$

where Z' is the observed outcome for the dependent variable, $\mathbf{Z}$ is a vector of explanatory variables, and $\boldsymbol{\beta}_k$ is the vector of regression coefficients in the kth run (corresponding to the kth class) of regression (Equation 10.16). Since $p(Z' = k)$ sums to 1 over all K candidate classes, the K sets of regression coefficients $\boldsymbol{\beta}_k$ are not unique. We may set $\boldsymbol{\beta}_K$ (of class K) to null (i.e., $\boldsymbol{\beta}_K = \mathbf{0}$) so that $p(Z' = K) = 1 / \left[2 + \displaystyle\sum_{j=1}^{K-1}\exp(\mathbf{Z\beta}_j)\right]$.

Thus, the regression coefficients represent the effects of the independent variables on the probability of choosing the kth class (i.e., alternative in a choice set) over the last alternative (i.e., class K). (See So and Kuhfeld [1995] for technical details about multinomial logistic regression.) The unknown parameters in each vector $\boldsymbol{\beta}_j$ are typically estimated by maximum *a posteriori* estimation. The solution is typically found using an iterative procedure such as iteratively reweighted least squares (Wedderburn 1974).

GLMs can be usefully combined with geostatistics and are valuable for spatial analysis and prediction, in particular those concerning non-normal data and categorical data and when spatial correlation is present, as discussed by Gotway and Stroup (1997). Brown et al. (2002) applied geostatistics and generalized additive models (GAMs) (a blending of GLMs and additive models that provide more flexibility than the linear predictor $\mathbf{Z}\boldsymbol{\beta}$ in GLMs) for simulation of land cover change. Guisan et al. (2002) reviewed the use of GLMs and GAMs for ecological modeling, discussed some of their related statistics used for predictor selection, model diagnostics, and evaluation, and presented new approaches applicable to GLMs and GAMs, such as the methods for the identification of interactions by a combined use of regression trees. In terms of area-class mapping applications, logistic regression models based on topography and other environmental factors have been used to predict vegetation types and species distributions (Rupprecht et al. 2011). As Cairns (2001) discussed in the context of predictive vegetation modeling, it is important to use a suite of predictive models to help understand the environment–vegetation relationships.

Consider an elaboration of Equation 10.12 in the context of change categorization. A compound classification according to Bayes's theorem is performed where each pair of pixels (coregistered) is analyzed with the aim of finding the most likely pair of class labels. This Bayesian method is useful as it allows for tabulation of to–from transitions and accommodates temporal dependence, as shown below.

Let $\mathbf{Z}^1_{(x)}$ and $\mathbf{Z}^2_{(x)}$ be random vectors representing the observations at times t_1 and t_2, respectively. Suppose $\{v_i, i = 1, \ldots, K_1\}$ and $\{\omega_j, j = 1, \ldots, K_2\}$ are the sets of possible classes at times t_1 and t_2, respectively. The compound classification is written as

$$\hat{C}_1(x) = v_k \quad \text{and} \quad \hat{C}_2(x) = \omega_l \quad \text{if and only if}$$

$$p(v_k, \omega_l | \mathbf{Z}^1(x), \mathbf{Z}^2(x)) = \max_{i,j} p(v_i, \omega_j | \mathbf{Z}^1(x), \mathbf{Z}^2(x)) \tag{10.17}$$

where $\hat{C}_1(x)$ and $\hat{C}_2(x)$ are the predicted class labels for location x at times t_1 and t_2, respectively. A change categorization rule may be expressed in terms of a class-conditional density function and prior probabilities in Bayesian classification. To relax the requirement for large numbers of training samples in estimation of joint class-conditional density $p(\mathbf{Z}^1(x)), \mathbf{Z}^2(x) | v_i, \omega_j)$, class-conditional independence is usually assumed as a reasonable approximation (Swain 1978). It is then possible to replace the quantity in the right-hand side of Equation 10.17 with

$$p(v_i, \omega_j | \mathbf{Z}^1(x), \mathbf{Z}^2(x)) \approx p(\mathbf{Z}^1(x) | v_i) p(\mathbf{Z}^2(x) | \omega_j) p(v_i, \omega_j)$$

$$= \frac{p(v_i | \mathbf{Z}^1(x)) p(\omega_j | \mathbf{Z}^2(x)) p(v_i, \omega_j)}{p(v_i) p(\omega_j)} \tag{10.18}$$

where $p(v_i|\mathbf{z}^1(x))$ and $p(\omega_j|\mathbf{z}^2(x))$ are single-data *a posteriori* probabilities, which may be estimated using kernel methods in $\mathbf{Z}$ space, and $p(v_i)$, $p(\omega_j)$, and $p(v_i, \omega_j)$ are *a priori* probabilities and joint probability $p(v_i, \omega_j)$, respectively, which may be estimated in a data-driven manner (Bruzzone et al. 1999; Zhang et al. 2009).

10.2 SPATIAL SCALES AND PATTERNS IN AREA CLASSES

Geostatistical tools, such as variograms, and their extensions, such as various auto-correlation indices, have been described in Chapters 4 and 5. These models and other scale descriptors have been further discussed at length in the context of terrain analysis in Chapter 9. As shown previously in this chapter, area-class maps represent variability through patches that are internally homogeneous (in relative terms) and exhibit abrupt interpatch transition. They are inherently better equipped for computing many spatial pattern indices (Gustafson 1998). Also, pattern quantification is essential to relate patterns (e.g., area classes) to underlying processes (e.g., discriminant variables), to identify spatial structure, and to detect changes in spatial heterogeneity through time. Moreover, pattern–process relationships should be studied at multiple scales (Levin 1992; Scheiner 1992), and pattern analysis is crucial for understanding the scale dependence of area classes. The emphasis here is thus placed on various scale descriptors based on spatial pattern indices, which are applicable for area-class maps, as alternatives to geostatistical models. This section provides a review of how spatial patterns may be quantified through various pattern indices and variance-based descriptors.

Spatial pattern indices fall into two general categories: those that evaluate the composition of the map without reference to spatial characteristics, and those that evaluate the spatial configuration of properties with spatial information incorporated. Spatial configuration indices are either patch oriented or neighborhood oriented. Patch-oriented indices are calculated considering only a single patch and its edges, whereas neighborhood-oriented indices are calculated using spatial neighborhoods and may consider all patches within the neighborhood or only neighboring pixels (Gustafson 1998).

Composition is readily quantified and is typically described by (1) the number of classes in the map, (2) the proportion of each class relative to the entire map, and (3) diversity. The number of classes and their proportions are generated by simple counting algorithms. Diversity can be measured by various methods, as shown by Gustafson (1998).

Entropy (Shannon 1948) can be used as a measure of diversity and is calculated as

$$H = -\sum_{i=1}^{K} p_i \ln p_i \qquad (10.19)$$

where H is the entropy, p_i is the proportion of cover type i, and K is the number of classes (e.g., land cover types or patch types) observed. The larger the value of H, the more diverse the landscape. Entropy-related concepts, such as H and that in Equation 10.22, will be described in more detail in Chapter 11.

Diversity measures typically combine two components of diversity: richness, which refers to the number of classes present, and evenness, which refers to the distribution of area among the classes. Romme's relative evenness index (Romme 1982), EV, is

$$EV = 100 \times \left[-\ln\left(\sum_{i=1}^{K} p_i^2 \right) \right] \Bigg/ \ln K \tag{10.20}$$

where K is the total number of patch types and p_i is the probability of occurrence for type i. In Equation 10.20, the quantity $-\ln\left(\sum_{i=1}^{K} p_i^2 \right)$ is actually the modified Simpson's index, as explained by Romme (1982). The evenness responds to the number of patch types and their proportions in a landscape, as reviewed by Gustafson (1998).

Dominance is the complement of evenness (dominance = 1 – evenness), indicating the extent to which the map is dominated by one or a few classes (O'Neill et al. 1988). It has been used widely in landscape ecology research. The measure of dominance, DMN, is calculated as the deviation from the maximum possible diversity:

$$DMN = H_{\max} - \left(-\sum_{i=1}^{K} p_i \ln p_i \right) = \ln K + \sum_{i=1}^{K} p_i \ln p_i \tag{10.21}$$

where, as in Equation 10.19, p_i is the proportion of the landscape in cover type i and K is the number of cover types observed (O'Neill et al. 1988). $H_{\max} = \ln K$ is the maximum diversity when all classes are present in equal proportions. Inclusion of $H_{\max}$ in Equation 10.21 normalizes the index for differences in numbers of landcover types between different landscapes. Large values of DMN indicate a landscape dominated by one or a few cover types, and small values indicate a landscape with many classes represented in approximately equal proportions (O'Neill et al. 1988). However, the index is not useful in a completely homogeneous landscape (i.e., $K = 1$) because DMN then equals zero.

For spatial configuration, there are patch-based indices, such as size, shape, density, connectivity, and fractal dimension; and pixel-based indices, such as contagion and lacunarity, as reviewed by Gustafson (1998). Patch-based measures of pattern can be calculated for all classes together or for a particular class of interest. Most shape indices use a perimeter–area relationship (Forman and Godron 1986), whereas contiguity (LaGro 1991) and elongation and deformity indices (Baskent and Jordan 1995) have also been proposed. The interspersion and juxtaposition index developed by McGarigal and Marks (1995) measures the extent to which patch types (patches of different classes) are interspersed. This patch-based index is conceptually similar to the pixel-based contagion index (see below), evaluating patch adjacencies (Gustafson 1998).

The most commonly used pixel-based neighborhood index is contagion (O'Neill et al. 1988; Li and Reynolds 1994; Riitters et al. 1996). Contagion ignores patches *per se* and measures the extent to which cells of similar class are aggregated (Gustafson 1998). The index is calculated using the frequencies with which different pairs of classes occur as adjacent pixels on the map, as described below.

The measure of contagion, or the adjacency of land-cover types, *CTG*, is calculated from an adjacency matrix, **P**, in which p_{ij} is the proportion of cells of type i that are adjacent (diagonals are excluded) to cells of type j, such that

$$CTG = H'_{max} + \sum_{i=1}^{K} \sum_{j=1}^{K} p_{ij} \ln p_{ij} \qquad (10.22)$$

where $H'_{max} = 2K \ln K$ and is the absolute value of the summation of $p_{ij} \ln p_{ij}$ when all possible adjacencies between cover types occur with equal probabilities (O'Neill et al. 1988; Gustafson 1998). H'_{max} normalizes landscapes with differing values of K and causes *CTG* to be zero when $K = 1$ or all possible adjacencies occur with equal probability. When $K \geq 2$, large values of *CTG* will indicate a landscape with a clumped pattern of cover types. As the contagion index measures the extent to which classes are aggregated or clumped, large values indicate large deviations from the equiprobable maximum (thus, large and contiguous patches are found on the landscape), whereas small values mean that the landscape is dissected into many small patches (O'Neill et al. 1988).

The relative contagion index *RCTG* is calculated as

$$RCTG = 1 + \sum_{i=1}^{K} \sum_{j=1}^{K} p_{ij} \ln p_{ij} \Big/ (2K \ln K) \qquad (10.23)$$

where K is the total number of patch types in a landscape mosaic and p_{ij} is the probability that two randomly chosen adjacent pixels belong to patch types i and j. *RCTG* in Equation 10.23 is clearly a normalized (by $H'_{max} = 2K \ln K$) version of *CTG* in Equation 10.22. As mentioned earlier, the contagion measures the extent to which patches are aggregated (i.e., their spatial arrangement) (O'Neill et al. 1988).

Based on Romme's formula for relative patchiness (Romme 1982), we may calculate relative patchiness index *RPI* as

$$RPI = 100 \sum_{i=1}^{K} \sum_{j=1}^{K} E_{ij} D_{ij} \Big/ N_b \qquad (10.24)$$

where K is the total number of patch types in a landscape mosaic, E_{ij} is the number of edges between patch types i and j, D_{ij} is the dissimilarity value for patch types i and j, and N_b is the total number of edges of pixels (i.e., each pixel has four edges), as shown by Li and Reynolds (1994). The patchiness index measures the contrast of neighboring patch types in a landscape mosaic and may reflect indirectly the spatial arrangement (Farina 2006).

The fractal dimension is an index for quantifying the dependence of geographic measures on the sampling density and can measure the complexity of shapes in the landscape (Goodchild 1980a). If the landscape is composed of simple geometric shapes like squares and rectangles, the fractal dimension will be small, approaching 1.0. If the landscape contains many patches with complex and convoluted shapes, the fractal dimension will be large (O'Neill et al. 1988). Fractal dimension, $D_{fractal}$, of a patch's perimeter P is related to its area A through a formula $P \sim \sqrt{A}^{D_{fractal}}$ (Lovejoy 1982) and is calculated from regressing a linear model:

$$P_k = l\sqrt{A_k^{D_{fractal}}} \tag{10.25}$$

where $D_{fractal}$ is the fractal dimension of patch shape in a landscape, l is a constant, and A_k and P_k are the area and the perimeter of patch k, respectively; an $A\sim P$ relation formula equivalent to Equation 10.25 was prescribed by Li and Reynolds (1994). $D_{fractal}$ is estimated by a regression of log P_k on log A_k for all individual patches and is twice the slope of the regression line. The fractal dimension derived can be used to measure the irregularity of overall patch shape in a landscape (Krummel et al. 1987; O'Neill et al. 1988), in particular the wiggliness or degree of contortion (Lovejoy 1982) of patch perimeters. Using this area–perimeter relation, Lovejoy (1982) investigated the fractal geometry of satellite- and radar-determined cloud and rain areas between 1 and 1.2×10^6 km^2, and found no characteristic horizontal length scale between 1 and 1000 km. Also, changes in the fractal dimension through time can be used to measure the degree to which human activities (e.g., agriculture or urbanization) are disturbing and simplifying the landscape patterns, regardless of the specific land uses, as discussed by O'Neill et al. (1988).

Spatial heterogeneity is generally defined as the complexity and variability of a property (e.g., cover type) in space. For categorical maps, spatial heterogeneity is reflected in the number of patch types, the proportion of each type, the spatial arrangement of patches, patch shape, and contrast between neighboring patches, as discussed by Gustafson (1998). Various indices, such as contagion, evenness, patchiness, and fractal dimension, as shown previously, were developed to quantify spatial heterogeneity in categorical maps (Li and Reynolds 1994).

Lacunarity analysis is a multiscale method used to determine the heterogeneity (or texture) of a property represented as a binary response in one, two, or three dimensions (Plotnick et al. 1993, 1996). It provides insights into scale-dependent variation in the spatial structure of a landscape, or patterns in the alternation of patches and gaps, thereby facilitating an understanding of the underlying processes that operate at different scales (Plotnick et al. 1996). For calculating lacunarity, a conceptually and computationally simple gliding box approach was proposed (Allain and Cloitre 1991; Plotnick et al. 1993). This algorithm involves exhaustive scanning of a whole landscape. First, place the gliding box of a given size (*pix*, the side length of a square box) at one corner of the problem domain and determine the box mass $n(pix)$ (i.e., the number of cells containing a species of interest). The box is then systematically moved through the domain one cell at a time and at each step the box mass is determined. Then, the lacunarity ($LCN(pix)$) for box size pix is defined to be the second moment of $n(pix)$ normalized by its first moment squared and can be calculated by adding one to the ratio of the variance and the squared mean of the box mass:

$$LCN(pix) = \frac{E\{[n(pix)]^2\}}{[E[n(pix)]]^2} = 1 + \text{var}[n(pix)]\Big/\big(mean[n(pix)]\big)^2 \qquad (10.26)$$

as explained by Plotnick et al. (1993).

As the lacunarity is a measure of spatial heterogeneity, a more heterogeneously distributed species will lead to a higher lacunarity value, which implies that the species tends to be more clustered with frequent, extensive gaps around clusters. Lacunarity is thus very indicative of landscape fragmentation over time (e.g., Wu et al. 2000). The lacunarity index also provides information about the domains of the scale of spatial variation. When a sampling scale, or the box size, is smaller than the scale of species pattern, the variance of the box mass among different locations in the plot is large, increasing the lacunarity index. If the box size captures the scale of the pattern, the variance approaches zero, making the index approach one, as reviewed by Kim et al. (2009).

We describe two variance methods and compare them to wavelet-based descriptors for spatial scales and patterns, following Dale and Mah (1998). Hill's (1973) blocked quadrat variance method comes in two versions, one with two terms ($V_2(blk)$) and one with three ($V_3(blk)$), where blk is the block size. Suppose there is a belt transect of n quadrats, with z_i being the density of a particular species in the ith quadrat. Hill's blocked quadrat variance for scale blk is calculated as

$$V_2(blk) = \frac{1}{2blk(n+1-2blk)} \sum_{i=1}^{n+1-2blk} \left(\sum_{j=i}^{i+blk-1} z_j - \sum_{j=i+blk}^{i+2blk-1} z_j \right) \qquad (10.27a)$$

for two terms and

$$V_3(blk) = \frac{1}{8blk(n+1-3blk)} \sum_{i=1}^{n+1-3blk} \left(\sum_{j=i}^{i+blk-1} z_j - 2\sum_{j=i+blk}^{i+2blk-1} z_j + \sum_{j=i+2blk}^{i+3blk-1} z_j \right) \qquad (10.27b)$$

for three terms; the three-term version is generally recommended as superior to the two-term version. Peaks in the plot of variance as a function of block size indicate the scales comparable to corresponding block sizes blk used, as reviewed by Dale and Mah (1998) and Dale et al. (2002).

Hill's (1973) paired quadrat variance is calculated as

$$V_{paired}(blk) = \frac{1}{2(n-blk)} \sum_{i=1}^{n-blk} \left(z_i - z_{i+blk}\right)^2 \qquad (10.28)$$

Peaks in this variance plot are interpreted in the same way. In comparisons of these two methods, three-term quadratic variance is frequently recommended over paired quadrat variance because, if there is a pattern at scale *blk*, the latter produces nondiminishing resonance peaks at 3*blk*, 5*blk*, etc., of variance equal to the peak at *blk*, which can be confusing (Dale and Mah 1998).

Wavelets can be usefully explored for spatial pattern analysis (Dale and Mah 1998), as discussed in Chapter 9. Wavelet variance responds to the scale of the pattern, which is the average of the sizes of the patches and of the gaps, and can detect dominance patches when quantitative data on plant species occurrence are available or when using sample scores from an ordination analysis such as correspondence analysis. The position variance can be used to identify density peaks or density transitions in the data (Dai and Maarel 1997; Dale and Mah 1998).

The wavelet method seems to offer several advantages over Hill's methods of quadrat variance calculations. First, we can plot wavelet variance as a function of scale, in which the peak variance height is determined by pattern intensity and does not increase with scale (so, depending on the wavelet chosen, the position of the variance peak indicates the scale), as discussed in Chapter 9. Second, the method produces only faint resonance peaks, if any, and, by using several different wavelet forms, different characteristics of the pattern can be investigated. Lastly, the wavelet method is able to portray trends in the data clearly, when the pattern is nonstationary (Dale and Mah 1998; Dale et al. 2002).

There is yet another variance-based scale descriptor, scale variance, that may be usefully explored for area-class maps. As mentioned in Chapter 1, scale variance analysis, a hierarchical analysis first developed by Moellering and Tobler (1972), aims to determine level-wise variability in a known nested hierarchy and to evaluate each level's relative, independent contribution to the total variability of the whole system (Wu et al. 2000). To conduct scale variance analysis in area-class maps, one can systematically aggregate spatial data by increasing grain size (termed a scale level) progressively so that a nested data hierarchy is formed (Barnsley et al. 1997). On the basis of nested data hierarchies, scale variance analysis proceeds by computing the total sum of squares, partitioned sums of squares, and scale variance at each scale level. Finally, scale variance or the percent total sum of squares is plotted against scale levels, resulting in the scale variance plot. A peak in the plot implies that much variability occurs at the corresponding scale level, which is indicative of the average size of dominant patches in the landscape. The height of the peak reflects the relative contribution of that particular scale level to the total variability of the landscape (Wu et al. 2000).

The various pattern indices and variance-based scale descriptors described above are well suited for quantifying spatial scales of variation in area-class maps. As discussed in the first section, discriminant space provides a coherent framework for area-class mapping whereby the class $C(x)$ prevailing at individual location x can be predicted from class-controlling factors $\mathbf{Z}(x)$. Thus, $\mathbf{Z}$-based scale modeling can be associated with C-based pattern description. As many of the approaches to scale modeling for interval/ratio fields can be applied to $\mathbf{Z}$, the richness of scale modeling approaches in both $\mathbf{Z}$-fields and C-fields can be utilized so that process–pattern relationships (i.e., $\mathbf{Z}$ vs. C relationships) may be explored through discriminant models, thus shedding light on scale modeling in area-class information.

10.3   SCALING AREA-CLASS INFORMATION

10.3.1   Upscaling

Aggregation of a fine-resolution categorical map to a coarse resolution is a common approach to upscaling (Moody and Woodcock 1994; Turner et al. 1989). Several aggregation methods have been used, such as random selection, majority rule, and reclassification of coarsened images.

The random rule assigns class labels for an output cell by picking randomly a label from those of the input fine-resolution grid cells. Because the probability that a class is selected to represent an aggregated area is proportional to its presence in the input fine-resolution data, this aggregation method is less likely to create data distortion than the majority rule, except under extreme conditions, such as when the input map is aggregated to only a few pixels (He et al. 2002).

The majority rule is used widely (Turner et al. 1989; Moody and Woodcock 1994, 1995). The majority rule finds the majority class label (i.e., the most frequently occurring or mode class) from the input fine-resolution grid cells and records that label in the corresponding aggregated coarse-resolution cell on the output grid, as implemented by He et al. (2002). Specifically, this is done by overlaying a coarse-resolution output grid on the fine-resolution input grid. Each cell in the output grid is assigned to the majority class of the fine-resolution grid cells falling within it. When two or more major classes occur with the same frequency (i.e., there is a tie), the output grid cell is assigned randomly to one of the major classes. Because of filtering by majority, dominant classes are expected to become more dominant and minor classes decrease in proportion in the aggregated map, as shown by He et al. (2002).

The measurements of landscape indices used by He et al. (2002) indicated that random-rule-based aggregation maintains spatial patterns better than majority-rule-based aggregation as the latter tends to alter the spatial patterns inherent in the original area-class maps. Random-rule-based aggregation is thus more suitable for studies requiring accuracy in spatially explicit representations of spatial patterns.

The degradation–reclassification method is an upscaling method that works by image degradation and then classification. In this approach, the same training data and classification procedures are often used for all resolution levels (Yang and Merchant 1997). Such upscaled maps are not directly comparable to those obtained from different sensors or satellites because they do not account for differing sensor characteristics, viewing geometries, and atmospheric conditions. Rather, they are simulations similar to those of the original fine-resolution data except for the pixel size, as discussed by Yang and Merchant (1997). However, such images allow investigation of scaling issues without the influence of other factors among different sensors.

The degradation–reclassification method assumes implicitly that the response function of the sensor is a square-wave function: all radiances are averaged equally within a coarse pixel (but anything outside excluded). However, a more accurate representation of an image obtained with a coarser-resolution sensor would involve modeling the transfer function between different resolutions to derive spatial filters for simulating coarse-resolution data (Justice et al. 1989), as explained by Yang and Merchant (1997).

Similar to the degradation–reclassification method, a statistical finite mixture model was used to classify land-cover classes after aggregation of Landsat TM spectral values to coarser-resolution images (Ju et al. 2005). Comparison with results from majority-based aggregations of the 30-m land-cover classification indicated that the finite mixture model approach performed consistently more accurately, with the lowest error at spatial resolutions between 60 and 150 m.

The discriminant models and geostatistics may be valuable in designing upscaling procedures, which can be built on the work of Gotway and Stroup (1997). The key to this is to construct a discriminant space from data space (i.e., the space spanned by the input dataset, which is not necessarily amenable for class definition, interpretation, and analysis); aggregation is then performed in the discriminant space, and the upscaled discriminant variables are inserted into classification procedures to derive area classes. The categorical information thus obtained should be more conformal to the definitions of area classes at corresponding scales to which the discriminant variables are aggregated.

Effects of upscaling on area-class information can be checked through the changing spatial patterns of area-class maps with aggregation. The aggregation of raw spectral values in the degradation–reclassification method tends to result in an increasing proportion of pixels with a mixed spectral response and, thus, misclassification. Retrieving area estimates from coarse-spatial-resolution maps is therefore generally inaccurate owing to the effect of spatial aggregation on class proportions (Mayaux and Lambin 1995).

The proportion estimation error due to spatial aggregation is defined as (Moody and Woodcock 1994)

$$\varepsilon_i = (P_{Ci} - P_{Fi})/P_{Fi} \tag{10.29}$$

where ε_i represents the proportion by which class i is over- or underestimated, and P_{Ci} and P_{Fi} are the proportions of class i at the coarse and fine resolutions, respectively. The magnitude of the errors in the estimation of the fine-resolution proportions depends on the spatial resolution of the map, the fine-resolution proportions of different classes, and the spatial arrangement of fine-resolution area classes (this reasoning is also important for discussion of alternative approaches to downscaling in the next subsection). Thus, error ε_i can be expressed as

$$\varepsilon_i = f(\Delta, P_{Fi}, S_{Fi}) \tag{10.30}$$

where Δ is the spatial resolution of the map, P_{Fi} is the proportion of the class i at fine resolution, and S_{Fi} is a descriptor of the spatial pattern of the class i at fine resolution, as shown by Mayaux and Lambin (1995).

10.3.2 DOWNSCALING

As reviewed by Mayaux and Lambin (1995), it is theoretically possible to predict the loss of information in the scaling process if the degree of adjacency or contiguity of land-cover types is known. As Moody and Woodcock (1995) and Mayaux and

Lambin (1995) explained, there are logical and consistent relationships between the fine-resolution spatial characteristics of the cover types and the proportional errors at coarse resolution. Thus, by modeling the errors associated with coarse-resolution mapping, it is possible to perform backward scaling, that is, to retrieve fine-resolution class proportions from the coarse-resolution data, at least to some extent. Below, we review some of the methods for downscaling area estimates based on coarse-resolution image classification, following Mayaux and Lambin (1995). Then, a description is provided of methods for downscaling species data (McPherson et al. 2006), which can be usefully explored for downscaling area-class data from coarse-resolution pixels to fine-resolution pixels.

Inversion of Equation 10.30 suggests

$$P_{Fi} = f(P_{Ci}, S_{Fi}) \tag{10.31}$$

where the spatial arrangement at a fine resolution, measured by S_{Fi}, needs to be quantified for completion of the inversion of P_{Fi} from P_{Ci} (Mayaux and Lambin 1995). However, since the spatial pattern of a class is not preserved when scaling up, information about spatial arrangement is only available at coarse resolution (Turner et al. 1989). Equation 10.31 may be resolved either by modeling spectral mixtures at the coarse-pixel level or by modeling the spatial patterns of classes for coarse pixels (Mayaux and Lambin 1995), as described below.

Mixture modeling is a direct approach that seeks to establish a relationship between the cover types at a fine resolution (fine-resolution class proportion to be more accurate) and the spectral response measured for the corresponding coarse-resolution pixels (Iverson et al. 1989; Cross et al. 1991; Quarmby et al. 1992):

$$P_{Fi} = f(Z_C(x)) \tag{10.32}$$

where $Z_C(x)$ is the spectral value of the pixel x at a coarse resolution. Equation 10.32 may be considered as the inversion of Equation 7.27 where $RFL_b(x)$ actually represents $Z_C(x)$ in Equation 10.32. This approach assumes that spectral response measured by the coarse-resolution data (e.g., AVHRR data) is proportional to the number of candidate cover types within the sensor's instantaneous field of view, as reviewed by Mayaux and Lambin (1995). In Equation 10.32, the estimated proportion of cover types at coarse resolution is not included, nor is the spatial pattern of the cover types.

Equation 10.32 can be determined empirically or on a physical basis, as reviewed by Mayaux and Lambin (1995). Iverson et al. (1989) estimated the function f empirically using Landsat TM data, applying the concept of double sampling with regression estimators. The empirical procedure is very sensitive to the sample selection and sample size of the fine-resolution data used for calibration. Cross et al. (1991) developed an explicit linear mixture model. Such a model is based on the identification, at the coarse resolution, of the spectra of pure pixels corresponding to "end members." A model based on these extreme spectra was then applied to a large region to estimate the proportions of different classes by decomposing the spectral values of mixed pixels using the end-member components. The physically based procedure is highly sensitive to the determination of the end-member spectra (Mayaux and Lambin 1995).

In addition to the mixture modeling approaches, we may relate the fine-resolution cover data to the coarse-resolution cover data directly (rather than the coarse-resolution spectral data) using regression (Mayaux and Lambin 1995). Without including the spatial pattern as an independent variable in the model, Equation 10.31 is simplified to

$$P_{Fi} = f(P_{Ci})$$
(10.33)

In the experiment by Nelson (1989), the independent variable P_{Ci} is the estimated forest cover percentage at a coarse resolution, measured in a block of pixels. However, regression results based on Equation 10.33 are unstable owing to the exclusion of S_{Fi} measuring the spatial pattern of cover types at fine resolution, as the regression parameters depend heavily on the cover type analyzed (Mayaux and Lambin 1995).

This could be fixed by including covariates S_{Fi} quantifying spatial pattern of land-cover types at a coarse spatial resolution to control the relationship between the dependent and the independent variables in Equation 10.33 (Mayaux and Lambin 1995). The relationship between class proportions at fine and coarse resolutions (i.e., Equation 10.33) can be rewritten as

$$P_{Fi} = f(P_{Ci}, S_{Ci})$$
(10.34)

where S_{Ci} is a descriptor of the spatial pattern of the class i at the coarse resolution. Equations 10.33 and 10.34 are known as inverse calibration models (Mayaux and Lambin 1995). This model was demonstrated to be more reliable than classical calibration to correct remotely sensed areal estimates for misclassification bias (Czaplewski and Catts 1992; Walsh and Burk 1993), as discussed by Mayaux and Lambin (1995).

The aforementioned methods for downscaling class areal estimates can recover their finer-resolution counterparts, although with some error. They are all oriented to map-wide estimation of class proportions, except for the mixture modeling approach in Equation 10.32, which operates at the pixel level. We review below two methods for pixel or grid cell-level downscaling described by McPherson et al. (2006), which seek to predict species' distributions (i.e., area classes) at a finer resolution from occurrence records at coarser resolution.

First, we consider the *direct* approach. Assume availability of environmental conditionals (i.e., controlling factors) at fine resolution, whereas the species occurrence data are at coarse resolution. According to McPherson et al. (2006), the direct approach works in three steps: (1) pixel amalgamation to calculate average environmental conditionals for each larger pixel (coarse resolution), (2) model parameterization based on the statistical relationship between species occurrences and averaged environmental data, and (3) application of parameter values to environmental data at smaller pixels, yielding fine-resolution predictions (Collingham et al. 2000; Araújo et al. 2005). Unfortunately, predictions by the direct approach may be unreliable because of extrapolation- and interpolation-induced uncertainty as models parameterized at coarser resolution may not be applicable directly to predicting class occurrences at finer resolution.

Second, we consider the *iterative* approach, which seeks to circumvent the unreliability issue associated with the direct approach. It works by trying to minimize the discrepancy between the environmental values encountered during calibration and prediction. It moves from coarse to fine resolution incrementally, doubling the spatial resolution of predictions at each step (McPherson et al. 2006). Model calibration starts at the original resolution of the coarser-resolution data (e.g., 16 km). If performing satisfactorily at that resolution, the model's parameters are used to make predictions at the next-finer resolution (e.g., 8 km). These predictions are classified as presence or absence using a predicted probability of occurrence of 0.5 as the threshold. The outcome serves as training data for a new model, recalibrated at half the starting resolution (i.e., 8 km), and the process begins anew, as described by McPherson et al. (2006). This approach assumes that models calibrated at coarse resolution make decent predictions at finer resolution as long as the zoom factor (resolution gap) is modest. As noted by McPherson et al. (2006), because the dominance of environmental factors controlling species' distributions shifts with spatial scale, cross-scale predictions remain a challenging task.

Gardner et al. (2008) proposed a robust method for rescaling land-cover data, which allows map resolution (grain size) to be either increased or decreased while holding the total proportion of land-cover types constant. The method uses a weighted sampling net of variable resolution to sample an existing map and then selects randomly from the frequency of cover types derived from this sample to assign the cover type for the corresponding location in the rescaled map.

The procedures described previously for downscaling class-proportional data often determine only the fractions of candidate classes within a coarse pixel without locating them in space, with the exception of mixture modeling and the method proposed by Gardner et al. (2008). More importantly, downscaling should go further to turn these proportions into a fine-resolution map of class labels (Boucher 2007). Subpixel mapping is underdetermined in that many different fine-resolution maps can lead to an equally good reproduction of the available coarse fractions. Thus, the unknown fine-resolution land-cover map is regarded as a realization of a random set. Boucher (2007) proposed simulation-based approaches to downscaling. At any pixel along a path visiting all fine-scale pixels, a class label is simulated from a local probability distribution made conditional on the coarse class-fraction data, any simulated land-cover classes at fine pixels previously visited along that path, and a prior structural model.

The downscaled area-class maps, generated through stochastic simulation as above, all honor the coarse proportion data and any fine-scale data available, and exhibit the spatial patterns called for by the input structural model. A useful adaptation to this would be to replace indicator kriging-based simulation by discriminant-variable-based simulation. The simulated, downscaled discriminant variables are then submitted to classification procedures to derive simulated, downscaled area-class maps.

10.4   DISCUSSION

So far, the discussions have focused on spatial scales. The interactions between spatial scale and thematic detail are important for categorical mapping and should be dealt with. The interaction between the resolution of images used, the thematic detail observable, and the size of patches detected can give rise to some unwanted consequences in spatial modeling (Ju et al. 2005). As the image resolution becomes coarser, there is a tendency to detect fewer of the small patches of land cover. Also, as the image resolution becomes coarser, the ability of the analyst to distinguish thematic classes may be reduced and, depending on the classes and the type of images, a point may be reached at which the number of classes must be reduced. This means that discussion of spatial scale is only informative if the classification scheme addresses thematic resolution as well as spatial resolution. Moreover, Buyantuyev and Wu (2007) showed that changing thematic resolution may significantly affect landscape metrics and, in turn, their ability to detect landscape changes. Thus, the effects of thematic resolution must be considered in area-class mapping and landscape pattern analysis.

Area-class mapping often needs to operate with remote-sensing data of different resolutions. As Strahler et al. (1986) suggested, we may consider H-resolution and L-resolution models in remote sensing in a relative sense. H-resolution models are characterized by resolution cells smaller in size than scene objects (making identification of individual objects possible), whereas L-resolution models have resolution cells larger than scene objects (tending to result in mixels). In studies of urbanization and urban sprawl, fine-grained urban fabrics can be imaged effectively by fine-resolution aerial photographs and satellite sensor images. The class "urban" refers to a land use, but it is likely a mixture of fine-grained patterns of different land cover types. A coarse-resolution image sees only a mixture of land cover types in urban areas and that mixture is commonly interpreted as a land-use class.

Investigators often select an appropriate spatial scale for image-based land-cover classification using statistics such as local variance and scale variance. However, where such statistics vary locally over the region of interest, their use in selecting a single spatial resolution may be undermined. Furthermore, differences observed between the red and near-infrared wavelengths have implications for users who wish to select a single spatial resolution for multispectral images. Therefore, for a mixed land type (e.g., as is present in most remotely sensed images), or a land type that contains a variety of different land-cover objects (e.g., urban), it may be difficult to select a single appropriate spatial resolution (Atkinson and Aplin 2004).

Scale is an issue that should be considered in close association with uncertainty. Discriminant models facilitate process-based strategies not only for scale-related research, but also for uncertainty modeling. In discriminant space, a linear model may be prescribed for measurement vector $\mathbf{Z}$ at location x: $\mathbf{Z}(x) = m_Z(x) + \delta\mathbf{Z}(x)$, which states that $\mathbf{Z}(x)$ is a version of the mean vector $m_Z(x)$ corrupted by an error vector of $\delta\mathbf{Z}(x)$, which may have zero mean. Discriminant models assume that there exists a mean value $m_Z(x)$ and hence a mode class label $C(x)$ at every location, and measurement $\mathbf{Z}(x)$ containing error $\delta\mathbf{Z}$ leads to error-prone area-class $C^*(x)$. This modeling approach is advantageous as it overcomes problems typically resulting from indicator-kriging-based alternatives, such as nonreplicability in statistics (e.g., mean and variance) computed on the basis of indicator kriging-simulated area-class maps, as demonstrated by Goodchild et al. (2009).

In addition to measurement-error-induced uncertainty in $C^*(x)$, uncertainty exists owing to class semantics. It is important to recognize that empirically built discriminant models do not always support a one-to-one correspondence between $m_Z(x)$ and $C(x)$. Thus, data classes $C_d(x)$ associated with $m_Z(x)$ by a data-specific membership function $f_k(m_Z)$ should be discerned from Z space to ascribe a one-to-one relationship between Z and C_d. Data classes can be used to bridge the gap between measurement and information classes, denoted above by C, which pertain to the classification scheme designed for the specific application. The differentiation between data and information classes will be useful for objective quantification of uncertainty (Foody 2010), which may be due to measurement error in Z (variance) and model imperfection regarding $C_d - C$ association (bias), respectively. The issue of uncertainty will be the topic for Chapter 12.

11 Information Content

When discussing scale in spatial information and analysis, it is natural to ask about the meaning of information therein and how spatial information is measured and communicated, especially in relation to scale. There are various types of carriers or media for spatial information, such as texts, maps, graphs, images, animations, and audio/video records. The concept of information has come a long way from daily expressions of various messages to Shannon's (1948) classic information theory (Cover and Thomas 2006). This chapter will introduce informational analysis of spatial information sources and their flows through geo-processing to information products, offering information-theoretic perspectives on scale issues, in order to complement the geostatistical and statistical methodology of previous chapters.

Remotely sensed images, both optical and microwave (radar), have become major sources of spatial information. As described in Chapter 3, these images are formed by recording the reflected radiance or energy from a scene. A typical digital image consists of a two-dimensional array of pixels, each representing the average reflectance, emittance, or backscattering of the surface within the sensor's instantaneous field of view, as described in Chapter 3. The images are used to detect the presence of certain phenomena, map their spatial extents, and estimate certain biophysical variables, such as leaf-area index and surface temperature. Usually, the raw images are processed using various operations, such as filtering, compression, enhancement, transformation, feature extraction, thematic classification, and others, where the analyst seeks to maximize the information content in the images for certain applications (Lee 1980; Bell 1988; Price 1994; Price 1997; O'Sullivan et al. 1998; Narayanan et al. 2002; Peng et al. 2005). The information extracted from images is usually depicted in the form of raster or vector maps. It is important to be able to quantify amounts of information provided through the processes of measurement, geo-processing, and representation of spatial entities and distributions, so that we can increase the information potential of a dataset for a particular application (Barnsley et al. 1997; Goodchild 2003; Harrie and Stigmar 2010).

Scale is clearly an important metric for spatial information and stands as a key factor in information content and flows in images, maps, and geo-processing. Intuitively, images at finer (spatial) resolution and maps at finer scale often contain more information content than otherwise, although the relationships are rarely linear or simple, as mentioned in Chapter 1 when this chapter was outlined. Models of scale reviewed in earlier chapters have been seen to be geostatistical, variance based, fractal, pattern indicator based (e.g., surface ruggedness and area-class shape indices), and so on. These models would be complemented usefully by information theoretic perspectives. Methods for scaling and analyzing scale effects would also benefit from information theory. For example, information theory offers insightful views of scale issues, such as entropy rates and their use in quantification of spatial correlation, limits to super-resolution mapping, and sampling designs based on informational measures, as will be discussed in this chapter. In summary, the purpose of this chapter is to provide a primer for informational analysis of scales and change of scale through a review of key concepts and methods in information theory so that further research on scale issues can build on the initial informational framework laid by this chapter.

Various methods have been proposed to characterize the information content in remotely sensed data (Dowman and Peacegood 1989; Kalmykov et al. 1989; Oliver 1991; Blacknell and Oliver 1993; Metternicht and Zinck 1998). Such methods often attempt to relate information content to parameters and variables in image capture and processing, such as sensor characteristics, image resolution (both radiometric and spatial), variability of the properties of interest, and others (Huck et al. 1981, 1996). Information theory provides the basis for characterizing the information content in images and other spatial data (Shannon 1948; Cover and Thomas 2006). For example, entropy and mutual information, as important concepts in information theory, measure the information in a random variable and the amount of information in common between two random variables, respectively. The former quantity indicates how easily messages can be compressed, as applied by Aiazzi et al. (2009) to hyperspectral image data, whereas the latter can be used to find the communication rate across a channel (Sheppard and Larkin 2003). Below, we list some of the methods for information quantification in images, both optical and radar.

Through examining the effect of spatial resolution on radar image interpretation, it was proposed that the interpretability or information content can be determined by the so-called spatial-gray-level resolution volume, which is the product of the range resolution, the azimuth resolution, and a gray-level resolution defined as the ratio of the 90% level to the 10% level on the fading distribution for the samples independently averaged for each pixel (Moore 1979). Later, statistical models were applied to study an imaging radar as a noisy communication channel with multiplicative noise, and an information-theoretical approach was used to quantify the information content conveyed by the radar image (Frost and Shanmugam 1983). Shannon's information theory was used to quantify the information content of Landsat-4 TM and MSS image data (Price 1984). With visible and near-IR band data over five agricultural scenes, it was found that the Landsat TM images provide more information-gathering capability as expressed in terms of bits per pixel, as their six reflective bands actually acquire 18 bits of information per pixel out of a possible $6 \times (8 \text{ bits}) = 48$ bits, whereas the four MSS bands actually obtain 10 bits of information per pixel out of a possible $4 \times (7 \text{ bits}) = 28$ bits (Price 1984). It was then argued that the TM and MSS are equally efficient in gathering information (with information efficiency calculated as the ratio of actual to possible information amount, i.e., $18/48 \approx 10/28$). Blacknell and Oliver (1993) applied both intuitive and information-theoretic approaches to characterize the information content of coherent images contaminated with additive or multiplicative noise. They analyzed the effects of both noise types on the trans-information (i.e., mutual information that measures the information amount conveyed by a received message about a source) of a coherent imaging system, as a particular form of communication channel, by relating information-theoretic concepts to intuitive descriptions of information based on physical models. Information theory was also used to compute the informational redundancy, or mutual information, of L- and C-band Shuttle Imaging Radar-C (SIR-C) images, thereby increasing the accuracy of land-cover classification by performing fusion of the two bands of data (Le Hégarat-Mascle et al.

1997). Narayanan et al. (2002) proposed a simple mathematical model to relate information content in images, evaluated based on classification accuracy, to spatial resolution, based on the reasoning that loss of information occurs if pixels in the image are wrongly classified. Using this approach, they separately developed negative-exponential models relating information content to image parameters, such as spatial and spectral resolution.

Clearly, the aforementioned methods can be categorized, as also reviewed in Narayanan et al. (2002). There is the interpretability model based on visual identification of targets at degraded spatial resolutions, which may be used as an indicator of information content (Moore 1979). The interpretability model was superseded by a method based on classification accuracy (Narayanan et al. 2002). The classic concept of entropy can be used to evaluate information in terms of image-gathering capability (Narayanan et al. 2002), which is useful for efficient data compression and transmission (Chen et al. 1987; Cover and Thomas 2006). The mutual information model measures the information about spatial entities or distributions from images, which is more applicable in the design of imaging systems in which accuracy is important for inversion of surface properties (Bell 1988).

Entropy, mutual information, and other informational measures have been employed not only for image assessment but also for system design (Fellgett and Linfoot 1955; Huck et al. 1999). Information theory has become a systematic methodology for informational analysis of spatial information processes. Taking the processes of image gathering, encoding, reconstruction, and display individually and as a whole, Huck et al. (1996) provided an information-theoretic framework for visual communication, whose assessment can be based on such information measures as the rate of transmission of information (i.e., mutual information) that the image-gathering system produces for the radiance field residing within its field of view, and the information rate of the observed image that the image-restoration system produces from the received information on an image-display medium. They showed that the best possible picture can be produced from a visual channel at the lowest data rate only if the image-gathering device produces the maximum-realizable information rate and the image-restoration algorithm properly accounts for the critical limiting factors that constrain the visual communication.

Spatial data that are not in image formats are often involved in image processing. Also, end products of image data are often represented as maps. For example, data from the Moderate Resolution Imaging Spectroradiometer (MODIS) instrument on the NASA Terra satellite launched in 1999 (complemented by another MODIS on the Aqua satellite in 2002) were used to provide various information products in support of global change research and natural resource management with spatial resolutions of 250 m to 1 km, such as land cover dynamics and land surface temperature (Justice et al. 2002). Information products are also derived from remote-sensing images of finer spatial resolution, which are more useful for local applications requiring finer-grained images for cartographic and topographic mapping (Toutin 2011). Thus, it is important to make the link between information content in image sources and that in maps so as to characterize information flows in remote sensing, geographic information systems (GIS), and spatial analysis (Mendicino and Sole 1997; Niedda 2004). To some extent, maps are representations of the entities extracted or variables mapped from remote measurements. It is interesting to investigate informational dynamics as measurements are used to create various information products (Bishop et al. 2001; Rodgers and Connor 2003).

Information theory has clear relevance and great potential for research and applications involving geographic information, as briefly shown above and continually documented in the literature (Marchand 1972; Batty 1974; Davis and Dozier 1990; Rossi and Posa 1992; Chen et al. 2008; Guo 2010). Scale is an important dimension of geographic information and analysis, as has been discussed in previous chapters of this book. There seems to be merit in making the links between information theory and scale. Although variograms have conventionally been used for modeling scales (Meisel and Turner 1998), entropy and other information-theoretic quantities may be applied as complementary tools for describing scales and analyzing their effects on information transmission in spatial information processing and applications. For example, entropy over lagged bivariate distributions was defined as a measure for spatial disorder by Journel and Deutsch (1993), which is an example of early work on integrating information-theoretic and geostatistical approaches to the modeling of spatial variability. As shown by Johnson et al. (2001), the entropy of spatial pattern associated with a particular pixel resolution, conditional on the pattern of the next coarser resolution, may be calculated and plotted as a function of resolution, resulting in a two-dimensional plot called a conditional entropy profile, which provides a means for visualization of multiscale landscape fragmentation patterns. It is well known that spatial entropy is affected by the number of geographic units over which it is computed. As a result, the size and number of areal units, such as census tracts, stand as intervening factors in direct and meaningful comparisons between features or regions with differing geographic dimensions. Relative entropy measures adjusted for scale effects are advocated by researchers (Heikkila and Hu 2006). A thorough understanding of information theory and its links with scale is of mutual benefit for scale-related research and further developments of information theory in the context of geographic information science (GIScience).

This chapter will first describe the fundamental concepts of information theory. Then, methods for quantifying the information content in remotely sensed images will be reviewed; images with additive and multiplicative noise are described separately. This is followed by a section on the association between resolution and information, and about the effects of resolution on information content on the one hand and information-theoretic perspectives on super resolution on the other. After descriptions about informational characteristics in image sources, information content in maps will be discussed, while accounting for spatial patterns. Lastly, we summarize the key points presented in this chapter and provide an outlook for further research and development in the light of information theory.

11.1 INFORMATION THEORY

Information theory plays an important role in communication and other fields of study including geography via such quantities as entropy and mutual information (Marchand 1972; Finn 1993; Singh 1997; Goodchild 2003). As mentioned previously, *entropy* is a measure of the uncertainty associated with a random variable, whereas *mutual information* is a measure of the amount of information one random variable contains about another. Mutual information is a special case of a more general quantity called *relative entropy*, which is a measure of the distance between two probability distributions (Cover and Thomas 2006). We describe these and other related concepts below.

As a measure of the uncertainty of a random variable, entropy quantifies the amount of information required on average to describe the random variable. Let S be a discrete random variable with probability mass function $p(s) = \Pr\{S = s\}$, $s \in \mathbb{S}$, $\mathbb{S}$ being the set of states that S can take. We denote the probability mass function by $p(s)$ rather than by $p_S(s)$, for convenience. Thus, if $p(s)$ stands for the probability distribution of S, the expected value of the random variable $g(S)$ is written as $E_p g(S) = \sum_{s \in \mathbb{S}} g(s) p(s)$ or more simply as $Eg(S)$, with E denoting expectation, when the probability mass function is given implicitly by the context (Cover and Thomas 2006). The *entropy* $H(S)$ of a discrete random variable S is defined by

$$H(S) = E[I(s)] = E\left[\log_2 \frac{1}{p(s)}\right] = -\sum_{i=1}^{n} p(s_i) \log_2 p(s_i) \tag{11.1}$$

where $I(s)$ denotes the entropy contribution of an individual message s, which is also known as the self-information of s, as shown later, with n being the number of states for S in the set $\mathbb{S}$. With logarithms in base 2, as in Equation 11.1, the entropies are measured in bits (by default, all logarithms in this book are to base 2 unless specified otherwise); if the base of the logarithm is e, the entropy is measured in *nits* or *nats* (1 nat $= \log_2 e \approx 1.443$ bits). We will use the convention that $0 \log 0 = 0$, which is easily justified by continuity since $s \log s \to 0$ as $s \to 0$ (Cover and Thomas 2006). The concept of entropy was related to spatial variability by Mendicino and Sole (1997) and to spatial heterogeneity by Aiazzi et al. (2004).

For continuous random variables, the concept of differential entropy should be used, which is similar in many ways to entropy for discrete random variables. Let S be a continuous random variable with probability density function $pdf(s)$. We can define its differential entropy through an integral:

$$H(S) = -\int_{-\infty}^{\infty} pdf(s) \log_2 pdf(s)\, ds \tag{11.2a}$$

The differential entropy $H(S)$ of a Gaussian random variable S (with $pdf(s) = \dfrac{1}{\sqrt{2\pi\sigma^2}} e^{-\frac{(s-\mu)^2}{2\sigma^2}}$ where μ and σ represent the mean and standard deviation of S, respectively) can be calculated as

$$
\begin{aligned}
H(S) &= -\int_{-\infty}^{\infty} pdf(s) \log_2 pdf(s)\, ds \\[2mm]
&= -\int_{-\infty}^{\infty} pdf(s) \log_2 \frac{1}{\sqrt{2\pi\sigma^2}} e^{-\frac{(s-\mu)^2}{2\sigma^2}}\, ds \\[2mm]
&= \int_{-\infty}^{\infty} pdf(s)\left(\log_2 \sqrt{2\pi\sigma^2}\right) ds + \int_{-\infty}^{\infty} pdf(s)(\log_2 e)\left[\frac{(s-\mu)^2}{2\sigma^2}\right] ds
\end{aligned}
\tag{11.2b}
$$

As $\int_{-\infty}^{\infty} p(s) \frac{(s-\mu)^2}{2\sigma^2} ds = \frac{1}{2} E(S^2) = \frac{1}{2}$ (noting the unit variance of a standardized normal distribution $E(S^2)$), we have

$$H(S) = \log_2 \sqrt{2\pi\sigma^2} + \frac{1}{2}\log_2 e = \frac{1}{2}\log_2 2\pi e\sigma^2 \qquad (11.2c)$$

There are similar derivations for the differential entropy of a normal distribution in many studies including Cover and Thomas (2006).

The *joint entropy* $H(S, Z)$ of a pair of discrete random variables (S, Z) with a joint distribution $p(s, z)$ is defined as $-E \log p(S, Z)$, that is,

$$H(S,Z) = \sum_{i=1}^{n}\sum_{j=1}^{m} p(s_i,z_j) I(s_i,z_j) = -\sum_{i=1}^{n}\sum_{j=1}^{m} p(s_i,z_j) \log_2 p(s_i,z_j) \qquad (11.3)$$

where $I(s_i, z_j)$ is the entropy contribution of an individual message coded as the alphabet pair $\{s_i, z_j\}$, and n and m represent the numbers of states of S and Z, respectively (Cover and Thomas 2006). $H(S, Z)$ can be seen as an extension to univariate $H(S)$ or $H(Z)$, and can be extended to modeling the entropy of sequences of random variables (i.e., stochastic processes), as shown later in this section. This is useful for information quantification in images or maps, which can be modeled as stochastic processes generating messages via reflectances or map symbols.

We also define the conditional entropy of a random variable given another as the expected value of the entropies of the conditional distributions, averaged over the conditioning random variable. If $(S, Z) \sim p(s, z)$, the *conditional entropy* $H(S|Z)$ is defined as

$$H(S|Z) = -E \log p(S|Z) \qquad (11.4a)$$

This quantity is computed as follows. The entropy of S conditional to z_j, $H(S|z_j)$, is

$$H(S|z_j) = \sum_{i=1}^{n} p(s_i|z_j) I(s_i|z_j) = -\sum_{i=1}^{n} p(s_i|z_j) \log_2 p(s_i|z_j) \qquad (11.4b)$$

Taking expectation over all z_j $(j = 1, 2, \ldots, m)$, we have

$$H(S|Z) = \sum_{j=1}^{m} p(z_j) H(S|z_j) = -\sum_{j=1}^{m}\sum_{i=1}^{n} p(z_j) p(s_i|z_j) \log_2 p(s_i|z_j)$$

$$\qquad (11.4c)$$

$$= -\sum_{j=1}^{m}\sum_{i=1}^{n} p(s_i, z_j) \log_2 p(s_i|z_j)$$

as is deduced in many textbooks including that of Cover and Thomas (2006). Clearly, in computing the conditional entropy, the weighting assigned to each information term is the joint probability of the states involved. Although entropy $H(S)$ is regarded as a measure of uncertainty about a random variable, $H(S|Z)$ is a measure of the amount of uncertainty remaining about S after Z (the conditional) is known. Similarly, conditional entropy $H(Z|S)$ can also be defined. Both $H(S|Z)$ and $H(Z|S)$ are useful constructs and will be discussed in the context of mutual information later in this section.

The joint entropy of a pair of random variables is related to the entropy of either variable and to the conditional entropy of the other. As shown by Cover and Thomas (2006), we have

$$
\begin{aligned}
H(S,Z) &= H(S) + H(Z|S) \\
&= H(Z) + H(S|Z)
\end{aligned}
\tag{11.5}
$$

which says that the entropy of a pair of random variables is the entropy of one plus the conditional entropy of the other. The concepts of joint and conditional differential entropy can be defined in a way similar to differential entropy in Equation 11.2a, but will not be elaborated here.

The *relative entropy* or Kullback–Leibler distance is a measure of the distance between two distributions (Kullback and Leibler 1951). In statistics, it arises as the expected logarithm of the likelihood ratio (Cover and Thomas 2006). The relative entropy $D(P\|Q)$ between two probability mass functions $P = \{p(s_i)\}$ and $Q = \{q(s_i)\}$ ($s_i \in \mathbb{S}$, $\mathbb{S}$ being the set of values for a discrete random variable S defined earlier with respect to Equation 11.1) is defined as

$$
D(P\|Q) = \sum_{i=1}^{n} p(s_i) \log \frac{p(s_i)}{q(s_i)}
\tag{11.6}
$$

where s_i is the value of underlying discrete random variable S and n is the number of states for S (Kullback and Leibler 1951; Cover and Thomas 2006). In the above definition, we use the convention that $0 \log \frac{0}{0} = 0$ and the convention (based on continuity arguments) that $0 \log \frac{0}{q} = 0$ and $p \log \frac{p}{0} = \infty$ (for $p, q > 0$). Thus, if there is any symbol s_i such that $p(s_i) > 0$ and $q(s_i) = 0$, then $D(P\|Q) = \infty$ (Cover and Thomas 2006). Relative entropy can also be defined for continuous random variables through integrals, although not described further here.

We now introduce mutual information, which, as mentioned previously, is a measure of the amount of information that one random variable contains about another random variable. Consider two random variables S and Z with a joint probability mass function $p(s, z)$ and marginal probability mass functions $p(s)$ and $p(z)$. The mutual information $I(S, Z)$ is the relative entropy between the joint distribution $p(s, z)$

and the product distribution $p(s)p(z)$ when assuming independence between S and Z (Cover and Thomas 2006), and is defined as

$$I(S,Z) = D(p(s,z)\|p(s)p(z))$$

$$= \sum_{i=1}^{n}\sum_{j=1}^{m} p(s_i,z_j)\log_2 \frac{p(s_i,z_j)}{p(s_i)p(z_j)} \tag{11.7a}$$

For continuous random variables, summation in Equation 11.7a is replaced with a definite double integral:

$$I(S,Z) = \int_{-\infty}^{\infty}\int_{-\infty}^{\infty} pdf(s,z)\log\left(\frac{pdf(s,z)}{pdf(s)pdf(z)}\right)dsdz \tag{11.7b}$$

where $pdf(s, z)$ is the joint probability density function of S and Z, and $pdf(s)$ and $pdf(z)$ are the marginal probability density functions of S and Z, respectively (Cover and Thomas 2006).

We can derive alternative expressions for mutual information based on Equation 11.7a. For example, we may write

$$I(S,Z) = \sum_{i=1}^{n}\sum_{j=1}^{m} p(s_i,z_j)\log_2 \frac{1}{p(s_i)} + \sum_{i=1}^{n}\sum_{j=1}^{m} p(s_i,z_j)\log_2 \frac{1}{p(z_j)}$$

$$- \sum_{i=1}^{n}\sum_{j=1}^{m} p(s_i,z_j)\log_2 \frac{1}{p(s_i,z_j)} \tag{11.8a}$$

$$= H(S) + H(Z) - H(S,Z)$$

Substituting Equation 11.5 for $H(S, Z)$ above, we have

$$I(S,Z) = H(S) - H(S|Z)$$

$$= H(Z) - H(Z|S) \tag{11.8b}$$

Thus, S says as much about Z as Z says about S, explaining the meaning of mutual information as the amount of information that knowledge of either variable provides about the other (Cover and Thomas 2006).

According to Equation 11.8b, the mutual information $I(S, Z)$ is the reduction in the uncertainty of S due to the knowledge of Z (i.e., the difference between the entropy of the information source $(H(S))$ and the equivocation of the information source $(H(S|Z))$. If we consider S to be the input to a communication channel and Z to be the output, we can see that Equation 11.8b defines the mutual information transferred between them, which is the change of uncertainty about S when we are

given the output Z (Cover and Thomas 2006). When the total information is trans-
ferred, $H(S|Z) = 0$ and $I(S, Z) = H(S)$. At the other extreme, when Z does not contain
any information relatable to S, $H(S|Z) = H(S)$ and therefore $I(S, Z) = 0$ (i.e., there is
no mutual information).

A special occasion is the mutual information of a random variable with itself. We
note that

$$I(S,S) = H(S) - H(S,S) = H(S) \tag{11.9}$$

because $H(S|S) = 0$. Thus, the mutual information of a random variable with itself
is the entropy of the random variable. This is the reason why entropy is sometimes
referred to as *self-information* (Cover and Thomas 2006), as mentioned previously.

In summary, mutual information can be equivalently expressed as

$$\begin{aligned}
I(S,Z) &= H(S) - H(S|Z) \\
&= H(Z) - H(Z|S) \\
&= H(S) + H(Z) - H(S,Z) \\
&= H(S,Z) - H(S|Z) - H(Z|S)
\end{aligned} \tag{11.10}$$

where $H(S)$ and $H(Z)$ are the marginal entropies, $H(S|Z)$ and $H(Z|S)$ are the condi-
tional entropies, and $H(S, Z)$ is the joint entropy of S and Z. Since $H(S) \geq H(S|Z)$, this
characterization is consistent with the non-negativity property.

In communications, *encoding* is the process by which information from a source
(e.g., place names) is converted into symbols (e.g., ASCII) to be communicated,
whereas *decoding* is the reverse process by which these symbols are converted back
into information a receiver can understand. A *communication channel* is a system in
which the output depends probabilistically on its input. It is characterized by a prob-
ability transition matrix $\{p(Z|S)\}$ that determines the conditional distribution of the
output given the input. Channel capacity, which is the upper limit for mutual information transferrable through the channel (Cover and Thomas 2006), is an important
concept in information theory. For a communication channel with input S and output
Z, we can define the capacity C by

$$C = \max_{p(s)} (I(S,Z)) \tag{11.11}$$

where the maximization is over the probability distribution of input S.

The ultimate limit on the rate of communication of information over a channel
is given by the channel capacity. The channel coding theorem shows that this limit
can be achieved by using codes with a long block length (Shannon 1948; Cover and
Thomas 2006). In information theory, a code is usually considered as an algorithm
that maps symbols from some source alphabet to some target alphabet. In practical
communication systems, there are limitations on the complexity of the codes that
can be used and, therefore, it may not be possible to achieve capacity, as shown in the
context of imaging by Huck et al. (1996).

The concept of mutual information may be generalized to more than two random variables. For instance, the *conditional mutual information* (Cover and Thomas 2006) of random variables S and Z given Z_0 is defined by

$$I(S, Z | Z_0) = H(S | Z_0) - H(S | Z, Z_0) \qquad (11.12)$$

which is the reduction in the uncertainty of S due to the knowledge of Z when Z_0 is given. Bell (1988) studied waveform design to maximize the conditional mutual information conveyed by radar systems.

The concepts of mutual information can be extended to multispectral situations by letting the variables S and Z become multidimensional vectors $\mathbf{Z}$ and $\mathbf{Z}'$, with $\mathbf{Z} = (Z_1, ..., Z_{N_Z})$ and $\mathbf{Z}' = \left(Z_1', ..., Z_{N_{Z'}}'\right)$. Usually, $N_{Z'} \le N_Z$. The information transfer achieved by the communication channel is used here in a general sense to represent both simple selections of spectral band subsets and more complex transformations, such as the Tasseled Cap transformation (Malila 1985). The selection of subsets of spectral bands is a special case of the mutual information expression,

$$I(\mathbf{Z}, \mathbf{Z}') = H(\mathbf{Z}) - H(\mathbf{Z} | \mathbf{Z}') \qquad (11.13)$$

where $\mathbf{Z}'$ is a subset of $\mathbf{Z}$, as described by Malila (1985). High-dimensional generalization that maximizes the mutual information between the spectral bands and the target variables (e.g., class indicators) is found to be useful in feature selection (Peng et al. 2005).

To work with multiple random variables, we need to explore entropy further. Regular entropy is defined for a single random variable and quantifies the number of bits needed, on average, to represent the value of the random variable. It is established that $nH(Z)$ bits suffice on average to describe n independent and identically distributed (i.i.d.) random variables $\{Z_1, ..., Z_n\}$ (Cover and Thomas 2006). However, we would like to know what happens when the random variables are dependent or form a stationary process, as in remotely sensed images where reflectances in neighboring pixels usually exhibit spatial dependence (i.e., they are rarely i.i.d.).

As an extension of the concept of entropy, the entropy rate refers to situations where we have a sequence of random variables forming a random process. In this case, the entropy rate is the number of bits that is needed to encode each new element in the sequence, given all the previous ones (Cover and Thomas 2006). In other words, the entropy rate is the amount of new information (on average) contained in each new element in the random process. Again, consider pixel digital numbers (DNs) in remotely sensed images. The amount of new information in each new pixel DN tends to be different for images flown over homogeneous vs. heterogeneous terrain. We will also see the relevance of the concept of entropy rate toward the end of this section where information redundancy in images due to spatial dependence is analyzed. In the next section, we show how spatial dependence in remotely sensed images is accommodated in the computation of information entropy.

Dependence among the random variables in a stochastic process $\{Z_1, ..., Z_n\}$ is characterized by the joint probability distribution $p\{z_1, ..., z_n\} = \Pr[Z_1 = z_1, ..., Z_n = z_n]$.

A stationary stochastic process is one in which the joint distribution of any subset of the sequence of random variables is invariant with respect to shifts in the index, as discussed in Chapter 4 where geostatistical models of scale are introduced. A discrete stochastic process $Z_1, Z_2, \ldots$ is Markov if for $n = 1, 2, \ldots, \Pr[Z_{n+1} | Z_n, \ldots, Z_1] = \Pr[Z_{n+1} | Z_n]$ for all $z_1, \ldots, z_n, z_{n+1} \in \mathbb{Z}$. Furthermore, a Markov chain is space- or time invariant if the conditional probability $\Pr[Z_{n+1} | Z_n, \ldots, Z_1] = \Pr[Z_{n+1} = z' | Z_n = z] = \Pr[Z_2 = z' | Z_1 = z]$ for all $z, z' \in \mathbb{Z}$. Thus, a time-invariant Markov chain is characterized by its initial state and a probability transition matrix $\mathbf{P} = [p_{zz'}]$ $(z, z' \in \mathbb{Z})$, where $p_{zz'} = \Pr[Z_{n+1} = z' | Z_n = z]$, as discussed by Cover and Thomas (2006).

Given a sequence of n random variables, to determine how the sequence's entropy grows with n, we can define the entropy rate of a stochastic process $\{Z_1, \ldots, Z_n\}$:

$$H(\mathbb{Z}) = \lim_{n \to \infty} \frac{1}{n} H(Z_1, \ldots, Z_n) \tag{11.14}$$

when the limit exists (Cover and Thomas 2006). Thus, the entropy $H(Z_1, \ldots, Z_n)$ grows linearly with n at a rate $H(\mathbb{Z})$, the entropy rate of the process. For the i.i.d. stochastic process $\{Z_1, \ldots, Z_n\}$ mentioned previously, the entropy rate is simply the entropy of the i.i.d. random variable Z, $H(Z)$. A related quantity for entropy rate is

$$H'(\mathbb{Z}) = \lim_{n \to \infty} \frac{1}{n} H(Z_n | Z_{n-1}, \ldots, Z_1) \tag{11.15}$$

Thus, $H(\mathbb{Z})$ is the per-symbol entropy of the n random variables and $H'(\mathbb{Z})$ is the conditional entropy of the last random variable given the past (Cover and Thomas 2006). For a stationary stochastic process, both limits in Equations 11.14 and 11.15 exist and they are equal: $H(\mathbb{Z}) = H'(\mathbb{Z})$. Furthermore, by the definition of a stationary Markov process, we can add conditional entropy to this equality: $H(\mathbb{Z}) = H'(\mathbb{Z}) = H(Z_2 | Z_1)$ (for details, see Cover and Thomas 2006). As $H(Y_2 | Y_1) \leq H(Y_2)$, the entropy rate of a stochastic process is usually less than the regular entropy of a random variable at a location in space or time.

As seen above, entropy rate for a spatially dependent field (e.g., an intensity image) is often less than per-sample (i.e., per-pixel) entropy and, thus, reflects the information redundancy residing in the field. There is a growing literature on the computation of image entropy that provides a measure of spatial redundancy. Petrov and Li (2003) proposed a mathematical framework that permits the factorization of a joint probability distribution into its localized components for a two-dimensional array of pixels. The factorization was used to estimate the contribution to mutual information of two- and three-pixel correlations for a large ensemble of natural images, shedding light on the information redundancy in images. Chandler and Field (2007) described a technique for estimating the entropy of image patches based on the proximity distribution of pixel values. The advantage of this proximity-based technique over simple statistics, such as the power spectrum, is that the proximity distribution is dependent on all forms of redundancy. This function was used to estimate the redundancy of 3×3 patches of known entropy as well as of 8×8 patches of Gaussian white noise, natural scenes, and noise with the same power spectrum as natural scenes. It was found that the sample of 8×8 patches of natural scenes used in their study has less than half the entropy of 8×8 white noise and less than 60% of the entropy of noise with the same power spectrum. The approach proposed by Chandler and Field (2007) provided estimates of the redundancy and compressibility of natural image patches with good accuracy and, thus, a reference for the efficiency of any coding strategy aiming to reduce redundancy, although it depends on extrapolation for images larger than 3×3 pixels.

We will discuss the effects of spatial dependence on information entropy of images in the following section, where the entropy rate of an image is estimated using spatially dependent image models and can be used as an approximation for the per-sample (pixel) entropy accounting for interpixel spatial dependence. As scale reflects the domain and strength of spatial dependence, a description of relations between an image's spatial dependence and its information statistics will be helpful for elaborating the effects of scale on information content.

11.2 INFORMATION CONTENT IN REMOTELY SENSED IMAGES

11.2.1 INFORMATIONAL CHARACTERISTICS OF IMAGES

Spatial measurements are undertaken to obtain information about geographic entities and distributions, decreasing the *a priori* uncertainty about them, as discussed in the context of radar remote sensing by Bell (1988). Remote measurements acquired through various sensors are becoming increasingly important as primary sources of spatial information. We examine remote sensing as a measurement process in the light of information theory.

In a measurement system, there are fields or objects to be measured, a measurement mechanism, and an observer. Suppose that a random vector **S** refers to parameters characterizing what is measured. The measurement mechanism maps **S** to another random vector **Z**, often stochastically. The observer observes **Z** and determines the information about **S** from **Z** (Bell 1988). For example, a radar image is a spatial mapping of the scattering characteristics of the surface being illuminated by the radar transmitter (Bell 1988; Oliver and Quegan 2004), whereas an optical image records reflectance measurements related to the surface's physical properties, but subject to atmospheric effects and other factors (Holm et al. 1989). So, the image consists of intensities $z(i, j)$ at pixel coordinate pair (i, j); where intensity is a function of the scattering characteristics (for a radar) or reflectance (for an optical sensor) of the ground location corresponding to (i, j). Without causing confusion about image intensities, we use $z(i, j)$ to represent either radar or optical measurement at location (i, j), although radar images are complex (having in-phase and quadrature components, or amplitude and phase in polar form) as opposed to real in the case of optical images (Oliver and Quegan 2004).

Because of the inherent uncertainty in the measurement mechanism, the measurement may contain errors. Since **S** and **Z** are jointly distributed random vectors, they have mutual information $I(\mathbf{S}, \mathbf{Z})$ between them. $I(\mathbf{S}, \mathbf{Z})$ quantifies the amount of information that observation of **Z** provides about **S**. The greater this mutual information, the more information we are obtaining about **S** by measurement **Z**, as described in the previous section. We can determine the maximum number of equiprobable classes into which we can assign **S** by observation of **Z** based on knowledge of $I(\mathbf{S}, \mathbf{Z})$, as discussed by Bell (1988). The mutual information between source **S** and measurement **Z** will be discussed in the next two subsections, whereas this subsection will describe the information-theoretic characteristics of discrete images. The effects of interpixel spatial dependence in an image on its information content will also be examined. The analyses are also applicable for the underlying stochastic processes (informational sources) that have generated the observed images, although the latter is a sampled, discretized, and, often, nonlinearly related version of the former.

Consider an image **Z** as an $m \times n$ array of pixels, with an intensity value $z(i, j)$ at coordinates (i, j). The intensity level at each of the pixels may be modeled as a continuous variable. However, in most real imaging systems and those using digital processing, these levels will be quantized into discrete levels. For instance, remote sensors often produce signals that have a fixed maximum number of signal levels in each spectral band, usually expressed as a number of bits (e.g., 6 bits for 64 levels in Landsat MSS bands and 8 bits for 256 levels in Landsat TM bands) (Chen et al. 1987; Bell 1988). Denote the possible intensity values in each pixel by $z \in \mathbb{Z}_q = \{z_1, ..., z_K\}$. Thus, the image under study can be considered to be an $m \times n$ array, with each element of the array taking on one of the K elements of $\mathbb{Z}_q$. If we view the image as a noiseless communication channel, there will be $N_{\mathrm{MS}} = K^{mn}$ unique messages that can be conveyed by such a channel. In other words, there are N_{MS} distinguishable images that can be constructed. The amount of information I that can be conveyed by this image viewed as a noiseless channel is

$$I = \log N_{\mathrm{MS}} = mn \log K \tag{11.16}$$

Assume that we generate the image **Z** (a matrix) by assigning a value from $\mathbb{Z}_q$ to each of the $m \times n$ pixels *independently*, with each cell having probabilities $p_1, ..., p_K$ of taking on values $z_1, ..., z_K$ from $\mathbb{Z}_q$, respectively. The entropy of the image, denoted by $H(\mathbf{Z})$ (i.e., the amount of information it conveys when generated by such a source) is

$$H(\mathbf{Z}) = mnH(pix) \tag{11.17a}$$

where $H(pix)$, the entropy of an individual pixel, is

$$H(pix) = -\sum_{k=1}^{K} p_k \log p_k \tag{11.17b}$$

The maximum of $H(pix)$ occurs when $p_k = 1/K$ for all $k = 1, ..., K$, which is $\log K$, confirming the result shown in Equation 11.16 with the i.i.d. assumption (Bell 1988).

However, as shown previously, spatial dependence is inherent to many spatial forms and processes (Tobler 1970). Thus, there are usually significant similarities in the reflectance or scattering characteristics of the terrain surface for adjacent pixels, which imply information redundancy in the image **Z** as a whole, challenging the results obtained in Equations 11.17a and 11.17b.

Construct an $M = m \times n$ by 1 column vector $\mathbf{z}$, which is obtained from $\mathbf{Z}$ by column scanning. $\mathbf{z}$ can be thought of as the output of an information source, realized as any of the $N_{MS} = K^{mn}$ possible vectors $\mathbf{z}$, as explained by Bell (1988). Consider the *a priori* probability of occurrence for each of the N_{MS} possible images $\mathbf{z}_i$, which is $p(\mathbf{z}_i)$, the joint probability distribution of the individual pixel intensities, i.e., $p(\mathbf{z}_i) = \text{Pr}[\mathbf{z}_i(1), ..., \mathbf{z}_i(M)]$, where $i = 1, ..., N_{MS}$. The entropy of the image (Bell 1988) is

$$H(\mathbf{z}) = -\sum_{i=1}^{N_{MS}} p(\mathbf{z}_i) \log p(\mathbf{z}_i) \qquad (11.18)$$

We have the following inequality: $H(\mathbf{z}) \leq mn \log K$, with equality if and only if $p(\mathbf{z}_i) = K^{-mn} = 1/N_{MS}$, for $i = 1, ..., N_{MS}$.

We can express joint entropy $H(\mathbf{z})$ for the image $\mathbf{z}$ using the chain rule described by Cover and Thomas (2006):

$$H(\mathbf{z}) = H(\mathbf{z}(1), \mathbf{z}(2), ..., \mathbf{z}(M)) = \sum_{l=1}^{M} H[\mathbf{z}(l) | \mathbf{z}(l-1), ..., \mathbf{z}(1)] \qquad (11.19)$$

where the general term $H[\mathbf{z}(l) | \mathbf{z}(l-1), ..., \mathbf{z}(1)]$ can be interpreted as the average information provided by the pixel indexed at l when the values of the preceding $l-1$ pixels are known. Equation 11.19 provides a conditional entropy-based formula for computing the source entropy of the image (Bell 1988).

Assume that the general term $H[\mathbf{z}(l) | \mathbf{z}(l-1), ..., \mathbf{z}(1)]$ in Equation 11.19 approaches a nonzero limiting value, denoted as $H[\mathbf{z}(l) | \infty]$, as l becomes large, which is actually the entropy rate of the stochastic process that generated the image intensity field. This was elaborated by Anastassiou and Sakrison (1982), confirming that the conditional entropy of $\mathbf{z}(l)$ defined at the site l, given knowledge of the variables (i.e., $\mathbf{z}(l-1), ..., \mathbf{z}(1)$) in its infinite past, is the entropy rate of the underlying field, as introduced in Section 11.1. Furthermore, assume $\mathbf{z}$ is a stationary process, $H[\mathbf{z}(l) | \infty]$ is constant for all l, then, we have

$$H(\mathbf{z}) \approx \sum_{l=1}^{M} H[\mathbf{z}(l) | \infty] \approx MH[\mathbf{z}(l) | \infty] \qquad (11.20)$$

We must determine the limiting conditional entropy $H(\mathbf{z}(l) | \infty)$ of $\mathbf{z}(l)$ from a finite sequence of pixels. The approximation can be performed through an extension to Equation 11.4c:

$$H(\mathbf{z}(l) | \infty) \approx H(\mathbf{z}(l) | \mathbf{z}(l-1), ..., \mathbf{z}(l-j))$$

$$= \sum_{k_0=1}^{K} ... \sum_{k_j=1}^{K} p(\mathbf{z}(l) = k_0, ..., \mathbf{z}(l-j) = k_j) \log \frac{p(\mathbf{z}(l) = k_0, ..., \mathbf{z}(l-j) = k_j)}{p(\mathbf{z}(l-1) = k_1, ..., \mathbf{z}(l-j) = k_j)} \qquad (11.21)$$

where $p(\mathbf{z}(l)=k_0, \ldots, \mathbf{z}(l-j)=k_j)$ is the joint probability of pixel l and its past j pixels, and $p(\mathbf{z}(l-1))=k_1, \ldots, \mathbf{z}(l-j)=k_j)$ is the joint probability of pixel ls past j pixels (Bell 1988).

The joint distribution $p(\mathbf{z}(l), \ldots, \mathbf{z}(l-j))$ for a K-level image has K^{j+1} distinct sets of states at which the joint distribution must be estimated. Computation is intensively expensive even for relatively small j. Additional difficulty lies in the number of samples required for a reliable estimation of the joint distribution from histograms. Despite these, Bell (1988) found through hypothetical examples that the information contained in a group of pixels is often less than the sum of the individual, mutual information conveyed individually by the pixels. This makes sense as we note the inequality $H[\mathbf{z}(l)|\infty] \leq H[\mathbf{z}(l)|\mathbf{z}(l-1)] \leq H[\mathbf{z}(l)]$.

The problem of dimensionality encountered in the computation of image entropy calls for low-dimensional and closed-form models for pixel value distributions, such as Markovian models, as shown above. Spatially dependent image models that are amenable to easy computing are very important for quantification of image entropy and other information measures. A quadrilateral Markov random field model was introduced by Razlighi et al. (2009) to overcome the limitations with the usual Markov mesh random field models, such as strong diagonal dependency in neighborhood. The proposed Markovian model with a neighborhood size of two pixels is used to decompose an image prior into a product of two-dimensional joint pdfs, which are estimated using joint histograms under the homogeneity assumption. Hosseini et al. (2010) proposed the concept of multi-information rate and provided improved estimates of the multi-information rate of natural images for the sake of understanding spatial structures therein. Whereas the common approach is to estimate the multi-information for patches of increasing sizes and divide it by the number of pixels, Hosseini et al. (2010) showed that the limiting value of this sequence, the multi-information rate, can be better estimated by using another sequence (with a limiting value) by measuring the mutual information between a pixel and a causal neighborhood of increasing size around it. These methods may be usefully combined with geostatistics to flexibly accommodate spatial dependence in informational analysis.

We have discussed information-theoretic characteristics for a single-band image. However, multispectral images are often employed for spatial-information extraction. To quantify entropy in multispectral images, the quadrilateral Markov random field model described by Razlighi et al. (2009) may be usefully explored, treating per-pixel multispectral measurements as vectors. We may also try informational analysis in the feature space, as follows.

When the probabilities in the entropy equations are based on all possible combinations of image intensity quantization levels, absolute information measures will result. However, techniques are usually employed to utilize only relative amplitude information between signals from various scene elements. Thus, the information resides in the number of spectral cells occupied in the feature space and in the distribution of spectral values within them. Malila (1985) developed an expression that gives a relative entropy value, H_R, for multispectral image data, in terms of counts of occurrences of observations in cells of the feature space. It is repeated here (for six bands indexed b_1–b_6):

$$H_R(\mathbf{Z}_1, ..., \mathbf{Z}_6) = \log_2 N_{obs} - \left(\frac{1}{N_{obs}}\right) \sum_{b_1} ... \sum_{b_6} n_{b_1 b_2 b_3 b_4 b_5 b_6} \log_2 n_{b_1 b_2 b_3 b_4 b_5 b_6} \qquad (11.22)$$

where $n_{b_1 b_2 b_3 b_4 b_5 b_6}$ is the count of occurrences in the cell having level b_1 in $\mathbf{Y}_1$, level b_2 in $\mathbf{Y}_2$, etc., and N_{obs} is the total number of observations in the data set being analyzed.

Equation 11.24 can be written as

$$H_R(\mathbf{Z}) = H_{max} - H_{loss} \qquad (11.23a)$$

It is informative to divide the total information loss due to spectroradiometric concentration of signals into two components, one due to the reduced number of spectral cells that are occupied, and the other which occurs when the duplicate observations are not uniformly distributed among those cells, that is,

$$H_{loss} = L_{cell} + L_{unif} \qquad (11.23b)$$

where L_{cell} is the cell loss or loss in number of cells, that is,

$$L_{cell} = \log_2 N_{obs} - \log_2 N_{cells} = -\log_2\left(\frac{N_{cells}}{N_{obs}}\right) \qquad (11.23c)$$

and L_{unif} is the uniformity loss,

$$L_{unif} = H_{loss} - L_{cell} \qquad (11.23d)$$

as described by Malila (1985).

11.2.2 INFORMATION CONVEYED IN IMAGES WITH ADDITIVE NOISE

In the previous subsection, it was assumed that each pixel displayed an intensity that was quantized to one of K levels. So, the maximum amount of information that could be obtained per pixel was log K (bits). However, the actual information content of each individual pixel was not discussed. In other words, we did not examine the information that could be expected from the remote measurement of a pixel and the image, which was a persistent topic in the literature (Fuderer 1988; Gupta and Virdi 1989; Prasad 2000; Jiang and Rahman 2012). This is now addressed in this subsection.

An imaging system is a communication channel in which the message, the observed image, conveys information about the underlying scene (Blacknell and Oliver 1993). The communication channel is not perfect, since the imaging process is affected by errors (Huck et al. 1981). Thus, it is important to know to what extent the image provides information about the imaged scene. Next, we describe statistical models for images, referring to Huck et al. (1996).

There are two types of signals involved in remote sensing, coherent and incoherent, depending on whether the illumination used in an imaging system exhibits spatial coherence (Cox and Sheppard 1986). For coherent imaging, it is appropriate to describe the illumination as a spatial distribution of complex-valued field amplitude, typically in radar. For incoherent illumination, it is appropriate to describe it as a spatial distribution of real-valued intensity (Goodman 2005). Thus, we focus on the analysis of linear systems with real-valued images first, whereas complex-valued images are discussed briefly as a generalization afterward.

Fourier analysis is a useful mathematical tool for informational analysis of image gathering and processing. For a function G of X, Y in two-dimensional space, its Fourier transform (the Fourier spectrum or frequency spectrum) as represented by $\mathbb{F}_G$ is

$$\mathbb{F}_G(f_X, f_Y) = \int_{-\infty}^{\infty} \int_{-\infty}^{\infty} G(X,Y)\exp(-2I\pi(f_X X + f_Y Y))\,\mathrm{d}f_X\,\mathrm{d}f_Y \qquad (11.24)$$

where f_X, f_Y are referred to as frequencies, with $I^2 = -1$ (for details, see Huck et al. 1996; Goodman 2005). In the following text, we use one-dimensional location x (e.g., a pixel) and spatial frequency f_x (with units of cycles per pixel) to simplify description, without causing any confusion.

The imaging equation shown in Equation 3.19 may be rewritten as

$$Z(x) = [\mathbb{k}S(x)*PSF(x)]\ sampling\ (x) + N(x) \qquad (11.25a)$$

where $Z(x)$, $S(x)$, and $N(x)$ represent the pixel-specific output datum, input signal, and noise (assumed additive and includes photodetector and quantization noises), respectively, with $\mathbb{k}$ being the steady-state gain of the linear radiance-to-signal transformation, $sampling\ (x)$ being the spatial sampling lattice designed for the sensor, and PSF (x) being the imaging sensor's point spread function (Huck et al. 1996). Equation 11.25a can be expressed equivalently, in the spatial frequency domain, as

$$\begin{aligned} \mathbb{F}_Z(f_x) &= [\mathbb{k}\mathbb{F}_S(f_x)\mathbb{F}_{PSF}(f_x)] * \mathbb{F}_{sampling} + \mathbb{F}_N(f_x) \\ &= \mathbb{k}\mathbb{F}_S(f_x)\mathbb{F}_{PSF}(f_x) + \mathbb{F}_{N_a}(f_x) + \mathbb{F}_N(f_x) \end{aligned} \qquad (11.25b)$$

where $\mathbb{F}_Z$, $\mathbb{F}_S$, and $\mathbb{F}_N$ represent the Fourier transforms of the output datum, the input signal, and the noise, respectively, with $\mathbb{F}_{PSF}$ being the spatial frequency response (SFR) of the sensor. Also in Equation 11.25b, the quantity $\mathbb{F}_S(f_x)\mathbb{F}_{PSF}(f_x)$ is the blurred component of the digital signal and $\mathbb{F}_{N_a}$ actually represents the quantity of $[\mathbb{k}\mathbb{F}_S(f_x)$ $\mathbb{F}_{PSF}(f_x)]*sampling_{\text{sidebands}}(f_x)$ and is the aliased signal component that, with insufficient sampling, folds into the sampling passband, which is largely ignored in the digital image processing literature, as discussed in detail by Huck et al. (1996). Equation 11.25b can be reorganized as

$$\mathbb{F}_Z(f_x) = \mathbb{k}\mathbb{F}_S(f_x)\mathbb{F}_{PSF}(f_x) + \mathbb{F}_{NOISE}(f_x) \qquad (11.25c)$$

where $\mathbb{F}_{NOISE}(f_x) = \mathbb{F}_{N_a}(f_x) + \mathbb{F}_N(f_x)$ is the sum of the Fourier transforms of noise due to aliasing, use of a discrete photodetector, and quantization.

To quantify the trans-information $I(S(x), Z(x))$ between input signal $S(x)$ and output datum $Z(x)$ based on Equation 11.25a, we may apply the rule $I(S(x), Z(x)) = H(Z(x)) - H(Z(x)|S(x))$ at any one location x and then sum over all locations to yield the total information. However, quantification of trans-information from the continuous radiance field to the digital signal may be carried out in the frequency domain via Equation 11.25c whereby convolution in the spatial domain is replaced by multiplication.

Based on Equation 11.25c, we can write the mutual information between the measurement $\mathbb{F}_Z(f_x)$ and the source $\mathbb{F}_S(f_x)$ as

$$I(\mathbb{F}_S(f_x), \mathbb{F}_Z(f_x)) = H(\mathbb{F}_Z(f_x)) - H(\mathbb{F}_Z(f_x)|\mathbb{F}_S(f_x)) \tag{11.26a}$$

which says that the trans-information is the entropy of the received signal less the conditional entropy of the received signal given the radiance field, as also shown by Huck et al. (1996). Because the noise is treated as independent and additive, $H(\mathbb{F}_Z(f_x)|\mathbb{F}_S(f_x))$ does not depend on $\mathbb{F}_S(f_x)$ and is solely a function of the noise statistics, that is, $H(\mathbb{F}_Z(f_x)|\mathbb{F}_S(f_x)) = H(\mathbb{F}_{NOISE}(f_x))$ (Shannon 1948). A proof for this in the context of statistical optics is given by Frieden (1983). Thus, we can simplify Equation 11.26a to

$$I(\mathbb{F}_S(f_x), \mathbb{F}_Z(f_x)) = H(\mathbb{F}_Z(f_x)) - H(\mathbb{F}_{NOISE}(f_x)) \tag{11.26b}$$

where $H(\mathbb{F}_{NOISE}(f_x))$ is the entropy of the noise.

Assume $\mathbb{F}_{NOISE}(f_x)$ is Gaussian with zero mean. Then, the joint probability for complex $\mathbb{F}_{NOISE}$ is

$$p_N(\mathbb{F}_{NOISE}(f_x)) = \frac{1}{\pi PSD_{NOISE}(f_x)} \exp\left(-\left|\mathbb{F}_{NOISE}(f_x)\right|^2 \Big/ PSD_{NOISE}(f_x)\right) \tag{11.27}$$

where $PSD_{NOISE}(f_x)$ is the power spectrum for $\mathbb{F}_{NOISE}$ (Huck et al. 1996). Similarly, on the assumption of normality for the data in the frequency domain (i.e., $\mathbb{F}_Z$), we can write its probability density as

$$p_Z(\mathbb{F}_Z(f_x)) = \frac{1}{\pi PSD_Z(f_x)} \exp\left(-\left|\mathbb{F}_Z(f_x)\right|^2 \Big/ PSD_Z(f_x)\right) \tag{11.28}$$

where $PSD_Z(f_x)$ is the power spectrum for $\mathbb{F}_Z$ (Huck et al. 1996). A derivation is also given by Frieden (1983), although with simplification.

Applying the formula to compute the differential entropy of multivariate Gaussian distributions and using Equation 11.26b, Huck et al. (1996) showed that

$$I(\mathbb{F}_S, \mathbb{F}_Z) = \frac{1}{2}\log(PSD_Z/PSD_{NOISE}) \tag{11.29}$$

Hence, the information conveyed by the data is the logarithm of the ratio of the power spectrum of the image data to that of the noise.

The mutual information $I(\mathbb{F}_S, \mathbb{F}_Z)$ described above is for any one frequency f_X. We can sum over frequency to obtain the total information,

$$\begin{aligned}
I(\mathbb{F}_S, \mathbb{F}_Z)_{\text{total}} &= \frac{1}{2}\int_{-\Omega}^{\Omega} df_x \log[PSD_Z(f_x)/PSD_{NOISE}(f_x)] \\
&= \frac{1}{2}\int_{-\Omega}^{\Omega} df_x \log\left[1 + \frac{\Bbbk^2 |\mathbb{F}_{PSF}(f_x)|^2 PSD_S(f_x)}{PSD_{NOISE}(f_x)}\right]
\end{aligned} \tag{11.30}$$

where $PSD_S(fx)$ is the power spectrum of the underlying radiance field being imaged and the interval $(-\Omega, \Omega)$ stands for the passband, indicating that only those frequencies f_x within the passband will contribute to the information content (Huck et al. 1981, 1996; Frieden 1983). Classic work on the informational aspects of optical design seeking to maximize image information content and assessment of optical images includes that of Linfoot (1955) and Fellgett and Linfoot (1955), and also provides some details about the Fourier treatment of imaging processes.

As shown in Equation 11.30, all frequencies (i.e., f_x) contribute in like manner. In other words, high and low frequencies convey information in a similar way. Furthermore, the ratio $|\mathbb{F}_{PSF}(f_x)|^2/PSD_{NOISE}(f_x)$ in Equation 11.30 will be high for a given imaging system, even one with a low $|\mathbb{F}_{PSF}(f_x)|$, if the noise power spectrum $PSD_{NOISE}(f_x)$ is low enough. So, it is noise, and not system aberrations, that degrades the information content and sets a limit to restoring an image (Frieden 1983).

The theoretic upper bound of $I(\mathbb{F}_S, \mathbb{F}_Z)_{\text{total}}$ is the channel capacity for a bandwidth-limited system with an average power limitation and is given in Huck et al. (1996) as

$$[I(\mathbb{F}_S \mathbb{F}_Z)_{\text{total}}]_{\max} = \frac{1}{2}|\Omega| \log\left[1 + \frac{\Bbbk^2 \sigma_S^2}{\sigma_N^2}\right] \tag{11.31}$$

where σ_S^2 is the variance of the underlying radiance field, σ_N^2 is the variance of photodetector and quantization noise, and $|\Omega|$ stands for the area of the frequency passband (i.e., the product of bandwidths in the f_X and f_Y directions). This upper bound is achieved only when $PSD_S(f_x) = \sigma_S^2$ and $|\mathbb{F}_{PSF}(f_x)| = 1$ for f_x within the passband (both are zeroes otherwise).

Frieden (1983) assumed a simplified imaging equation with additive noise: $Z = S + N$, where Z, S, and N stand for measurement, source, and noise, respectively, and derived a formula similar to Equation 11.30 for computing information content in images. Based on such a simplified model, Blacknell and Oliver (1993) elaborated on the two extreme cases of information content in images: uncorrelated and fully correlated signals. In both cases, the noise is uncorrected and identically distributed for each sample. For the case of uncorrelated signals, the information content, which is a maximum, is

$$[I(S,Z)_{total}]_{max} = \frac{npix}{2} \log\left[1 + \frac{\sigma_S^2}{\sigma_N^2}\right] \tag{11.32a}$$

where σ_S^2 and σ_N^2 stand for the variance of the signal and the noise, respectively; $npix$ is the number of pixels, representing the number of samples discretizing the frequency bandwidth-limited domain. For the case of a fully correlated signal, the total information content of the image with $npix$ pixels is reduced to

$$[I(S,Z)_{total}]_{min} = \frac{1}{2} \log\left[1 + npix\, \frac{\sigma_S^2}{\sigma_N^2}\right] \tag{11.32b}$$

Clearly, the information content in a real image dataset will fall between these two extremes as a real dataset is usually spatially correlated, although not perfectly. Also, it can be seen that reduction in noise will increase the information measure (Blacknell and Oliver 1993). The effects of signal and noise spatial dependencies and their interaction on information content should be studied further.

Coherent images, such as radar images, are increasingly common in remote sensing, as described in the context of synthetic aperture radar (SAR) imaging by Oliver and Quegan (2004). Suppose that the spectral quantities $\mathbb{F}_Z$, $\mathbb{F}_S$, and $\mathbb{F}_{NOISE}$ are complex numbers, having real and imaginary parts (i.e., $\mathbb{F}_Z$ being a number couplet $\left(\mathbb{F}_{Z_{re}}, \mathbb{F}_{Z_{im}}\right)$, similarly for $\mathbb{F}_S$, and $\mathbb{F}_{NOISE}$). The real and imaginary parts of $\mathbb{F}_Z$, $\mathbb{F}_S$, and $\mathbb{F}_{NOISE}$ are assumed uncorrelated (at each frequency). For simplicity, it is further assumed that the real and imaginary parts of any spectral quantity (at a given frequency) are identically distributed. Thus, the power spectrum PSD_n is twice the power spectrum of either the real or the imaginary components, noting that $PSD_{NOISE} = E\left(\mathbb{F}_{NOISE_{re}}^2\right) + E\left(\mathbb{F}_{NOISE_{im}}^2\right) = 2PSD_{NOISE_{re}} = 2PSD_{NOISE_{im}}$. As the power spectra are ratioed in Equation 11.29, the information content of a coherent image may also be expressed in the same way, but without the factor of 1/2 (Fellgett and Linfoot 1955; Frieden 1983). A proof for the formula of mutual information between complex-valued input radiance and output signal ensembles was given by Fales and Huck (1991). The significance of complex signals in imaging and informational analysis necessitates a comprehensive study of complex random variables and information measures (Neeser and Massey 1993; Schreier and Scharf 2010).

11.2.3   INFORMATION IN IMAGES WITH MULTIPLICATIVE NOISE

As described in Chapter 3, radar sensors measure the radar reflectivity of the terrain by illuminating the scene with coherent electromagnetic radiation at a given wavelength and measuring the power returned in the echo signal. To increase the spatial resolution of the sensor in the azimuth direction, which is practically limited by the length of the physical antenna, a much larger antenna is synthesized by taking advantage of the motion of the radar platform so that azimuth resolution can be greatly increased. Imaging radars that perform this coherent processing of returns received at different locations of a real-aperture radar are called synthetic-aperture radars (Bell 1988; Oliver 1991). The radar image $\mathbf{Z}$ displaying a coherently processed signal as a function of the terrain's reflectivity $\mathbf{S}$ can be used to extract information about the nature of the terrain. Both $\mathbf{S}$ and $\mathbf{Z}$ are modeled as random processes. In this subsection, we quantify the average amount of information that can be extracted about the reflectivity $\mathbf{S}$ from the recorded image $\mathbf{Z}$, with the radar system modeled as a noisy communication channel, following Frost and Shanmugan (1983) and Bell (1988).

Although imaging systems are usually modeled as additive white Gaussian noise channels, especially those for incoherent imaging, as shown in the previous subsection, the coherent nature of radar illumination often introduces the equivalent of multiplicative noise into the images, known as speckle (Frost and Shanmugan 1983; Oliver 1991). As a phenomenon inherent to coherent imaging, speckle occurs because, when incident waves are scattered by randomly distributed scatters, the resultant scattered field contains random interference terms. Specifically, speckle noise results from random fluctuations in the return signal from an object no larger than a resolution cell for a conventional radar image, whereas, in a SAR image, speckle is caused by coherent processing of backscattered signals from multiple distributed targets (Tso and Mather 2009). As speckle noise is the result of interfering backscattering and varies from pixel to pixel, a radar image shows a typical speckled appearance (Oliver 1991; Oliver and Quegan 2004). Being different from additive noise, multiplicative noise should be studied to analyze its effects on information content (Corner et al. 2003).

The multiplicative model (for a pixel) can be written as

$$Z = SN \qquad (11.33)$$

where S represents the terrain reflectivity, Z represents the radar measurement, and N is the multiplicative noise process, with these three quantities referring to a particular pixel. The random processes S and N are statistically independent since the former depends only on the scene being imaged and the latter is strictly system dependent through the number of looks averaged (Frost and Shanmugan 1983; Bell 1988). Next, the average amount of information that can be extracted about S from Z (i.e., the mutual-information measure $I(S, Z)$) will be evaluated on a pixel-by-pixel basis. This is also useful as a basis for a discussion about the interaction between radiometric resolution and spatial resolution in the context of radar imaging in the next section.

An imaging radar illuminates areas of a terrain within its field of view and records the value of the power returned from resolution cells on the ground. An area on the ground covered by a resolution cell (pixel) is typically made up of a large number of scatterers. We can model the signal received by the radar (before detection) as a narrow-band Gaussian random process. Then, with a square-law detector, the value of the received power PW from a pixel has a pdf as below (Porcello et al. 1976; Frost and Shanmugan 1983; Bell 1988):

$$f_{PW}(pw) = \mu^{-1} \exp(-pw/\mu) \text{ for } pw \geq 0 \tag{11.34}$$

where μ is the mean of PW.

Often, several independent measurements of PW for each pixel are obtained and averaged to form the image intensity value Z for the pixel, $Z = \dfrac{1}{NL} \sum_{i=1}^{NL} PW$, with NL being the number of independent measurements or "looks." The pdf of Z may be modeled by the gamma distribution

$$pdf_Z(z) = \frac{z^{NL-1}(\mu/N)^{-NL}\exp(-zNL/\mu)}{\Gamma(NL)} \tag{11.35}$$

where $\Gamma(NL) = (NL-1)!$. As μ is proportional to the radar reflectivity S of the resolution cell, we can assume that $\mu = S$ and rewrite Equation 11.35 as a conditional pdf (Frost and Shanmugan 1983):

$$pdf_{Z|S}(z \mid s) = \frac{z^{NL-1}s^{-NL}\exp(-zNL/s)}{\Gamma(NL)NL^{-NL}} \tag{11.36}$$

from which it is inferred that $\sigma_{Z/S}^2 = S^2/NL$.

As Frost and Shanmugan (1983) and Bell (1988) show, without any *a priori* knowledge, a uniform distribution for the reflectivity S is assumed:

$$pdf_S(s) = (\beta_2 - \beta_1)^{-1} \text{ for } 0 \leq \beta_1 \leq x \leq \beta_2 \tag{11.37}$$

where β_1 and β_2 define the minimum and maximum values of S, respectively, and $10 \log_{10}(\beta_2/\beta_1)$ is the dynamic range of the target reflectivity. Then, the pdf of Z is computed as

$$pdf_Z(z) = \int_{-\infty}^{\infty} pdf_{Z/S}(z/s)\,pdf_S(s)\,ds$$

$$= \frac{NL}{(NL-1)(\beta_2 - \beta_1)}\left[\exp(-NLy/\beta_2)\sum_{r=0}^{NL-2}(NLy/\beta_2)^{NL-r-2}\Big/(NL-r-2)!\right.$$

$$\left. -\exp(NLy/\beta_1)\sum_{r=0}^{NL-2}(NLy/\beta_1)^{NL-r-2}\Big/(NL-r-2)!\right]$$

$$(11.38a)$$

whereas the mean and variance of Z are computed as

$$\mu_Z = (\beta_1 + \beta_2)/2 \tag{11.38b}$$

$$\sigma_Z^2 = (\beta_2 - \beta_1)^2/12 + \left(\beta_1^2 + \beta_1\beta_2 + \beta_2^2\right)/3NL \tag{11.38c}$$

as shown by Frost and Shanmugan (1983).

The conditional entropy $H(Z|S)$ can be evaluated on the basis of the conditional pdf specified in Equation 11.36. Frost and Shanmugan (1983) obtained the following formula:

$$H(Z|S) = \ln \Gamma(NL) + NL - (NL-1)\left[\sum_{j-1}^{NL-1}\frac{1}{j} - cnst\right]$$

$$+ \frac{1}{\beta_2 - \beta_1}(\beta_1 - \beta_2 + \beta_2 \ln \beta_2 - \beta_1 \ln \beta_1) - \ln NL \tag{11.39}$$

where $cnst$ is Euler's constant $(0.577\ldots)$. However, it is not possible to express $H(Z)$ in closed form, so it must be evaluated numerically. Nevertheless, it is possible to compute the limits to $H(Z)$ and hence $I(S, Z)$. Consider the upper bound on $H(Z)$. Of all continuous random variables with finite variance σ^2, a Gaussian random variable has the largest differential entropy, $\log_2(2\pi e\sigma^2)^{1/2}$ (Cover and Thomas 2006). As NL becomes large, we have $H_{upper}(Z) = \ln(\beta_2 - \beta_1) + 0.1765$ (nats) (Bell 1988). Also, with a large NL, $H_{lower}(Z)$ will have a lower bound equal to $H(S)$, that is, $\ln(\beta_2 - \beta_1)$, the entropy of a uniformly distributed source in the interval $[\beta_1, \beta_2]$, as proved by Bell (1988). From the lower and upper bounds on $H(Z)$, we can derive the lower and upper bounds on $I(S, Z)$, respectively.

Moore (1979) found that the information content of radar images of a terrain is very low because of the large variance of N. With NL increasing, the variance of N approaches 0 and $\lim_{NL \to \infty} I(S, Z) = \infty$, meaning that the radar system approaches a noiseless communication channel, that is, one with no loss of information when NL is extremely large, as discussed by Frost and Shanmugan (1983) and Bell (1988).

The discussion above has focused on per-pixel information content in radar imaging when subject to multiplicative noise. Per-pixel information content may be generalized to the whole image extent through a simple factor of the number of pixels when interpixel spatial dependence is ignored, as by Frost and Shanmugan (1983). It is, however, questionable that spatial dependence is negligible in radar images of fine spatial resolution. Thus, the effects of spatial dependence on information content in radar images should be examined, as in the preceding subsection where additive noise models are discussed. Also, interaction between additive and multiplicative noise would be usefully examined, as these two types of noise are common in radar imaging (Corner et al. 2003).

11.3 IMAGE RESOLUTION AND INFORMATION CONTENT

11.3.1 EFFECTS OF RESOLUTION ON INFORMATION CONTENT

In the previous section, we considered an image of $m \times n$ pixels, with each element of the array taking on one of the K discrete values. If we view the image as a noiseless communication channel, there will be $N_{ms} = K^{mn}$ distinguishable images that can be constructed. The amount of information that can be conveyed by this image viewed as a noiseless channel is $mn \log K$, as shown in Equation 11.16. For an image acquired over a given study area, the finer its spatial resolution, the more pixels there will be for recording radiance or reflectivity signals from the terrain surface, thus increasing information capacity. Similarly, increased radiometric resolution (i.e., with increased levels of quantization K) will also mean potentially more information.

However, interpixel spatial dependence in the image implies informational redundancy among its $m \times n$ pixels $\mathbf{Z} = \{Z_1, Z_2, ..., Z_{m \times n}\}$. The information content of the image should be evaluated objectively based on joint entropy $H(Z_1, Z_2, ..., Z_{m \times n})$, which may be simplified as $mnH(\mathbb{Z})$, where $H(\mathbb{Z})$ is the entropy rate of the stochastic process underlying the image. If the image is assumed to follow a multivariate normal distribution, we may calculate its entropy as

$$H(Z_1, ..., Z_{m \times n}) = \log \sqrt{(2\pi e)^{m \times n} |\text{cov}_\mathbf{Z}|} \qquad (11.40)$$

where e is the base of natural logarithms, $\text{cov}_\mathbf{Y}$ is the variance–covariance matrix of the $m \times n$ pixels, and $\| \|$ indicates the determinant of a matrix (Cover and Thomas 2006). For non-Gaussian distributions, the computation of joint entropy becomes complicated, although Markov-chain models may be applied for entropy calculation

(Bell 1988). Generalized Gaussian models may also be employed for entropy modeling in image data (Aiazzi et al. 1999), whereas geostatistics can be helpful for quantifying spatial dependence in entropy modeling (Journel and Huijbregts 1978).

Consider an image as a whole, with each pixel representing a possible realization of a single random variable Z, say, image intensity. The maximum entropy is equal to the logarithm of the number of pixels that discretize an image, when all pixel values are equally probable. After coarsening or upscaling the image, there will be a reduction in the number of pixels to describe the same study area, thus a reduction of the possible values that the pixels may take. We can therefore normalize the information content of an image with the maximum entropy, which is log [$m \times n$], to eliminate the variation in entropy due to the reduction of the number of pixels (i.e., image coarsening) (Mendicino and Sole 1997).

Images have been assumed to be noiseless communication channels in the discussion above. However, imaging processes are always subject to uncertainty. For images contaminated with additive noise, Equation 11.30 may be applied to examine the effect of image spatial resolution on the mutual information conveyed by pixel measurement. The mathematical models described by Huck et al. (1981) for line-scan imaging mechanisms can be explored usefully for modeling the effects of image coarsening, in addition to noise in imaging, on information content.

To explore the relationships between spatial resolution and information content, Frost and Shanmugan (1983) defined the radiometric resolution, ρ_r, for the imaging radar as the ratio of the dynamic range of reflectivity over the number of distinguishable levels, that is,

$$\rho_r = \frac{\text{dynamic range in decibels}}{\exp[I(S, Z)]} \tag{11.41}$$

where the denominator is the number of distinguishable levels of the target reflectivity S in the measurement Z. One approach to sharpening the radiometric resolution (i.e., to decrease ρ_r) is by averaging a large number (NL) of independent samples from a given target to increase $I(S, Z)$. However, this implies coarsening the spatial resolution.

According to Frost and Shanmugan (1983), as the scene dynamic range $10 \log_{10}(\beta_2)$ increases (with $\beta_1 = 1$), the mutual information can be approximated as

$$I(S, Z) \approx f(NL) + \text{constant} \tag{11.42}$$

where $f(NL)$ represents the NL-dependent terms on the right-hand side of Equation 11.47. Thus, for large β_2, $I(S, Z)$ does not depend on the image dynamic range. This implies that increasing the image dynamic range does not result in an increase in the information content of radar images owing to the nature of multiplicative noise. Therefore, the implications for imaging-radar system design are that there is a fundamental limitation on the information content set by NL, beyond which an increase in image dynamic range does not increase the information content or the radiometric resolution, and that increasing the transmitter power does not result in an improvement in the system performance (Frost and Shanmugan 1983; Bell 1988).

The discussion above concerning the relations between information resolution and resolution seems to be limited to the case of radar images contaminated with multiplicative noise. Further work is needed to develop methods that can handle image data of different types or with different noise characteristics. There is research undertaken to explore information-resolution relations using other approaches, as shown below. As argued by Narayanan et al. (2000), traditional entropy-based methods for assessing the effects of resolution are more suited for quantifying the information content of unclassified images, whose probability density functions can be determined more accurately than those of classified images. Narayanan et al. (2000) proposed using classification accuracy as an index for information content in an image, based on the rationale that a loss in information would occur whenever a pixel in the classified degraded image is misclassified in comparison to the same pixel in the ground-reference image.

The spatial resolution of an image is often degraded by averaging pixel values within a moving window. As the size of the window increases, more of the background signal is mixed with the target or object signal and the mean reflectance of the object approaches that of the background. The rate of mixing depends on the size of the object, with means for small-sized objects attaining the background mean faster than for large-sized ones. The contrast between the object and the background also has a major role in classification accuracy, with low-contrast cases having low classification accuracy and high-contrast cases having high classification accuracy (Narayanan et al. 2000, 2002).

As the spatial resolution is coarsened and the pixel size, a_{pix}, increases, the amount of information to delineate the spatial extent of the target, a, decreases. The information content reduces to zero at a pixel size of ∞ (i.e., no information) (Narayanan et al. 2002). By plotting the information content data from the simulated images against a_{pix}/a, Narayanan et al. (2002) found that the relationship between information content and spatial resolution followed an exponential decay. We can model the information content, I, as a function of pixel size, a_{pix}, and the target characteristic dimension, a, as

$$I = \exp\left(-\beta_1 \left(\frac{a_{pix}}{a}\right)^{\beta_2}\right)$$

(11.43)

where β_1 and β_2 are the best parameters related to the interpretability of the image, as well as the contrast between the target and the background (Narayanan et al. 2002). The information content is unity for $a_{pix} = 0$ and is zero for $a_{pix} = \infty$. With the use of simulations to provide data points, various best-fit values (in the sense of minimized mean squared errors in regression analysis) can be computed from the above equations.

An experiment by Narayanan et al. (2002) found that, although the Landsat TM image has higher information content than the SIR-C images at finer pixel sizes, the opposite is true at coarser pixel sizes. The transition occurs at a pixel size of about 720 m. This suggests that the Landsat TM images are more useful for thematic classification for applications requiring fine resolution, whereas the SIR-C sensor is advantageous for applications involving coarser resolutions. Thus, the model proposed is useful in comparing different image types for different applications.

The information content in an image is also dependent on the radiometric resolution (i.e., gray-scale levels) of the image. To develop a model for the information content as a function of radiometric resolution, various simulation experiments were performed at Landsat TM and SIR-C SAR wavebands (Narayanan et al. 2000). Contrast-enhanced 8-bit images were generated using a total of five classes of terrain type, for various spatial arrangements. For each image, the radiometric resolution was degraded and the image reclassified. The classification accuracy, used as a measure of information content, was plotted as a function of the number of gray levels. Based on the results of their simulation experiments, the following model for information content I was proposed (Narayanan et al. 2000):

$$I = \exp\left(-k_1\left(\frac{L-1}{l-2}\right)^{k_2}\right) \tag{11.44}$$

where L is the number of gray levels in the original image, l is the number of gray levels in the degraded image ($0 \leq l \leq L$), and k_1 and k_2 are best-fit sensor-specific constants. It was assumed, for the sake of modeling, that a bi-level ($l = 2$) image does not convey any information about the textural features of the scene (i.e., $I = 0$), indicating that the degraded image does not furnish any information about the textural aspects of the scene. For $l = L$, $I = 1$ (i.e., all the information is preserved) (Narayanan et al. 2000).

In model development by Narayanan et al. (2000), classification accuracy was used as a measure to quantify information content. In such applications as classification, the loss in information content as a result of degraded radiometric resolution may not be significant, because certain pixels may still be correctly classified after gray-level degradation. Narayanan et al. (2000) found that the loss in information is exponential with respect to the number of gray levels. This was seen to be applicable for both Landsat TM and SIR-C images. Using this model for the information content of images as a function of radiometric resolution, one can specify an "optimal" radiometric resolution for an image.

11.3.2 INFORMATION CAPACITY AND SUPER RESOLUTION

As reviewed by Sheppard and Larkin (2003), according to Shannon's concept of the channel capacity of a temporal communications system, for a signal of duration T and a channel of bandwidth B_T, the number of points needed to specify the system is

$$M = 2TB_T + 1 \tag{11.45}$$

where the unity term accounts for the fact that a direct current (DC) signal can be measured with a channel of zero bandwidth. The term is usually neglected, as the other term $2TB_T$ is much greater than unity. The term $2TB_T$ represents the sampling rate sufficient for the exact recovery of the signal bandwidth limited to B_T, according to the Shannon sampling theorem. If the detected signal has an average power PW_S and an additive noise power PW_N, assumed band limited and uncorrelated, the number of levels that can be distinguished is given by $[(PW_S + PW_N)/PW_N]^{1/2}$. Thus, the total number of possible message states is

$$N_{ms} = [(PW_S + PW_N)/PW_N]^{M/2} \tag{11.46}$$

as described by Cox and Sheppard (1986) and Sheppard and Larkin (2003).

Information capacity, I (in bits), is defined as $I = \log_2 N_{ms}$, where N_{ms} is the total number of possible message states as quantified in Equation 11.46. Thus, the information capacity for an information channel with N_{ms} possible message states is

$$I = \frac{1}{2}(2TB_T + 1)\log_2(1 + PW_S/PW_N) \tag{11.47}$$

The information capacity of a two-dimensional imaging system, neglecting the DC terms (i.e., 1s as in Equation 11.45), is

$$I = 2B_xL_x2B_yL_y \log_2(1 + PW_S/PW_N)^{1/2} \tag{11.48}$$

where B_x and B_y are the spatial frequency bandwidths (in units of pixels per unit distance, say, meter); L_x and L_y are the dimensions of the field of view in the X and Y directions, respectively; and the value at each point of the object is real for an incoherent system and complex for a coherent system (Fellgett and Linfoot 1955; Cox and Sheppard 1986; Sheppard and Larkin 2003). It is helpful to note that the components $2B_xL_x$ and $2B_yL_y$ in Equation 11.48 represent the numbers of samples (pixels) in X and Y coordinates, respectively, required for the exact reconstruction of the field of image intensity, in accordance with the Shannon sampling theorem. Also, it is meaningful to make the link between Equations 11.32a and 11.48, both of which assume spatial independence among pixels.

It has been found that, if the system's transfer function is known with sufficient accuracy and the noise level is low, super-resolution procedures can provide resolution beyond the classical limits.

With information theory, the relationships between resolution and degrees of freedom of images were established, as reviewed by den Dekker and van den Bos (1997). All super-resolution techniques can be explained by the theorem of invariance of the number of degrees of freedom (Lukosz 1966; Toraldo di Francia 1969). For an imaging system, its number of degrees of freedom, D_F, is

$$D_F = 2(2L_x B_x + 1)(2L_y B_y + 1)(2TB_T + 1) \tag{11.49}$$

where, again, B_x and B_y are the spatial bandwidths; L_x and L_y are the dimensions of the field of view in the X and Y directions, respectively; T and B_T are the duration of exposure and temporal bandwidth B_T, respectively; and the factor 2 accounts for two polarization states (den Dekker and van den Bos 1997; Sheppard and Larkin 2003).

Lukosz's (1966) invariance theorem does not take noise into account. However, the practical limits to the resolution of imaging systems with known transfer functions are set by noise (den Dekker and van den Bos 1997). Cox and Sheppard (1986) combined the results of Fellgett and Linfoot (1955) and of Lukosz (1966) to include the DC terms, time dependence, noise, and two independent states of polarization, so that the information capacity or degrees of freedom in Equation 11.49 are modified as

$$D_F = 2(2L_x B_x + 1)(2L_y B_y + 1)(2TB_T + 1) \log_2 (1 + PW_S/PW_N)^{1/2} \tag{11.50}$$

as described by den Dekker and van den Bos (1997). Thus, in an imaging system, it is the information capacity, rather than the resolution, that is invariant, as discussed by Sheppard and Larkin (2003).

According to this invariance theorem, it is possible to extend the spatial bandwidth at the expense of one of the other factors in Equation 11.50 if *a priori* knowledge concerning the object is available (den Dekker and van den Bos 1997; Sheppard and Larkin 2003). Take a one-dimensional object as an example. It is possible to extend B_x above the classical value while reducing B_y so that D_F remains constant. This means that resolution is increased in the X direction by sacrificing the unneeded resolution in the Y direction (Lukosz 1966). As another example, consider an object that does not vary with time. Its spatial-frequency information could be encoded as temporal-frequency information. At the receiver, this temporal information could be decoded, and a high-resolution image could be formed. Analytic continuation can be regarded as an attempt to increase the spatial bandwidth by reducing the signal-to-noise ratio (SNR) in the resulting image (Cox and Sheppard 1986). Using this result, Cox and Sheppard (1986) derived the maximum resolution improvement obtainable by analytic continuation.

11.4 INFORMATION CONTENT IN MAP DATA

Sections 11.2 and 11.3 have focused on information content in remotely sensed images. It is important to investigate the informational aspects of maps, which are either products derived from remote-sensing data or ancillary data often employed in image analysis and other spatial problem-solving scenarios, as mentioned previously. To quantify the information content in a map, we need to identify its information sources (i.e., the events and characteristics on which the entropy computations should be based). The second issue is how to incorporate spatial elements in entropy computation.

To compute the entropy of a map, we may first decompose it to different themes (i.e., map layers), such as roads and land parcels. This is followed by statistical sampling to select typical units from the map for computing the probabilities of map symbols in a theme and then the entropy. The map entropy is computed as the sum of the entropies of its different themes. The joint entropy of different information sources should be evaluated properly on the basis of correlation (Sukhov 1970).

With a nonspatial treatment, identical values of entropy would be obtained for binary images if their proportions of black and white pixels are the same, even if they have different spatial patterns. The spatial correlation between neighboring pixels of a raster map can be accommodated by replacing the values of the pixels by their differences (Gatrell 1977): if two neighboring pixels have the same color, their difference is defined to be positive; for pixels having different colors, their difference is defined as negative. The entropy of the resultant binary image $\mathbf{Z}$ (a raster map) is

$$H(\mathbf{Z}) = -p^+ \log_2 p^+ - p^- \log_2 p^- \tag{11.51}$$

where p^+ is the probability of black–black and white–white neighbors, whereas p^- is the probability of black–white and white–black neighbors. Gatrell (1977) proposed computing the entropy of a binary image as a weighted mean value of the entropy at different orders of neighborhood:

$$H(\mathbf{Z}) = \sum_{k=0}^{n} w(k) \cdot H(\mathbf{Z})_k \tag{11.52}$$

where $w(k)$ is a weight function, k is the order of the neighborhood, and a high value of k corresponds to a global neighborhood, whereas a small value corresponds to a local neighborhood.

Bjørke (1996) pointed out that there are three aspects of information: syntactic (the relationship among the symbols), semantic (the meaning of the symbols), and pragmatic (applications). The semantic aspect of information is very subjective (depending on individual map readers), whereas the pragmatic aspect of information is dependent on the purpose of the map. Quantifying these aspects of information is complex. However, the syntactic part of the information may be analyzed by focusing on the map objects themselves. One way to quantify the map information is simply to count the number of objects in the map. Another way is to express the amount of information as the number of object points in the map, which, being the total number of points representing the objects in the map, is usually greater than the number of objects being represented. Other ways are to calculate the map area proportion covered by map objects or the total line length of the objects (Harrie and Stigmar 2007). The rationale is that these methods seem to be the equivalent of those for quantifying information content that pixels can convey in images, as can be inferred from Equation 11.52 when pixels are assumed spatially independent with each carrying 1 bit of information (i.e., binary raster map cells).

The Shannon information theory is intended for message communication, by which the information content (entropy, H) in a sent message is calculated using Equation 11.1, where p_i are the probabilities for the messages or symbols i. In cartographic communication, the messages may be carried by different symbols (e.g., object types). To measure the information content in maps, we can compute the probabilities for the object types (i) in the map by

$$p_i = \frac{n_i}{n_{total}} \tag{11.53}$$

where n_i is the number of symbols for object type i and n_{total} is the total number of map symbols. Then, the entropy (H) is defined as $H = -\sum_{i=1}^{n_{type}} p_i \log(p_i)$, where n_{type} is the number of object types in the map.

However, as pointed out by Li and Huang (2002), this measure does not consider the spatial distribution of the objects. The entropy will be the same whether the objects are tightly assembled or more widespread, if measures for probabilities, such as number of objects, number of object points, object line lengths, and object areas, are the same. The spatial distribution of information is important for entropy calculation.

Spatially influenced measures are recommended (Li and Huang 2002). To identify the "region of influence," the empty spaces surrounding each map object, modeled as Voronoi regions, are used. Three measures are introduced: geometric, topologic, and thematic (Li and Huang 2002). The geometric measure calculates the entropy of the Voronoi regions. The probability for each object is calculated as the ratio between its Voronoi region size and the total map size: $p_{SD,i} = A_i/A$, where A_i is the Voronoi area of the map objects $i = 1, 2, \ldots, N_{obj}$ (N_{obj} being the total number of objects in the map) and A is the total map area. The total map entropy for spatial distribution (H_{SD}) is then calculated as

$$H_{SD} = -\sum_{i=1}^{n_{obj}} p_{SD,i} \log p_{SD,i} \tag{11.54}$$

Thus, the map entropy is dependent on two properties: the number of regions (the entropy is larger with fewer regions) and the size differences of the regions (the more equally sized the regions are, the larger the entropy, with the maximum being log N_{obj} when all Voronoi regions are of equal size).

To cope with maps that have different numbers of objects, the entropy value can be normalized with the maximal entropy value for the same number of objects (i.e., log N_{obj} as indicated above). Hence, we obtain the following index:

$$H'_{SD} = \frac{H_{SD}}{\log N_{obj}} \tag{11.55}$$

where H_{SD} is given by Equation 11.62 and N_{obj} is the number of objects (Li and Huang 2002). The index will be equal to 1 if all Voronoi regions are of the same size and will be smaller the more uneven the Voronoi region sizes are, as can be deduced from Equation 11.54.

Based on the Voronoi regions and by differentiating between the neighbor types, the thematic measure of entropy can also be calculated on the assumption that the complexity increases when the objects are mixed (i.e., with neighbors of contrasting types) (Li and Huang 2002). Also, we can measure the spatial distribution of points in terms of entropy based on the Voronoi regions. A normalized index for the point distribution $\left(H'_{pnt} \right)$ may be computed as

$$H'_{pnt} = \frac{H_{pnt}}{\max[H_{pnt}]} = \frac{-\sum_{i=1}^{NP} p_{size,i} \log p_{size,i}}{\log NP} \tag{11.56}$$

where $p_{size,i}$ is the relative size of the Voronoi region for point i over the total area of the mapped extent and NP is the number of points (Li and Huang 2002). Clearly, the denominator $\log NP$ in Equation 11.56 acts, in a way similar to that (i.e., $\log N_{obj}$) in Equation 11.55, to normalize the entropy of point distribution (H_{pnt}) by the maximal entropy value for the same number of objects.

We have reviewed the methods for quantifying information content in maps by properly measuring the probabilities of the object distributions depicted on maps. Spatial patterns are central to proper informational analysis of map symbols, because the information content of the same type and number of objects mapped would be the same regardless of how they are distributed on the maps. The methods reviewed so far in this section are, however, implicitly oriented to maps depicting discrete spatial entities, such as points and areas. For lines, such as linear features (i.e., coastlines and drainage networks), special treatment is necessary for proper quantification of entropy and other informational measures (Fiorentino et al. 1993; Battersby and Clarke 2003; Lawford 2006). There are other types of maps, widely produced and used in the GIScience community, whose informational analysis will need to be adapted for their own characteristics. Two types of field variables are often represented on maps: interval/ratio fields (e.g., digital elevation models) and categorical fields (area-class maps). Below, we review methods for measuring information content in these two types of maps.

Consider maps of interval/ratio fields. In Section 11.2, we described mutual information-based methods for measuring the amount of information conveyed by an image about the scene being imaged, for which system parameters in image gathering (e.g., point spread function (PSF) or SFR) need to be specified in addition to the Gaussian assumption. These methods may be applied to raster maps of interval/ratio fields. One of the difficulties in extending image informational analysis to raster maps concerns the processes involved in map production, which are often undocumented. This makes the mechanistically based methods originally proposed for image gathering not straightforwardly applicable to raster maps. However, these methods may be extended usefully to raster maps with simplification and approximation, leading to an empirical or, at best, semi-mechanistic solution. A strategy is to bound the estimates of map information content by referring to Equations 11.32a and 11.32b, and to refine the estimates by accounting for spatial correlation in the fields mapped. Clearly, this strategy is mostly data driven, depending on the normality assumption and empirical parameterization of SNR, spatial correlation, and other factors. For quantifying information in upscaled maps, we need to obtain a hierarchy of upscaled maps at aggregate scales. For this, methods for upscaling described in Chapter 9 may be usefully explored.

For area-class maps, the trans-information from class-controlling covariates to the end products seems more difficult to quantify than for maps of interval/ratio fields. This is because of the often nonlinear transformations implied in the mapping processes, not to mention the burden of probabilistic modeling of conditional or joint distributions, as in the case of interval/ratio fields. Suppose *a posteriori* probability vectors of class occurrences at individual pixels are derived from the discriminant $\mathbf{Z}$-space (Goodchild et al. 2009). It is easy to compute entropy measures for class probabilities at individual pixels. To compute entropy in upscaled area-class maps, we need to obtain probability maps at aggregated scales. For this, upscaling is preferably performed through discriminant $\mathbf{Z}$-space as shown in Chapter 10. Unfortunately, computing mutual information between area-class maps and their underlying truth maps is extremely difficult, if not impossible, as true class labels are subject to a host of inherent uncertainties, as shown in Chapter 10, although mutual information between an area-class map and its class-defining $\mathbf{Z}$ may be more tractable.

In addition to the empirical methods shown above, geostatistics may also be used for information analysis of raster maps of both continuous and categorical variables, provided they are derived directly or indirectly from kriging. Based on image information models incorporating additive noise as shown in Equation 11.26b, Bishop et al. (2001) proposed using kriging prediction and variance to evaluate information content in raster maps produced from kriging. Consider kriging over a problem domain of areal extent A discretized by N_b regular blocks. Suppose block v_i is predicted to be $z(v_i)$ with kriging variance $\sigma^2(v_i)$. Then, the information within the block, which is related to SNR (Fellgett and Linfoot 1985), is

$$I(v_i) = \log_2 \frac{z(v_i)}{\sigma(v_i)} \tag{11.57a}$$

where the concept of distinguishable levels is applied implicitly.

Then, the total information of the kriged map for variable Z is

$$I = \sum_{i=1}^{N_b} n_{effobs} I(v_i) \tag{11.57b}$$

where n_{effobs} is the number of independent observations. We can further normalize the information content I with respect to the areal extent of the study area:

$$I_A = I/A \tag{11.57c}$$

where I_A is a measure of information per unit area or information density (Bishop et al. 2001). Scale effects can be examined by plotting I or I_A against the block size in kriging.

There are two different routes whereby information in area-class maps can be assessed. One is to apply Equations 11.57a to 11.57c to the results from indicator kriging, which produces both the kriged class probabilities and the associated variance. A necessary adaptation is to have the kriged class probabilities collapsed into the maximum class probability, say, p_{max}. The information measure for block v_i is

$$I(v_i) = \log_2 \frac{p_{max}(v_i)}{\sqrt{p_{max}(v_i)[1 - p_{max}(v_i)]}} \tag{11.58}$$

The information measures obtained from Equation 11.58 can be inserted into Equations 11.57b and 11.57c to derive estimates for the total amount of information from the kriged area-class map and the information density of the map, respectively.

The other approach is to compute information measures for individual map layers of the class-defining $\mathbf{Z}$. A hurdle is how we may quantify the amount of information transmission through the class-mapping algorithms, which are often nonlinear. Suppose class probability vectors $\mathbf{P}$ are derived from the classifier and the $\mathbf{Z}$ vectors. Even if source entropy $H(\mathbf{Z})$ and receiver entropy $H(\mathbf{P})$ are estimated, there is no simple way to compute conditional or joint entropies for $\mathbf{P}$ and $\mathbf{Z}$. At a fundamental level, the methods described above for quantifying information content in kriged maps are only loosely framed in Shannon's classic information theory.

Despite developments in information analysis of images and maps, as reviewed so far in this chapter, there is only a modest amount of work directly linking scale and information content. Journel and Deutsch (1993) put forward a measure of spatial entropy built on discretized bivariate probabilities over lags, which may be extended to compute multiple-point entropy. We review the concept of spatial entropy in the interest of highlighting the affinity between entropic measures and variograms in describing spatial variability and hence scale.

Consider a stationary random function (over a domain D) $Z(x)$ that is discretized into K classes (e.g., K intervals of values). Suppose that the discrete $Z(x)$ has stationary marginal probabilities $\{p_k = \text{prob } [Z(x) \in \text{class } k], k = 1, ..., K\}$ and bivariate probabilities $\{p_{kk'} = \text{prob } [Z(x) \in \text{class } k, Z(x + h) \in \text{class } k'], k, k' = 1, ..., K\}$, both of which are independent of location x and lag h. From marginal class probabilities, we can compute the entropy of Zs distributions $H(P) = -\sum_{k=1}^{K} p_k[\log p_k]$.

The entropy associated with the set of bivariate probabilities is defined as

$$H(h) = -\sum_{k=1}^{K}\sum_{k'=1}^{K} p_{kk'}(h)[\log p_{kk'}(h)] \tag{11.59}$$

It can be seen that, for lag $h = 0$, $H(0) = H(P)$, and for lag $h = +\infty$, $H(+\infty) = 2H(P)$ (Journel and Deutsch 1993). For most well-behaved aperiodic pdfs, for which $p_{kk'}(h)$ decreases continuously as $|h|$ increases in any direction, the bivariate entropy measure $H(h)$ will increase with $|h|$. Thus, the entropy measure $H(h)$ will be nondecreasing in the interval $[H(0), H(+\infty)]$. Therefore, a standardized relative measure of bivariate entropy is

$$H_R(h) = \frac{H(h) - H(0)}{H(0)} \in [0,1] \tag{11.60}$$

The upper bound or maximum relative entropy $H_R(+\infty) = 1$ corresponds to the case of maximum unpredictability due to maximum spatial disorder. On the other hand, if a particular class of z values persists over large distance h, then the distribution of z in space presents a feature or spatial order that helps in predicting the unsampled locations, as discussed by Journel and Deutsch (1993).

We can define an average bivariate entropy over a block A of measure $|A|$ as

$$\begin{aligned}\bar{H}_R(A, A) &= \frac{1}{|A|}\int_A dx \int_A H_R(x - x')dx' \\ &\approx \frac{1}{n^2}\sum_{i=1}^{n}\sum_{i'=1}^{n} H_R(x_i - x_{i'})\end{aligned} \tag{11.61}$$

where the integral is approximated by averaging through n discretization points x_i, $i = 1, \ldots, n$. Note the similarity between averaged bivariate entropy in Equation 11.61 and dispersion variance of $Z(x)$ within A, as defined by semivariance in the work of Journel and Huijbregts (1978).

11.5 DISCUSSION

It has been shown that information theory is a unique language for information processes in remote sensing and GIS as it can offer valuable insights on the gathering, processing, analysis, visualization, and applications of geographic information. The breadth and depth of information theory implies that our review of it in the context of geospatial information and analysis is hardly comprehensive. Nevertheless, it is our hope that this chapter has met the modest goal of synthesis of informational concepts and methods that have been developed from within or outside of GIScience and that it will be valuable for applications in GIScience. Below, we discuss the areas where adaptations to information theory are necessary or further developments are essential for its added value in geography.

As discussed by O'Sullivan et al. (1998), information theory provides a rigorous framework for problems of image formation in terms of imaging problem definitions (i.e., appropriate models for the image data and suitable space of image reconstruction), optimality criteria for image estimation, algorithm developments based on these criteria, and the quality evaluation of image estimation or approximation. We have shown that information analysis of the imaging processes from image gathering to display is facilitated by the combined use of information measures such as entropy and mutual information and certain sensor parameters such as PSFs and SNRs, as described in detail by Huck et al. (1996, 1999). Useful extensions with adaptations are highly anticipated in information-theoretic assessment and design for imaging systems with fine spatial, spectral, and temporal resolutions, as they result from technological advance and pressing application needs.

With examples showing implicit use of information theory in many existing visualization techniques, Chen and Jäenicke (2010) sought to formulate concepts and measurements for quantifying visual information. As argued by them, the emphasis of some traditional applications of information theory, such as those in communication, may not always suit visualization, as traditional information theory typically focuses on the efficiency of communication channels, whereas visualization focuses on the effectiveness in aiding the perceptual and cognitive process for data understanding and knowledge discovery. This suggests that research be furthered for adopting and adapting information theory for visualization of geospatial information, which is characterized by spatial dependence on the one hand and prior information on the other (Goodchild 2003).

The vast data quantity and digital format of remotely sensed images of finer spatial, spectral, and temporal resolution require developments of efficient data compression techniques that minimize unnecessary consumption of resources for data transmission, storage, and handling. This becomes a key issue in the development of web-based and distributed computing environments (Kidner and Smith 2003). Data compression is a classic application of information theory in image acquisition and processing, but has not been elaborated earlier in this chapter. Data compression is meant to be information preserving and efficient, aiming to represent a dataset with a minimum number of bits by reducing redundancy in the data. When an image is modeled as an information source to be encoded, the pixels are considered to be observations of a random variable Z with a set of possible discrete values. The entropy of source Z is useful as the lower bound on the average bit rate needed to encode the source Z. The efficient codes are those with their average codeword lengths coming close to the entropy bounds (Lempel and Ziv 1986; Chen et al. 1987). As pixel values are often statistically correlated with neighboring pixels in an image, the image should be modeled as a random field governed by both the randomness and spatial dependence of the pixel values. The conditional entropy of the image source computed, for example, from a Markov model, is interpreted as the average rate of information from the image given local pixels and sets the lower bound on the average bit rate needed to encode the spatially correlated source (Chen et al. 1987). Both spatial (intraband) and spectral (interband) correlations should be dealt with when working with multispectral and hyperspectral image data (Memon et al. 1994; Liang et al. 2012).

Clearly, despite current high-throughput and high-performance computing and storage capacity, data compression remains important for efficiency in information gathering and processing, especially when dealing with the large quantity of spatial data in image and map formats, as shown above. The effectiveness of data compression prompted the question of whether the Nyquist–Shannon sampling theorem should always be followed in data acquisition, as it provides a sufficient condition for the sampling and reconstruction of a band-limited signal (and a necessary condition to avoid aliasing if reconstruction is done via the Whittaker–Shannon interpolation formula). Indeed, if there are extra assumptions made about the signal, then the Nyquist sampling rate may no longer be a necessary condition, as demonstrated by the recent successful applications of compressed sensing, which allow for full reconstruction with a sub-Nyquist sampling rate (Candès et al. 2006; Donoho 2006). In compressed sensing, provided the signal is sparse or compressible in some domain (i.e., containing many coefficients close to or equal to zero when represented in a certain domain), it can be perfectly reconstructed if it is sampled at a rate slightly greater than the effective bandwidth (but smaller than the conventional bandwidth), in which reconstruction is completed by the solution to an optimization program for an underdetermined system of linear equations (Candès et al. 2006). There are many examples of sparse or compressible signals in geospatial applications. For example, over a large area there may only be a few targets that are of large reflectivity or of significant interest to analysts in radar imaging. Compressed sensing can be exploited for minimum sampling rate, but maximum information rate and accuracy in object detection and mapping (Herman and Strohmer 2009). Information theory will play a central role in research on and applications of compressed sensing in remote sensing through exploring scene sparsity and determination of optimal sampling rate with limiting factors such as SNR and resolution (Reeves and Gastpar 2012).

In geographic information processes, from gathering through dissemination, scale is probably one of the most central dimensions and metrics for understanding information mechanisms in GIScience and for information-optimizing spatial problem solving. As hinted at previously, spatial dependence will have complex effects on the entropy rate of the processes underlying the image acquisition and analysis, since process scale is inherently related to and dependent on the characteristics of spatial dependence. The similarity between standardized variograms and relative spatial entropy and the links between their regularized versions suggest that we may well be able to use informational measures to model and interpret scale in spatial processes and forms. Localized scale metrics may be built on information measures computed over areas of relative homogeneity. The effects of upscaling and downscaling can be examined through informational analysis of the images and maps under study, and informational graphing will be usefully explored in data aggregation and disaggregation. For example, the concept of degrees of freedom may be used as an index for the finest resolution at which informative and meaningful downscaling may be performed. Conflation of heterogeneous spatial information sources, such as maps, will benefit from information-theoretic modeling and assessment of the sources and the fusion products (Price 1999). It is inadequate to use conventional measures to evaluate the performance of different image fusion and data conflation algorithms for merging coarse spatial resolution, but fine spectral resolution images with a fine spatial resolution panchromatic image (Van der Meer 1997; Thomas et al. 2008). Although the coverage on links between information theory and scale in this chapter is far from comprehensive, it is hoped that information-theoretic concepts and their spatialized extensions described will have laid a firm foundation for further explorations of scale-related informational quantities.

The notion that entropy quantifies the uncertainty involved in a probabilistic distribution suggests the potential role of information theory in dealing with uncertainty. It is constructive to extend information theory to the discussion of spatial uncertainty, which will be the focus of the next chapter. As uncertainty is inherent in geographic information, it is important to characterize it better for increased understanding of geographic information itself. We have described additive and multiplicative noise (i.e., uncertainty in measurement) in the language of information theory, and these were shown to affect trans-information in image gathering in different ways in Section 11.2. Their effects on information content are seen as related to, but differing from, their statistical properties. Map comparison can be performed on the basis of mutual information to concentrate on the consistency between mapped classes in different maps rather than on the correctness of the map being tested (Finn 1993). When viewed from the perspective of changing scales, we can provide meaningful characterizations of the similarities and differences between area-class maps, including their spatial configuration (Remmel and Csillag 2006). Surely, information theory will shed light on uncertainty characterization and complement statistical methods for the description and analysis of uncertainty, a theme that seems to have been dominant in the literature, as will be seen in the next chapter.

12 Uncertainty Characterization

As mentioned toward the end of Chapter 11, uncertainty is inherent in spatial information and analysis. Uncertainty should thus be characterized properly. Although the information theory discussed in Chapter 11 provides a special language of reasoning in the face of uncertainty, this chapter addresses the topic of uncertainty characterization, in particular with respect to its relation to scale issues, the theme of this book.

Spatial information is derived from measurement, and all measurements are uncertain, which means that they may contain errors. There are abundant examples of errors in measurements, such as angular and distance errors in land surveying for position fixing, errors in image matching in digital photogrammetry for automatic elevation generation, and radiometric and multispectral measurement errors for feature identification and biophysical inversion in remote sensing. Thus, positioning by field surveys or satellite technology is prone to positional errors. Digital elevation models (DEMs) acquired from digital photogrammetry are subject to errors, in particular those due to inaccuracy in image correlation. Remote-sensing information products about spatial entities, land cover, and many other variables are known to be of limited accuracy. We give further detail concerning error occurrences in remote-sensing thematic mapping below.

In remote measurements, errors occur because of sensor performance, stability of the platform, and conditions of imaging. There may be systematic errors in detector response, which may be corrected through calibration (Thome et al. 1997). As shown in Chapter 3, various types of blurring occur during imaging, which, compounded by the usual limited resolution of the observing systems, contribute to errors in remote measurements and hence the resultant pixel values. There are also errors due to misregistration of equivalent pixels among multiple images and maps. Hence, geometric corrections often seek to hold the geometric mismatch to a displacement of no more than one pixel. As discussed by Goodchild et al. (2009), land cover mapping, as a prediction process, is subject to errors owing to the imperfection in class discrimination based on remote measurements because many biophysical processes underlying land cover cannot be remotely monitored with adequate accuracy owing to the difficulties of separating one class from another when both have similar spectra and, in a changing environment, discriminating real changes from natural variation (Lambin and Strahler 1994; Stehman and Czaplewski 1998; Coppin et al. 2004; Liu et al. 2005; Tso and Mather 2009). Therefore, land cover information derived from remotely sensed data is often of limited accuracy as a combined result of measurement errors and various human–machine limitations in the processes of information extraction. As land cover mapping and change detection have been extended from local, regional, and national to global scales, it is required that error in remotely sensed land cover information and its propagation in derivative products be quantified and handled adequately (Zhang et al. 2009b).

The notion of error implies that there must be a "true" or reference datum available to test a given measured or calculated value. But there are no rigorously defined truth values for spatial data, especially those concerning areal classes where inter- and intra-class variability, semantic inconsistency, and subjectivity in class labeling imply a host of compounding factors in determining true and unambiguous class labels (Foody 2010; Zhang 2010). Sometimes, the mean or modal value is taken from repeated or multiple values as a surrogate truth value. For many inherently fuzzy concepts, classes, or entities, there are no singleton truth values, but there are intervals or continua of gradation, such as an ecotone. Thus, we should extend the concept of errors and use uncertainty to cover all error and vagueness-related phenomena, so that errors are described using probability distributions, whereas fuzzy concepts, quantities, and objects are handled with fuzzy sets rather than with 0–1 crisp sets. More critically, the notion of spatial uncertainty promotes a strategy of handling inaccuracy in spatial information from a spatial perspective such that, for example, multipoint characteristics can be useful in uncertainty analysis (Zhang and Goodchild 2002; Li et al. 2012).

There has been increasing research on uncertainty in spatial information and analysis. Classification accuracy assessment and area-class error modeling have long been topics of increasing research in their respective communities (Goodchild et al. 1992; Hunsaker et al. 2001; Foody 2002; Nusser and Klaas 2003; Steele et al. 1998). Uncertainty for area-class maps can be approached from inaccuracy in position (i.e., area-class boundary placement) and class labeling, respectively, with the former handled by identifying and removing sliver polygons and the latter by constructing error matrices and computing various accuracy metrics, such as the percent correctly classified (*PCC*) and the kappa coefficient of agreement (Chrisman 1989; Foody 2002; Van Oort et al. 2004; Liu et al. 2005; Goodchild et al. 2009).

DEM errors have also been addressed extensively in the literature. As reviewed by Wilson and Gallant (2000), several methods have been proposed for the detection of errors and estimation of the magnitude and spatial distribution of errors (e.g., Brown and Bara 1994; Holmes et al. 2000). Error quantification often relies upon using topographic maps (e.g., Skidmore 1989) and field measurements (e.g., Bolstad and Stowe 1994) as reference data. The development of methods for analyzing the effects of DEM errors on terrain derivatives has been a research focus (e.g., Jenson and Domingue 1988; Carter 1992; Lee et al. 1992; Hunter and Goodchild 1997). Scale effects in DEMs and terrain analysis have also been addressed (MacEachren and Davidson 1987; Zhang and Montgomery 1994; Hodgson 1995; Gyasi-Agyei et al. 1995; Molnar and Julien 2000; Schoorl et al. 2000; McMaster 2002; Deng et al. 2007; Chow and Hodgson 2009).

Product validation is becoming increasingly integrated in operational remote sensing and spatial information production. According to Morisette et al. (2002, 2006), MODIS Land (MODLAND) product quality is ensured by calibration, quality assurance, and validation. The MODIS land validation effort helps to establish standards and protocols through close coordination with the Committee on Earth Observation Satellites (CEOS) Land Product Validation (LPV) subgroup, under its

Working Group on Calibration and Validation. MODLAND uses several validation techniques to develop uncertainty information for its products, such as leaf area index (LAI) products. These include comparisons with *in situ* data collected over a distributed set of validation test sites, comparisons with data and products from other sensors (e.g., Advanced Very High Resolution Radiometer, Thematic Mapper/ Enhanced Thematic Mapper Plus), intercomparison of trends derived from independently obtained reference data, and analysis of process model results.

In addition to error description, sometimes termed "measurement error characterization," and accuracy assessment, it is also important to assess the impacts of errors on spatial analysis and applications. This is known as error propagation (Heuvelink 1998; Zhang and Goodchild 2002; Li et al. 2012). In digital terrain analysis, we might want to evaluate the error margins in slope and aspect calculated from DEMs, which we know contain errors (Wechsler and Kroll 2006). For area-class mapping based on image classification, it is important to assess errors in areal estimates, which are derived through summation of grid cells labeled with certain land cover types, and change detection, which may be obtained by post-classification comparison or differencing of fractional cover maps, given error margins in source data (i.e., single-date classification or fractional cover) (Liu et al. 2005; Zhang and You 2011). The issue of error propagation arises because variance in class k's areal estimation based on summation of pixels (coded as indicators of class k) is rarely the sum of per-pixel variance of class k's occurrence unless interpixel dependence in class occurrence is ignored. This reasoning also applies to change detection where both spatial and temporal dependence should be accounted for in quantification of variance in class-change statistics.

Variance and covariance propagation may be applied for error propagation to quantify error in derivatives given knowledge of the variance and covariance of the input variables and the functions specifying the derivation. For error modeling in geo-processing, however, simple applications of the law of variance and covariance propagation without accounting for spatial dependence in spatial processes and geo-processing will lead to biased quantification of errors in derived data and analysis results (Liu et al. 2005; Zhang et al. 2009b), as shown above in the context of land cover mapping and change detection. Therefore, the impacts of measurement errors on spatial predictions should be assessed while accommodating spatial dependence (Rossi et al. 1993; Jager and King 2004).

Stochastic simulation may be pursued, as it generates multiple realized maps, each reproducing the same spatial structure as quantified by certain histograms and variograms and thus each representing alternative representations of the unknown "possible reality." The set of equiprobable realizations provides the basis for characterizing uncertainty in the mapped surfaces. The underlying rationale for stochastic simulation is that, since the properties at unsampled locations are largely unknown, the single kriging estimates at individual locations should be replaced by sets of alternative models that honor any sample information and reproduce a pattern that is similar to that of the true or reference field (Zhang et al. 2009a). More importantly, one of the greatest benefits of stochastic simulation is spatial uncertainty characterization, because one is seldom interested in the determination of an estimate or uncertainty model of a single unsampled location, but in the *joint* determination of all the unsampled values on a specific regular or irregular grid (Caers 2003; Zhang et al. 2009a).

Consider propagation of uncertainty in spatial modeling. The uncertainty in model prediction originates from errors in model inputs, which include measurement and interpolation errors, and in models themselves owing to conceptual, logical, and technical imperfection (Brown and Heuvelink 2005). Propagation of uncertainties in model inputs and models, through a model to outputs, can be quantified using analytical approaches based on Taylor's theorem of expansion and Monte Carlo simulation. The latter is preferred as it invokes fewer assumptions and is nonparametric. Continued improvements in computer technology and sampling techniques have allowed stochastic simulation to be applied to increasingly complex models. Nevertheless, there are many issues remaining, such as analysis of statistical dependence within and between inputs and model parameters, and assessment of structural uncertainty in models (Brown and Heuvelink 2005). More importantly, in uncertainty analysis, one of the most crucial issues concerns scale, owing to discrepancies of scales among input data, models, and applications. As the two important dimensions of geospatial information and geo-processing, scale and uncertainty are often interwoven. On the one hand, scale discrepancies complicate uncertainty characterization in multisource data and geo-processing. On the other hand, quantification of spatial uncertainty often operates at different scales, and it is necessary to analyze the scale dependence of the input and model uncertainties.

Therefore, this chapter will begin by introducing some basics about uncertainty. Then, we will address the often vexing problem of cross-scale validation, in which point or quasi-point support reference data are upscaled to facilitate the computing of error statistics. The topic of uncertainty description is followed by that of uncertainty modeling, for which we describe both analytical and simulation-based methods. Although analytical approaches based on variance and covariance propagation implicitly assume a homogenized scale for model inputs and, if required, model parameters, geostatistical simulation is more robust to scale discrepancies and should be explored in any scale-dependent modeling of uncertainty and applications. Methods for geostatistical simulation are described, which can be applied for point-support simulation with area-support data and for block simulation (univariate and multivariate) with both point and block support data, in addition to the conventional methods in which realizations are generated over grid nodes of supports conformal to those of the conditional data (implicitly, points or quasi-points). We will conclude this chapter with some remarks about future research along these lines.

12.1 ACCURACY METRICS AND ASSESSMENT

12.1.1 ACCURACY IN INTERVAL/RATIO FIELDS

Precision and bias are related concepts in spatial uncertainty characterization. Precision can be defined as the degree of conformity among a set of observations that are characterized as realizations drawn from a random variable. As such, precision is a measure of the dispersion of the probability distribution associated with a set of measured values (in practice, a repeatable measurement). Bias, on the other hand, can be defined as how close a mean estimate is to its true value (see Environmental Systems Research Institute et al. 1994 and Goodchild et al. 1994 for examples).

The *precision* of measurements is controlled by *random errors*, which are inherent in the measurement process. Repeated measurements of the same element yield different measured values, and these inconsistencies cannot be removed by refining the functional model or by applying corrections (Butler et al. 1998). With measurements viewed as random variables, a *stochastic model* can be used to assess random effects and obtain estimates of precision (Cooper and Cross 1988). The precision of stereo matching is positively related to the number of pixels that are associated with a target and the signal-to-noise ratio of the imagery (Butler et al. 1998). Systematic errors, on the other hand, determine the *bias* with which measurements (e.g., gridded DEM points derived from photogrammetric plotting) can be used to estimate the quantity sought (e.g., terrain elevation). Take DEMs as an example. A DEM's bias is evaluated and summarized by computing the mean of height discrepancies between the DEM-reported and reference elevations (Torlegård et al. 1986; Brown and Bara 1994; Nikolakopoulos et al. 2006). Last but not least, there is yet another type of error referred to as *gross errors*, which occur as blunders or mistakes, as discussed by Cooper and Cross (1988).

Ideally, estimates should be both precise and unbiased. In other words, an estimate or measurement should be as close as possible to its true value. At the same time, the degree of variability in the estimated mean value arising from repeat measurements (sampling) should be small. However, it is possible to have highly precise but biased estimates, and unbiased but imprecise estimates; the former are affected mostly by systematic errors, whereas the latter are considered to be subject to relatively large random errors alone. In accuracy assessment, both the degree of precision and the bias of the estimates derived from the sample will determine whether accuracy requirements can be considered to have been met. Usually, a representative sample of test points is used to check the data. For example, a minimum number of 28 test points per DEM is required: 20 interior points and 8 edge points, as specified in the National Map's DEM standards (http://nationalmap.gov/standards/demstds.html).

Let the true value of a field variable $Z(x)$ be $z(x)$ and define its measured or computed value as $\hat{z}(x)$. We can define error as $\varepsilon(x) = z(x) - \hat{z}(x)$. Suppose there is a sample of size n, related to n sampled locations $\{x_i, i = 1, \ldots, n\}$. The sample of mapped values can be denoted as $\hat{\mathbf{z}} = (\hat{z}(x_1), \ldots, \hat{z}(x_n))^{\mathrm{T}}$, whereas the sample of errors can be denoted as $(\varepsilon_1, \ldots, \varepsilon_n)$. Based on these, we can compute a variety of error descriptors for the mapped surface $\hat{z}(x)$ ($x \in D$). They include *RMSE* (root mean squared error), *RRMSE* (relative root mean squared error), *ME* (mean error), mean absolute error (*MAE*), mean absolute percentage error (*MAPE*), and *SD* (standard deviation), as follows:

$$RMSE = \sqrt{\sum_{i=1}^{n} \varepsilon_i^2 \Big/ n} \qquad (12.1a)$$

$$RRMSE = \sqrt{\sum_{i=1}^{n} \left(\frac{\varepsilon_i}{z_i}\right)^2 \Big/ n} \qquad (12.1b)$$

$$ME = \sum_{i=1}^{n} \varepsilon_i \Big/ n \qquad\qquad (12.1c)$$

$$MAE = \sum_{i=1}^{n} |\varepsilon_i| \Big/ n \qquad\qquad (12.1d)$$

$$MAPE = \sum_{i=1}^{n} \left|\frac{\varepsilon_i}{z_i}\right| \Big/ n \qquad\qquad (12.1e)$$

$$SD = \sqrt{\sum_{i=1}^{n} (\varepsilon_i - ME)^2 \Big/ n} \qquad\qquad (12.1f)$$

as also described by Zhang (2010).

The error measures above are explained below. By definition, *RMSE* is the square root of the mean squared error (*MSE*), which equals the sum of the variance (i.e., SD^2) and squared bias (i.e., ME^2). For *RRMSE* in Equation 12.1b, there are other variants: *RMSE* divided by the range of observed value (i.e., $RMSE/(\max(\hat{z}) - \min(\hat{z}))$, or *RMSE* normalized to the mean of the observed values (i.e., $RMSE/\mathrm{mean}(\hat{z})$). *SD* is the square root of the variance of the underlying population (hence expressed in the same units as the data) and is also called the standard deviation of the sample (which is the maximum-likelihood estimate of the true *SD* for a Gaussian population). But this estimator, when applied to a small-sized sample, tends to underestimate the true value, because it is a biased estimator. The unbiased estimator for the population standard deviation is the adjusted version, $\sqrt{\sum_{i=1}^{n} (\varepsilon_i - ME)^2 \Big/ (n-1)}$, as described in numerous statistical textbooks (e.g., Walpole et al. 1998).

As described in Chapter 6, kriging seeks to estimate the variable under study at an unsampled location given sample observations at nearby locations. It estimates the kriging variance as a byproduct of kriged surfaces. This is really valuable for error analysis. However, kriging estimates and the kriging variance do not lend themselves to quantification of spatial uncertainty (preferably through joint probability distributions) in the mapped fields, unless normality and spatial independence are assumed for the random function (RF) under study (Zhang et al. 2009a).

In statistics, uncertainty is quantified through a statistical distribution, describing the frequencies of outcomes expected (Bohling 2005). In the spatial domain, distribution functions depend on the number of sample data, their spatial configuration, the data values, and the specific spatial property under study. With distributions simulated, probability intervals or other measures of uncertainty can be calculated, as discussed by Zhang et al. (2009a). This can be taken care of by geostatistical simulation, which will be discussed later in this chapter.

12.1.2 ACCURACY IN AREA CLASSES

Many of the techniques for measuring uncertainty in area classes were developed in remote sensing, where accuracy assessment of land classifications is a key concern, as discussed by many researchers, such as Mayaux et al. (2006) and Lea and Curtis (2010). In accuracy assessment concerning area-class maps, it is common practice to select a set of sampled locations and to compare the class assigned to each location with some source of higher accuracy, usually ground reference data obtained by direct observation in the field or photo-interpretation based on large-scale aerial photographs. The results are then tabulated in the form of an error or misclassification matrix, also referred to as a contingency or confusion matrix, as will be shown below (Congalton 1991; Stehman and Czaplewski 1998; Congalton and Green 2009).

Suppose K is the total number of classes, n being the sample size. We can record the list of map and reference classes. An error matrix $\{e_{ij}, i, j = 1, ..., K\}$ can then be constructed by tabling the list and summarizing the resultant error matrix. In this error matrix, the columns define the classes in the reference data, and the rows define the classes in the data being evaluated for accuracy. The values in the cells in the table indicate the occurrences of map classes with respect to the reference classes. The diagonal elements of the matrix indicate the number or proportion of correct classifications. Therefore, an overall measure of accuracy is the percentage of cases that lie on the diagonal, known as PCC pixels (Rosenfield and Fitzpatrick-Lins 1986; Zhang and Goodchild 2002; Edem and Duadze 2004; Congalton and Green 2009). In other words, overall accuracy can be expressed as follows:

$$PCC = \frac{\sum_{i=1}^{K} e_{ii}}{e_{..}} \qquad (12.2)$$

where the subscript "." of e stands for summation across rows (if "." takes the first position) or columns (if "." takes the second position) or both as above.

To account for chance agreement between mapped and the reference classes, since a certain number of correct classifications will occur by chance, an often preferred index is the kappa index (Congalton and Green 2009). Kappa is computed as follows:

$$\kappa = \frac{PCC - P_{chance}}{1 - P_{chance}} = \frac{\dfrac{\sum_{i=1}^{K} e_{ii}}{e_{..}} - \sum_{i=1}^{K} \dfrac{e_{i.}}{e_{..}} \dfrac{e_{.i}}{e_{..}}}{1 - \sum_{i=1}^{K} \dfrac{e_{i.}}{e_{..}} \dfrac{e_{.i}}{e_{..}}} \qquad (12.3)$$

where PCC is the proportion of correctly classified entries and P_{chance} is the proportion of samples that could be expected to be classified correctly by chance, which is computed as $\sum_{i=1}^{K} \dfrac{e_{i.}}{e_{..}} \dfrac{e_{.i}}{e_{..}}$, (i.e., the sum of the products of the proportion of total

entries that are in row i and the proportion of all entries that are in column i), and K is the total number of rows or columns (Foody 2002; Zhang and Goodchild 2002).

In addition to an overall statement of accuracy, per-class accuracies can also be extracted from the error matrix, which can be differentiated into producers' and users' accuracy (also referred to as errors of omission and errors of commission, respectively). Producers' accuracy is an important measure of classification accuracy because the producers are interested in knowing how well area classes can be mapped. Producers' accuracy calculates the probability that a reference sample has been classified correctly. This quantity is computed by dividing the number of samples that have been classified correctly by the total number of reference samples in that class: $e_{ij}/e_{.j}$. Users' accuracy, on the other hand, is important for users who are interested in knowing how well an area-class map actually represents what exists in the real world. Users' accuracy indicates the probability that a mapped location is correctly labeled (relative to a ground reference). This type of accuracy is computed by dividing the number of correctly classified samples by the total number of samples that are classified as belonging to that class: $e_{ij}/e_{i.}$ (Edem and Duadze 2004; Congalton and Green 2009).

Ideally, both users' and producers' accuracy should be similar for all classes. However, users' and producers' accuracy may differ considerably among classes. Thus, both should be reported to give as complete a description of the error properties of individual classes as possible for both the users and the producers of thematic data (Edem and Duadze 2004).

Error matrices are often derived from samples that are assumed to be representative for an entire dataset. Thus, the accuracy measures computed from them are global as opposed to local, summarizing what is meant to be applicable for the whole area. But in reality, the chances of misclassification vary not only by class, but also across map extents, because misclassification errors are more likely in some areas than in others, as is the case for class-dependent accuracies (Li et al. 2012). Steele et al. (1998) advised the use of methods for estimating and interpolating classification accuracy, which they showed to be spatially varied. As shown by Van Oort et al. (2004), spatial variability in classification accuracy can be properly modeled, leading to improved estimation of pixel-level accuracy. In other words, accuracy assessment can be refined by taking such spatial heterogeneity into account.

Although error matrices and misclassification probabilities provide valuable information about classification accuracies, they may not be adequate on their own for quantifying uncertainty in derivative information and spatial queries involving the use of area-class maps. It is, however, possible to simulate error using Monte Carlo simulation and to propagate errors in area-class maps into information products obtained from error-prone maps (Zhang and Goodchild 2002; Liu et al. 2005). For instance, Goodchild et al. (1992) demonstrated the utility of simulation for error propagation based on probability vectors and spatial autocorrelation parameters with soil map examples, leading to effective description of generic problems of within-polygon heterogeneity and cross-boundary transition, typically involved in area-class maps, their spatial queries (areal estimates), and geo-processing (map overlay).

We will provide more details about geostatistical simulation that can be utilized for uncertainty characterization in area-class maps and their derivative products (Zhang et al. 2009b; Zhang and You 2011).

12.1.3 Positional Accuracy

Two-dimensional position is commonly represented by the easting and northing of a pair of Universal Transverse Mercator coordinates or of longitude and latitude coordinates and, thus, can be thought of as a combination of measurements, say, X and Y, respectively. Each measurement is subject to error. Analysis of measurement error is important for positional accuracy assessment, and many of the error measures described in Section 12.1.1 are useful here.

For example, repeated measurements made of the coordinates of a landmark on a topographic map would show variation in both X and Y coordinates, which is likely to follow a normal distribution or bell curve (Environmental Systems Research Institute et al. 1994), with most measurements clustered around the landmark's probable position. For the sake of simplifying error analysis, we may assume that errors in X and Y are uncorrelated: the direction and amount of error in one coordinate are independent of those in the other so that errors in X and Y can be treated separately and, if necessary, combined easily. The errors in both coordinates can be visualized through three-dimensional distribution curves. As mentioned previously, the dispersion of the measured coordinates reflects the imprecision of the measurements. On the other hand, the closeness of the average of all repeated measurements from the true position indicates the bias of these positional measurements: a small enough bias measure implies that the measured position is nearly unbiased.

In two dimensions, the circular Gaussian function is the distribution function for uncorrelated variates X and Y having a bivariate normal distribution and an equal standard deviation $\sigma = \sigma_x = \sigma_y$,

$$pdf(X,Y) = \frac{1}{2\pi\sigma^2} \exp\left(-[(X - \mu_X)^2 + (Y - \mu_Y)^2]/(2\sigma^2)\right) \qquad (12.4)$$

where μ_X and μ_Y represent the mean position in the X and Y directions, respectively.

Several statistics of the circular normal distribution are in common use to describe positional accuracy. Perhaps the most commonly used is the Circular Map Accuracy Standard, or CMAS, which forms the basis of the current National Map Accuracy Standard for positional accuracy (US Bureau of the Budget 1947), as also described by the accuracy assessment guidelines of the US National Biological Survey (NBS) and the National Park Service (NPS) in the context of vegetation mapping (Environmental Systems Research Institute et al. 1994). The most recent information about accuracy standards can be found on the FGDC (Federal Geographic Data Committee) website (http://www.fgdc.gov/). CMAS is defined as the 90th percentile of the circular normal distribution, or 2.146 times its standard deviation (ASPRS 1990):

$$\text{CMAS} = 2.146 \times \sigma_x \text{ or CMAS} = 2.146 \times \sigma_y \qquad (12.5)$$

where σ_x and σ_y are the standard deviation in each coordinate direction, respectively. In other words, CMAS forms a circle about the true location of the point, within which the observed location is expected to lie 90% of the time (Goodchild et al. 1991).

Positional accuracy can also be reported as a RMSE in each coordinate direction; it is a measure of inaccuracy that includes the effects of both bias and random error (Environmental Systems Research Institute et al. 1994). As in Equation 12.1f, it is computed as

$$RMSE_X = \left[\frac{1}{n} \sum_{i=1}^{n} (X_i - X_0)^2 \right]^{1/2} \tag{12.6}$$

where $RMSE_X$ is the standard error in the X-coordinate direction, n is the sample size, X_i is the measured coordinate location, and X_0 is the true coordinate location as determined in the source of higher accuracy. The RMSE for the Y-coordinate direction, $RMSE_Y$, can be computed similarly.

Positional accuracy of points can be evaluated in relative terms. For this, CMAS can be used. The accuracy with which two points can be positioned relative to each other depends on the positional accuracy of both. On the assumption of independence between errors in two points, relative positional accuracy $CMAS_{12}$ can be calculated by taking the square root of the sum of the squares of the two measures of positional accuracy $CMAS_1$ and $CMAS_2$:

$$CMAS_{12} = \sqrt{CMAS_1^2 + CMAS_2^2} \tag{12.7}$$

For example, if one point has a positional accuracy of 0.6 mm and a second point has a positional accuracy of 0.8 mm, a rough estimate of the error in their relative positions is 1.0 mm or 24 m on the ground for a 1:24,000 scale map (ASPRS 1990; Environmental Systems Research Institute et al. 1994).

However, positional accuracy in linear features is far more complicated than CMAS suggests (Tveite and Langaas 1999; Hughes et al. 2006; Lawford and Gordon 2010). Probabilistically formulated measures (e.g., Goodchild and Hunter 1997) may be more useful than conventional error descriptors for linear features, such as error band models. Furthermore, as for continuous and categorical fields, we definitely need to accommodate spatial dependence in positional data (Zhang and Goodchild 2002), especially those derived from sources and techniques whereby spatial correlation is induced during data reduction, such as semiautomatic feature extraction from fine-resolution imagery.

12.1.4 Confidence Intervals and Hypothesis Testing

As described in many statistics textbooks such as Walpole et al. (1998), a point estimate (e.g., of a mean or variance) requires an associated confidence interval to be truly useful. A confidence interval is an interval within which the true value of an estimate lies with a specific level of confidence. For example, interpretation of classification accuracy will be different between a case where 15 pixels are confirmed to be correctly classified in a test sample set of 20 pixels and another where 750 pixels out of a total of 1000 pixels are correctly classified, although both cases indicate a *PCC* measure of 75%. This is because we tend to have more confidence in the latter case as a result of its much larger test sample size. The concept of confidence interval is extremely useful, as developed in this subsection. A confidence interval may be two sided or one sided (also referred to as two tailed or one tailed). The width of a confidence interval is affected by the sample size used to derive the point estimate and by the confidence level itself. Larger sample sizes will result in a narrower confidence interval, as will lower confidence levels. Conventionally used confidence levels (denoted α in the next paragraph) are 90%, 95%, or 99% (Environmental Systems Research Institute et al. 1994; Goodchild et al. 1994). The concept of confidence intervals is very useful for accuracy assessment, as shown below.

For a population with unknown mean μ and standard deviation σ, to determine a confidence interval for the population mean, based on a simple random sample of size n, $\{z_1, ..., z_n\}$, we can use the distribution of the sample mean $\bar{z} = \dfrac{1}{n} \sum_{i=1}^{n} z_i$, which follows the t distribution with mean μ and standard deviation $SD/\sqrt{n}$, with $n - 1$ degrees of freedom, where SD is the sample standard deviation, as calculated through Equation 12.1f. Hence, we have the confidence interval for μ as

$$\bar{z} \pm t_{(1-\alpha)/2,n-1} \frac{SD}{\sqrt{n}} \tag{12.8}$$

where $t_{(1-\alpha)/2,n-1}$ is the upper $(1 - \alpha)/2$ critical value for the t distribution with $n - 1$ degrees of freedom and n is the sample size (Snedecor and Cochran 1967; Environmental Systems Research Institute et al. 1994). The sample set $\{z_1, ..., z_n\}$ mentioned above is relevant for data of continuous nature (i.e., interval/ratio fields and positional data) and is adaptable for categorical data if they are indicator transformed, for instance. However, a more useful statistical tool for accuracy assessment is hypothesis testing (Walpole et al. 1998), as follows.

The purpose of hypothesis testing is to determine whether observations (sample data) support the assumption (hypothesis) that the sample has been drawn from a population with specified parameter values, such as a normal distribution with a given mean or standard deviation. Thus, hypothesis testing can be performed to confirm whether or not a dataset conforms to a specific accuracy requirement. In a hypothesis test, we need to formulate a null hypothesis (H_0) and an alternative hypothesis (H_A). H_0 specifies the set of theoretical distribution parameters of a random variable against which the parameters estimated from the sample are to be compared. If the parameters computed from the sample are considered to be in agreement with the theoretical expectations on statistical grounds, one may conclude that H_0 is accepted as true (e.g., a dataset conforms to an accuracy criterion). Otherwise, H_0 is rejected as false and H_A is accepted with the stated confidence (e.g., a dataset fails a prescribed accuracy testing) (Environmental Systems Research Institute et al. 1994; Goodchild et al. 1994).

Consider t-testing for the difference between mean elevations in a DEM and a reference (Weih 2010). As the reference is assumed to be error free, the difference between mean elevations is also the mean of errors in the DEM relative to the reference. This can be easily seen from the data model below

$$z(x_i) = z_0(x_i) + \varepsilon(x_i) \tag{12.9}$$

where $z(x_i)$ and $z_0(x_i)$ are the recorded DEM elevation and the reference elevation at location x_i, respectively, and $\varepsilon(x_i)$ is the error in the DEM at location x_i.

Suppose the difference between means has been estimated from a sample of 25 checkpoints to be 3 m, and the sample standard deviation (i.e., SD) is 6 m. Thus, the aim of the hypothesis test is to determine whether or not a difference of 3 m in mean elevation is significantly different from a difference of 0 m in mean elevation. For this, consider the following null hypothesis H_0 and alternative hypothesis H_A.

H_0: The mean elevations of the DEM dataset and of the reference are the same.
H_A: The mean elevations of the DEM dataset and of the reference are different.

Under H_0, the random variable comparing the means is the t-statistic

$$t = \frac{\bar{z} - \bar{z}_0}{SD/\sqrt{n}} = \frac{\bar{\varepsilon}}{SD/\sqrt{n}} \tag{12.10}$$

which follows a t-distribution with $n - 1$ degrees of freedom, as also described by Environmental Systems Research Institute et al. (1994). In Equation 12.10, n is the sample size, $\bar{z}_0$ is the mean of the reference elevation, $\bar{z}$ is the sample mean of the DEM elevation, and $\bar{\varepsilon}$ is the mean of errors in the DEM dataset.

If the calculated absolute value for t exceeds the threshold of 1.71 set for a two-tailed hypothesis test at a 90% confidence level, H_0 should be rejected. Otherwise, H_0 should be accepted. For the given example, t is calculated as 2.5. H_0 should therefore be rejected. However, the same sample will not fail a two-tailed hypothesis test of H_0 at a 99% confidence level, as the threshold is then 2.80.

Consider determining conformance to an accuracy requirement for thematic accuracy, say, 80%. Suppose classification accuracy for a specific class has been estimated from a sample of 25 points, 18 (72%) of which have been properly identified. We can apply the hypothesis test to determine whether or not 72% is sufficiently low to consider it to be significantly different from the required accuracy of 80%, as described in the guidance for accuracy assessment in NBS/NPS Vegetation Mapping (Environmental Systems Research Institute et al. 1994), a collaborative program aimed to produce detailed and computerized maps of the vegetation for 250 national parks across the United States by processing Airborne Visible and Infrared Imaging

Spectrometer (AVIRIS) imagery along with ground sampling references (Xie 2008). For this, consider the null hypothesis H_0 and the alternative hypothesis H_A below.

H_0: The estimated accuracy and the required accuracy are the same.
H_A: The estimated accuracy and the required accuracy are different.

Under H_0, the random variable comparing the estimated and claimed accuracy

$$t = \frac{\hat{p} - p}{\sqrt{\dfrac{p(1-p)}{n}}} \tag{12.11}$$

will follow a t-distribution with $n - 1$ degrees of freedom (Environmental Systems Research Institute et al. 1994). In the equation, n is the sample size, p is the required accuracy, and $\hat{p}$ is the estimated accuracy.

If the calculated value for t exceeds the threshold of 1.71 set for a two-tailed hypothesis test at a 90% confidence level, H_0 should be rejected. Otherwise, H_0 should be accepted. For the given example, t is calculated as 1.0. Thus, H_0 is accepted, meaning that the classification accuracy of 72% meets the requirement for an accuracy of 80%. Although it is relatively easy to meet accuracy requirements using two-sided hypothesis testing, especially with small sample sizes, it is more difficult to satisfy the requirement that the required classification accuracy be met or exceeded (one-sided) at a given level of confidence (Environmental Systems Research Institute et al. 1994; Goodchild et al. 1994).

It is often necessary to determine the conformance of a map to some positional accuracy standards (i.e., whether or not its positional accuracy measure is equal to or exceeds the standards). For example, the ASPRS Accuracy Standards for Large-Scale Maps requires that 1:20,000 products have a standard error (in X or Y coordinate) of no more than 6 m for class 1 products (12 m for class 2 products) (Environmental Systems Research Institute et al. 1994). With a dataset of unchecked positional accuracy, it is important to test its positional accuracy by reference to standards. For this, hypothesis testing is again usefully employed.

To determine whether the estimated standard error in either the X or the Y coordinate direction (denoted SD_X or SD_Y, respectively) meets the accuracy requirement (for SD, which is the square root of variance), the null hypothesis H_0 and the alternative hypothesis H_A are formulated as follows:

H_0: The theoretical (required) standard error and the estimated standard error for a given coordinate direction are the same.
H_A: The estimated standard error is different from the required standard error.

Under H_0, the random variable defined below

$$\chi^2 = \left[\frac{(n-1)}{\sigma_X^2} SD_X^2 \right] \tag{12.12}$$

follows a chi-squared distribution with $n - 1$ degrees of freedom. In Equation 12.12, σ_X is the required accuracy standard and SD_X is the estimated standard error. Details of the statistical testing concerning positional accuracy are also described in the guidance for accuracy assessment in NBS/NPS Vegetation Mapping (Environmental Systems Research Institute et al. 1994). In addition to the χ^2 testing of positional error of random nature (based on SD), Vieira et al. (2004) applied the t-test approach for possible biases in position (i.e., systematic positional errors).

The data are accepted as meeting the accuracy requirement (i.e., H_0 is accepted) if the χ^2 statistic computed by Equation 12.12 is no greater than the threshold: $\chi^2 \leq \chi^2_{n-1,\alpha}$, with $\chi^2_{n-1,\alpha}$ being the theoretical χ^2 statistic at confidence level α with $n - 1$ degrees of freedom. For a 95% confidence level (5% significance level) and a sample size of 30 points, the maximum value that the computed χ^2 statistic may assume is approximately 43 (Walpole et al. 1998).

As implied in the descriptions above, accuracy should be assessed on a scale-specific basis (i.e., map accuracy standards are scale dependent). This is because phenomena shown on maps tend to be represented differently at different scales, and the number of different features and the detail with which they are represented decrease at smaller scales (i.e., coarser spatial resolution). The USGS, for example, guarantees that the mapped positions of 90% of well-defined points shown on its topographic map series at scales smaller than 1:20,000 will be within 0.02 inches of their actual positions on the map (see http://nationalmap.gov/gio/standards/). For maps at a scale of 1:20,000 or larger, 90% of all points will fall within 0.015 inch. For elevation data, the standards state that 90% of all contours will lie within ½ contour interval (CI), with none more than 1 CI.

In the discussions above, scale issues related to the sampling process have not been addressed. However, in practice, the samples are of finite support and will be irregularly shaped or nonuniformly spaced. The implicit requirement for the reference samples to be aligned with the data to be tested is often not met. We should advise methods to "homogenize" the data and use known techniques, such as upscaling, to make them comparable. This will be addressed in the next section.

12.2 GEOSTATISTICAL APPROACHES TO VALIDATION

We begin by introducing the MOLAND validation strategy for LAI (Morisette et al. 2006). There is a hierarchy of data layers for validation, individual measurements, elementary sampling units (ESU), and sites, upward to regional and global coverage. At the ESU level, individual measurement averaging is performed to estimate the ESU values (e.g., 10–100 measurements/ESU). Transfer functions are properly calibrated via fine spatial resolution images to estimate values at the site level (there are typically 20–100 ESUs per site). Correlation analysis between the global coarse-resolution products and the reference data is then performed to support global validation (with 50–100 sites globally).

Specifically, a two-stage nested sampling approach was adopted by Morisette et al. (2006), in which ESUs were used to capture the variability across the site extent and repeat measurements (field reference LAI measurement) within the ESU were used to capture the variability within the fine spatial resolution imagery. The smallest site extent was defined as the minimum area compatible with the resolution of the satellite product to be validated. For MODIS LAI products produced every 8 days globally at a spatial resolution of 1 km, a minimum site size is about 1 km^2, although this size is perhaps too small, considering the point spread function (PSF) and geolocation uncertainty of these sensors (Morisette et al. 2006). If the sites are relatively homogeneous, these issues can be ignored.

Once the field data have been obtained and converted to green LAI values, upscaling is performed to associate the field measurements with the spectral values from fine-resolution imagery. Upscaling is mainly based on the calibration of empirical transfer functions that establish a relationship between the average LAI values from each ESU and the multispectral values. The site-specific empirical relationships are usually established based on linear regression with vegetation indices or on multiple linear regression with top-of-the-atmosphere or top-of-canopy reflectances when available (Morisette et al. 2006). The result of regression-based upscaling is fine-resolution LAI maps estimated from fine-resolution images of 15- to 30-m spatial resolution, which is relatively consistent with the ESU extents.

Next, fine-resolution LAI maps were aggregated to match the MODIS product resolution to facilitate the comparison between the ground-based and MODIS LAI maps over a consistent statistical support area, as described by Morisette et al. (2006). Before upscaling, the satellite product and the LAI fine-resolution map were coregistered by reprojecting the fine-resolution images to match the moderate-resolution product. The LAI map was aggregated according to the apparent PSF of the MODIS product. Various processing steps, such as resampling and temporal compositing, mean that it is difficult to track how the sensor PSF relates to the ultimate product's effective PSF and that the validation process would be subject to approximation because of the often limited knowledge about the product's effective PSF.

Finally, global validation took place from the network of sites. The comparison between the aggregated fine-resolution LAI map and the satellite product should be performed over an area as large as possible containing an ensemble of sites. The sites should represent the variability and range of LAI and canopy types as observed over the earth's surface. Correlation analysis between the global-resolution products and the reference data was performed to support global validation (with 50–100 sites globally). Further detail is provided by Morisette et al. (2006).

In what follows, we will consider techniques based on block cokriging that can directly estimate the errors (i.e., the difference between the data and the assumed reference) and associated variance on a properly upscaled support.

Consider RFs $Z_1(x)$ and $Z_2(x)$, where Z_1 and Z_2 represent the same property, say, terrain elevation, but sampled by different means, with x referring to a location in the spatial domain. Suppose spatial covariance functions $\text{cov}_{Z_1}(x)$ and $\text{cov}_{Z_2}(x)$, and cross-covariance function $\text{cov}_{Z_1 Z_2}(h)$ can be defined. The aforementioned locations x refer implicitly to point or quasi-point supports. Many spatial data, such as elevation and slope, are actually defined on supports of finite size, which are assumed to be generated from finite-support random variables defined as a convolution of point-support variables. Denote such a block-support variable by $Z(v_x)$, with v_x standing for a finite support centered at x. This notation applies also to point-support variables, when a block v_x is reduced to a point x. Thus, block-support variables and their corresponding data or realizations are used as a general setting in the following text.

As shown previously (e.g., Chapter 8), the data on a block support v_α (i.e., $z(v_\alpha)$) can be considered as integrals of point-support values within their respective supports, that is, $z(v_\alpha) = \dfrac{1}{|v_\alpha|} \displaystyle\int_{v_\alpha} z(x)\,dx \approx \sum^{n_\alpha}_{s=1} b_\alpha(x_s)z(x_s)$, where $b_\alpha(x_s)$ is v_α's sampling kernel with n_α points discretizing the extent of the finite support of v_α (of area $|v_\alpha|$). The integrals may be approximated by simple averaging (i.e., $\sum^{n_\alpha}_{s=1} z(x_s)\big/n_\alpha$, with $b_\alpha(x_s) = 1/n_\alpha$, as in Chapters 6 and 8. Thus, it is possible to define the attribute mean and average covariance of Z on block supports:

$$E[Z(v_\alpha)] = \sum^{n_\alpha}_{s=1} b_\alpha(x_s)m_Z(x_s) \tag{12.13a}$$

$$\bar{c}_Z(v_\alpha, v_\beta) = \sum^{n_\alpha}_{s=1}\sum^{n_\beta}_{s'=1} b_\alpha(x_s)b_\beta(x_{s'})\,\mathrm{cov}_Z(x_s, x_{s'}) \tag{12.13b}$$

where n_α and n_β points are used to discretize the extents of the finite support v_α and v_β, respectively, and $\bar{c}_Z(v_\alpha, v_\beta)$ is the matrix populated by the average covariance between blocks v_α and v_β. In Equation 12.13b, the quantity $\mathrm{cov}_{Z_1 Z_2}(x_s, x_{s'})$ may be substituted for $\mathrm{cov}_Z(x_s, x_{s'})$ in the calculation of $\bar{c}_Z(v_\alpha, v_\beta)$, the average covariance between v_x and v_α:

$$\bar{c}_{Z_1 Z_2}(v_\alpha, v_\beta) = \sum^{n_\alpha}_{s=1}\sum^{n_\beta}_{s'=1} b_\alpha(x_s)b_\beta(x_{s'})\,\mathrm{cov}_{Z_1 Z_2}(x_s, x_{s'}) \tag{12.13c}$$

Now, consider bivariate cases whereby primary data generated from Z_1 and secondary data generated from Z_2 are involved in validation. Suppose Z_1 and Z_2 are sampled over data supports $\mathbf{V}_1$ and $\mathbf{V}_2$, which are denoted as $\mathbf{V}_1 = \{v_\alpha \in \mathbf{V}_1\colon Z_1(v_\alpha)$ known$\}$ and $\mathbf{V}_2 = \{v_\beta \in \mathbf{V}_2\colon Z_2(v_\beta)$ known$\}$, respectively, with the former usually being on smaller supports than the latter. The sets $\mathbf{V}_1$ and $\mathbf{V}_2$ are vectors of length n_1 and n_2, respectively, and, in general, different. The sets of block-support data sampled from $\mathbf{V}_1$ and $\mathbf{V}_2$ are denoted by $\mathbf{z}_1(\mathbf{V}_1) = \{z_1(v_\alpha), a = 1, \ldots, n_1\}$ and $\mathbf{z}_2(\mathbf{V}_2) = \{z_2(v_\beta), \beta = 1, \ldots, n_2\}$ (vectors of length n_1 and n_2), which are considered as realizations of RFs Z_1 and Z_2 over sets of data supports $\mathbf{V}_1$ and $\mathbf{V}_2$, respectively.

Assume $Z_1(x)$ and $Z_2(x)$ are nonstationary point RFs in the studied zone, having, as nonstationary expectations, drifts $m_1(x)$ and $m_2(x)$, respectively. Only algebraic (in) dependence between drifts $m_1(x)$ and $m_2(x)$ is considered below (Chilès and Delfiner 1999). The drifts are supposed to be linear combinations of known basis drift functions $\mathbf{tf}_1(x) = \{tf_{1_l}(x), l = 1,\ldots,L\}$ and $\mathbf{tf}_2(x) = \{tf_{2_l}(x), l = 1,\ldots,L\}$, respectively, that is,

$$m_1(x) = \sum_{l=1}^{L} wt_{1_l} tf_{1_l}(x) = \mathbf{wt}_1^T \mathbf{tf}_1(x) \tag{12.14a}$$

$$m_2(x) = \sum_{l=1}^{L} wt_{2_l} tf_{2_l}(x) = \mathbf{wt}_2^T \mathbf{tf}_2(x) \tag{12.14b}$$

where $\mathbf{wt}_1 = \{wt_{1_l}, l = 1,\ldots,L\}$ and $\mathbf{wt}_2 = \{wt_{2_l}, l = 1,\ldots,L\}$ are vectors of unknown coefficients of the known basis drift functions, with L being the index of typically monomials, as described by Chilès and Delfiner (1999). Note that, by convention, $tf_{1_l}(x)$ and $tf_{2_l}(x)$ are taken as 1 so that the constant-mean case is included in the model. Clearly, $tf_{1_l}(v_x)$ and $tf_{2_l}(v_x)$ may be obtained by averaging $tf_{1_l}(x)$ and $tf_{2_l}(x)$, respectively, over the extent of v_x.

The difference between primary and secondary means may be estimated based on the sample data, which may be written as

$$\hat{\varepsilon}(v_x) = \hat{m}_1(v_x) - \hat{m}_2(v_x)$$

$$= \sum_{\alpha=1}^{n_1} \lambda_1(v_\alpha) z_1(v_\alpha) - \sum_{\beta=1}^{n_2} \lambda_2(v_\beta) z_2(v_\beta) \tag{12.15}$$

$$= \lambda_1^T \mathbf{z}_1(\mathbf{V}_1) - \lambda_2^T \mathbf{z}_2(\mathbf{V}_2)$$

where λ_1 and λ_2 stand for the weight vectors assigned to data sets $\mathbf{z}_1(\mathbf{V}_1)$ and $\mathbf{z}_2(\mathbf{V}_2)$ in the neighborhood of v_x, respectively. It is customary for a kriging-based solution to require unbiasedness and minimum variance in the resultant estimator. The universal cokriging system for optimal weights in Equation 12.15 is (Papritz and Flühler 1994; Chilès and Delfiner 1999; Zhang, Kyriakidis, and Kelly 2009):

$$\begin{pmatrix} \bar{\mathbf{C}}_{Z_1(\mathbf{V}_1,\mathbf{V}_1)} & \bar{\mathbf{C}}_{Z_1 Z_2(\mathbf{V}_1,\mathbf{V}_2)} & \mathbf{TF}_1 & \mathbf{0}_{n_1 \times L} \\ \bar{\mathbf{C}}_{Z_2 Z_1(\mathbf{V}_2,\mathbf{V}_1)} & \bar{\mathbf{C}}_{Z_2(\mathbf{V}_2,\mathbf{V}_2)} & \mathbf{0}_{n_2 \times L} & \mathbf{TF}_2 \\ (\mathbf{TF}_1)^T & \mathbf{0}_{L \times n_2} & \mathbf{0}_{L \times L} & \mathbf{0}_{L \times L} \\ \mathbf{0}_{L \times n_1} & (\mathbf{TF}_2)^T & \mathbf{0}_{L \times L} & \mathbf{0}_{L \times L} \end{pmatrix} \begin{pmatrix} \lambda_1 \\ -\lambda_2 \\ -\psi_1 \\ -\psi_2 \end{pmatrix} = \begin{pmatrix} \mathbf{0}_{n_1} \\ \mathbf{0}_{n_2} \\ \mathbf{tf}_1(v_x) \\ -\mathbf{tf}_2(v_x) \end{pmatrix}$$

$$\tag{12.16}$$

where $\overline{\mathbf{C}}_{Z_1(\mathbf{V}_1,\mathbf{V}_1)}, \overline{\mathbf{C}}_{Z_2(\mathbf{V}_2,\mathbf{V}_2)}$, and $\overline{\mathbf{C}}_{Z_1 Z_2(\mathbf{V}_1,\mathbf{V}_2)}$ are auto- and cross-covariance matrices for Z_1 and Z_2 over data supports $\mathbf{V}_1 = \{v_\alpha\}$ and $\mathbf{V}_2 = \{v_\beta\}$, respectively; λ_1 and λ_2 represent the weight vectors assigned to Z_1 and Z_2 data, respectively; $\mathbf{TF}_1(\mathbf{V}_1) = [tf_{1_l}(\mathbf{V}_1)]$ and $\mathbf{TF}_2(\mathbf{V}_2) = [tf_{2_l}(\mathbf{V}_2)]$ are the matrices of drift functions arranged by columns l and rows of $\mathbf{V}_1$ and $\mathbf{V}_2$, respectively; ψ_1 and ψ_2 are vectors of length L, consisting of Lagrange multipliers for conditions of unbiasedness, respectively; and the $\mathbf{0}$'s are either vectors or matrices with proper dimensionality specified by subscripts.

From the solution of Equation 12.16, one can compute the difference between the means of the primary and the secondary RFs through Equation 12.15. Statistical testing of the difference between means of Z_1 and Z_2 may be performed using a t-test, as shown in Equation 12.11, with cokriged $\hat{\varepsilon}(v_x)$ and associated kriging variance (Zhang, Kyriakidis, and Kelly 2009). This geostatistical strategy for cross-scale data validation will be a useful extension to the existing practice in MODIS product validation described by Morisette et al. (2006). Since the set of blocks v_x being analyzed imply a certain support (usually in between the supports $\mathbf{V}_1$ and $\mathbf{V}_2$), there is the issue of balancing accuracy and resolution (spatial) in implementing cokriging for validating Z_2 against Z_1, as discussed in the context of downscaling in Chapter 9.

For categorical information, the validation process based on cokriging needs to be adapted. One way is to apply an indicator transform to both the reference categorical data and the data to be validated. The other way is to extend the cokriging algorithms for univariate datasets to those dealing with vector indicators. The vectors may also be probabilistic vectors, which are commonly involved in proportional data of individual classes within pixels.

12.3 ANALYTICAL APPROACHES TO ERROR PROPAGATION

Geo-processing involves certain operations on input fields or objects. As input datasets are often error prone, the output from geo-processing will also be contaminated with errors. This is referred to as error propagation, which is concerned about the effect of input variable uncertainties on the uncertainty of a function based on them (Endreny and Wood 2001; Hunsaker et al. 2001; Brown and Heuvelink 2005; Gonga-Saholiariliva et al. 2011). In statistics, this is often computed through the law of variance and covariance propagation.

Algebraically formulated geo-computing can be written as $G = f(Z)$, where G stands for output and Z for input variables. For example, slope can be calculated as proportional rise, $slope = dZ/dX$, where dZ stands for the difference between elevations along the maximum gradient and dX the corresponding horizontal distance. By the law of variance and covariance propagation, variance in slope can be calculated as $\sigma_{slope}^2 = 2\sigma_e^2(1-\rho)/dX^2$, where σ_{slope}, σ_e, and ρ stand for standard error of the slope, standard error of elevation, and correlation coefficient in elevation errors, respectively (Goodchild 1995; Zhang and Goodchild 2002). For nonlinear functions, we can linearize G using Taylor series expansion, which seeks to approximate $G(\cdot)$ using $\overline{G} + G'dZ$, where $\overline{G}$ and G' are, respectively, the mean and differential (against Z) of G. We can then apply the law of variance and covariance propagation to linearized functions. We describe this below.

Consider a set of linear functions $\mathbf{G}$ (a column vector of length L) that are related to a set of measurements

$$\underset{L1}{\mathbf{G}} = \underset{Ln}{\mathbf{G}_1} \underset{n1}{\mathbf{Z}} + \underset{L1}{\mathbf{G}_0} \tag{12.17}$$

where $\mathbf{Z} = (Z_1, \ldots, Z_n)^\mathrm{T}$ is the measurement vector, and $\mathbf{G}_1$ and $\mathbf{G}_0$ are, respectively, properly dimensioned matrix (gradients) and vector (intercepts) specifying the linear model in Equation 12.17.

The mean vector and variance–covariance matrix of $\mathbf{G}$ are

$$E(\mathbf{G}) = E(\mathbf{G}_1\mathbf{Z} + \mathbf{G}_0) = \mathbf{G}_1\boldsymbol{\mu}_\mathbf{Z} + \mathbf{G}_0 \tag{12.18a}$$

$$\begin{aligned}
\underset{LL}{\mathbf{C}_{\mathbf{GG}}} &= E\left[\left(\mathbf{G} - E(\mathbf{G}) \right)\left(\mathbf{G} - E(\mathbf{G}) \right)^\mathrm{T} \right] \\
&= \mathbf{G}_1 E\left[(\mathbf{Z} - \boldsymbol{\mu}_\mathbf{Z})(\mathbf{Z} - \boldsymbol{\mu}_\mathbf{Z})^\mathrm{T} \right] \mathbf{G}_1^\mathrm{T} \\
&= \underset{Ln}{\mathbf{G}_1} \underset{nn}{\mathbf{C}_{\mathbf{ZZ}}} \underset{nL}{\mathbf{G}_1^\mathrm{T}}
\end{aligned} \tag{12.18b}$$

where E denotes mathematical expectation, and $\boldsymbol{\mu}_\mathbf{Z}$ and $\mathbf{C}_{\mathbf{ZZ}}$ are the mean vector and variance matrix of measurement vector $\mathbf{Z}$, respectively.

Now, consider a nonlinear function of measurement vector $\underset{n1}{Z} = (Z_1, \ldots, Z_n)^\mathrm{T}$

$$G = f(Z_1, Z_2, \ldots, Z_n) \tag{12.19}$$

As above, the variance and covariance matrix of $\mathbf{Z}$ are $\mathbf{C}_{\mathbf{ZZ}}$. We would like to compute the matrix of variance of G $\mathbf{C}_{GG}$. Suppose we have an initial value set $Z^0 = \left[Z_1^0 Z_2^0 \ldots Z_n^0 \right]^\mathrm{T}$ for the measurement vector $\mathbf{Z}$, where the superscript 0's indicate approximations to the underlying variables. With the use of Taylor series expansion, G can be linearized at approximation $\mathbf{Z}^0$ to first order:

$$\begin{aligned}
G &= f\left(Z_1^0 Z_2^0 \cdots Z_n^0 \right) + \left(\frac{\partial f}{\partial Z_1} \right)_0 \left(Z_1 - Z_1^0 \right) + \left(\frac{\partial f}{\partial Z_2} \right)_0 \left(Z_2 - Z_2^0 \right) + \cdots \\
&\quad + \left(\frac{\partial f}{\partial Z_n} \right)_0 \left(Z_n - Z_n^0 \right) \\
&= \left(\frac{\partial f}{\partial Z_1} \right)_0 Z_1 + \left(\frac{\partial f}{\partial Z_2} \right)_0 Z_2 + \cdots \\
&\quad + \left(\frac{\partial f}{\partial Z_n} \right)_0 Z_n + f\left(Z_1^0 Z_2^0 \ldots Z_n^0 \right) - \sum_{i=0}^{n} \left(\frac{\partial f}{\partial Z_i} \right)_0 Z_i^0
\end{aligned} \tag{12.20}$$

where $\left(\dfrac{\partial f}{\partial Z_i}\right)_0$ is the partial differential of G with respect to variable Z_i (representing measurement), evaluated at Z_i^0 ($i = 1, \ldots, n$) (the initial approximation of Z_i).

If we use the following notations:

$$\mathbf{G}_1 = \left[\left(\frac{\partial f}{\partial Z_1}\right)_0 \left(\frac{\partial f}{\partial Z_2}\right)_0 \cdots \left(\frac{\partial f}{\partial Z_n}\right)_0\right]$$

$$G_0 = f\left(Z_1^0 Z_2^0 \ldots Z_n^0\right) - \sum_{i=0}^{n} \left(\frac{\partial f}{\partial Z_i}\right)_0 Z_i^0$$

Equation 12.20 can be expressed as

$$G = \mathbf{G}_1 \mathbf{Z} + G_0 \tag{12.21}$$

According to Equation 12.18b, the variance of G is computed as

$$\mathrm{var}_G = \mathbf{G}_1 \mathbf{C}_{\mathbf{ZZ}} \mathbf{G}_1^{\mathrm{T}} \tag{12.22}$$

We can apply variance and covariance propagation for error propagation in geo-processing, such as positional error propagation in polygonal areas and perimeters (Zhang and Goodchild 2002). Here, we discuss a model, which was employed for map error propagation by Haining and Arbia (1993), for error propagation in map subtraction or image differencing. This model is reviewed as it permits explicit accounting for scale effects in error propagation if adapted properly. However, it will become clear that analytical error propagation in the spatial domain is often demanding, with many spatial error statistics and parameters to account for, unless spatial independence and other simplifications are assumed, rendering it less suitable for uncertainty analysis.

In a problem domain D, let $\mathbf{S} = \{s(x), x\, D\}$ denote an $n \times 1$ column vector of (fixed) ground-reference values, if D is discretized into n locations x_i ($i = 1, \ldots, n$). Thus, $s(x_i)$ is the true (univariate) value associated with location x_i. Let $\mathbf{Z} = \{z_i\}$ denote the column vector ($n \times 1$) of corresponding observed values. As discussed by Haining and Arbia (1993) and Arbia et al. (1998), the Geman and Geman (1984) degradation model can be employed:

$$\mathbf{Z} = \mathbf{HS} + \mathbf{e} \tag{12.23}$$

where $\mathbf{H} = \{h_{i,j}\}$ is a fixed $n \times n$ matrix with row sums equal to unity and $h_{ij} \geq 0$ for all i and j (Ripley 1988). The vector $\mathbf{e}$ is an $n \times 1$ column vector of measurement errors.

If e is a multivariate normal random variable with mean vector μ and dispersion matrix $\sigma^2\sum$, we can write $e \sim \text{MVN}[\mu, \sigma^2\sum]$. Systematic error is represented by $\mu \neq 0$; σ^2 is a scalar measuring the size of random error, and $\sum$ here denotes an $n \times n$ correlation matrix in which the nonzero values in the off-diagonal cells of row i are associated only with the near neighbors of x_i. The structure of $\sum$ is defined by the structure of the point-spread function. Thus, the mean vector and covariance matrix of the map $\mathbf{Z}$ are

$$E[\mathbf{Z}] = \mathbf{HS} + \mu$$
$$E\left[(\mathbf{Z} - E[\mathbf{Z}])(\mathbf{Z} - E[\mathbf{Z}])^\mathrm{T}\right] = \sigma^2\sum \qquad (12.24)$$

where E, as usual, denotes mathematical expectation (Haining and Arbia 1993).

Haining and Arbia (1993) investigated error propagation in map addition, ratioing, and overlaying. We can apply this method to investigate map subtraction. Specify two maps, $\mathbf{S}_1$ and $\mathbf{S}_2$, and construct two new (observed) maps, $\mathbf{Z}_1$ and $\mathbf{Z}_2$, using Equation 12.23. e_1 and e_2 are mutually independent, and they are also assumed to be independent of the ground-reference maps. Perform the map operation on $(\mathbf{S}_1, \mathbf{S}_2)$ and also on $(\mathbf{Z}_1, \mathbf{Z}_2)$ and compare the results. To perform this, we have

$$\mathbf{Z}_\mathrm{d} = \mathbf{Z}_1 - \mathbf{Z}_2 = (\mathbf{H}_1\mathbf{S}_1 - \mathbf{H}_2\mathbf{S}_2) + (e_1 - e_2) \qquad (12.25a)$$

If we assume $\mathbf{H}_1 = \mathbf{H}_2 = \mathbf{H}$, we can rewrite Equation 12.25a

$$\mathbf{Z}_\mathrm{d} = \mathbf{H}(\mathbf{S}_1 - \mathbf{S}_2) + (e_1 - e_2) = \mathbf{HS}_\mathrm{d} + e_\mathrm{d} \qquad (12.25b)$$

where $\mathbf{S}_\mathrm{d} = \mathbf{S}_1 - \mathbf{S}_2$ and $e_\mathrm{d} = e_1 - e_2$.

The error vector is

$$e_{z_\mathrm{d}} = \mathbf{Z}_\mathrm{d} - \mathbf{S}_\mathrm{d} = (\mathbf{Z}_1 - \mathbf{Z}_2) - (\mathbf{S}_1 - \mathbf{S}_2)$$
$$= (\mathbf{H}_1 - \mathbf{I})\mathbf{S}_1 - (\mathbf{H}_2 - \mathbf{I})\mathbf{S}_2 + (e_1 - e_2) \qquad (12.26a)$$

where $\mathbf{I}$ is an $n \times n$ identity matrix, which can be simplified as

$$e_{z_\mathrm{d}} = (\mathbf{H} - \mathbf{I})(\mathbf{S}_1 - \mathbf{S}_2) + e_\mathrm{d} \qquad (12.26b)$$

when $\mathbf{H}_1 = \mathbf{H}_2 = \mathbf{H}$. Thus, the error for location x_i is given by

$$e_{z_\mathrm{d}}(x_i) = \left(z_1(x_i) - z_2(x_i)\right) - \left(s_1(x_i) - s_2(x_i)\right)$$
$$= \sum h_{ij}s_1(x_j) - \sum h_{ij}s_2(x_j) - (s_1(x_i) - s_2(x_i)) + e_\mathrm{d}(x_i) \qquad (12.26c)$$

The size of the error for location x_i is a function of the variance of $e_d(x_i)$ and the spatial correlation in the ground-reference maps (**S**), as discussed by Haining and Arbia (1993).

If the spatial scale $\{h_{ij}\}$ does not exceed the spatial scale at which the ground reference data vary (i.e., $\sum h_{ij}s_1(x_j) \approx s_1(x_i)$ and $\sum h_{ij}s_2(x_j) \approx s_2(x_i)$), then the presence of error due to **H** will not greatly influence the size of error at x_i, as can be seen from Equation 12.26c. Thus, maps (S_1, S_2) showing large spatial correlation will tend to have smaller errors than maps with small spatial correlation. If $\mathbf{H} = \mathbf{I}$ (i.e., when there are no errors due to **H**), $\mathbf{e}_{z_d}$ is independent of S_1 and S_2. Then the spatial properties of $\mathbf{e}_{z_d}$ are independent of the spatial properties of the ground reference data and depend only on the spatial properties of $\mathbf{e}_d$ (Haining and Arbia 1993; Arbia et al. 1998).

If the maps are independent, the variance–covariance matrix of $\mathbf{e}_d$ can be written as

$$\Sigma_d = \begin{bmatrix} \Sigma_1 & \mathbf{0} \\ \mathbf{0} & \Sigma_2 \end{bmatrix} \tag{12.27}$$

where Σ_1 and Σ_2 represent the variance and covariance matrices of $\mathbf{e}_1$ and $\mathbf{e}_2$, respectively. Furthermore, if we assume normality in $\mathbf{e}_1$ and $\mathbf{e}_2$, we can derive the properties for the error map $\mathbf{e}_{z_d}$. This is done by setting $\mathbf{D} = [\mathbf{I}: -\mathbf{I}]$ and $\mathbf{m} = \left[(\mathbf{H}-\mathbf{I})S_1 : (\mathbf{H}-\mathbf{I})S_2 \right]^T_{1\times2n}$. Then $\mathbf{e}_{z_d}$ is distributed as MVN[$\mathbf{Dm}$, $\sigma^2\mathbf{D}\Sigma_d\mathbf{D}^T$], where Σ_d is defined in Equation 11.27 (Haining and Arbia 1993; Arbia et al. 1998). So

$$E\left[e_{z_d}(x_i) \right] = E\left[z_1(x_i) \right] - E\left[z_2(x_i) \right] = m_i - m_{n+i}, i = 1,\dots,n$$
$$\mathrm{var}\left[e_{z_d}(x_i) \right] = \mathrm{var}\left[z_1(x_i) \right] + \mathrm{var}\left[z_2(x_i) \right], i = 1,\dots,n \tag{12.28}$$

It is likely that $\mathbf{e}_1$ and $\mathbf{e}_2$ are cross-correlated, and so are S_1 and S_2. To complicate this further, $\mathbf{e}_1$ and $\mathbf{e}_2$ may not follow Gaussian distributions. Then, analytical approaches to error propagation will become extremely complicated in applications where spatial autocorrelation and cross-variable correlation are present and should be taken care of. Furthermore, geo-processing is often formulated as logical and inferential rather than as algebraic. For example, we might want to inquire about the areal extent of arable land in a region with certain soil types. We can first derive the set of arable land (say, A_1) from the land cover map of the region. Similarly, the set of a particular soil type (denoted A_2) can be derived from the map depicting soil distribution. The answer to the question asked (A) is provided by logical AND operated on sets A_1 and A_2: $A = A_1 \cap A_2$. As A_1 and A_2 may contain various errors, we need to assess the level of uncertainty in the resultant set A. If the confidence levels for A_1 and A_2 are 85% and 80%, respectively, what level of confidence can we expect for A? To answer this, we need to formulate error calculus.

Last but not the least, geographic information systems (GIS) data are often used for spatial modeling. In predictive vegetation modeling (Van Niel et al. 2004), spatial distributions of needle leaf and broad leaf forests are correlated to topography. As topographic variables are usually derived from digital elevation data, which are themselves derived from field surveys, photogrammetric mapping, and interferometric synthetic aperture radar that are of limited accuracy, predicted vegetation distributions will be uncertain to some extent. Stochastic simulation will be necessary for uncertainty characterization in spatial modeling as well as in geo-processing, as follows.

12.4 GEOSTATISTICAL SIMULATION

12.4.1 SIMULATION OVER A POINT SUPPORT

Geostatistical simulation is a spatial extension of the concept of Monte Carlo simulation, a method for "sampling the unknown" using constraints (e.g., histograms and variograms inferred from existing data). Geostatistical simulations seek to honor or reproduce the spatial variability of data, usually characterized by a variogram model, in addition to histograms. If the simulations also honor the data themselves, they are said to be "conditional simulations" (Chilès and Delfiner 1999; Lantuejoul 2001; Bohling 2005). Instead of providing a single map as in kriging, geostatistical simulation aims to provide a set of alternative maps, termed *realizations*, that quantify the uncertainty about the joint outcomes of the unknown grid node values (Caers 2003; Zhang et al. 2009a). In the same sense as a set of outcomes drawn from a univariate distribution model represent the uncertainty about a random variable, a set of stochastically generated and equiprobable surfaces $\{z(x)^{(l)}, x \in D, l = 1, 2, \ldots, nsim\}$ represent the uncertainty about the unknown true RF $\{Z(x), x \in D\}$, where $nsim$ is the number of simulations and D is the problem domain. In this subsection, we describe the general methodology of geostatistical simulation over point supports with spatial structural models and, if necessary, conditional data that are also prescribed over point supports. Simulation of interval/ratio RFs is discussed first, followed by indicator approaches.

Recall that kriging estimates both the mean and the standard deviation of the variable at each grid node. Assume that the random variable at each grid node x follows a normal (Gaussian) distribution. Rather than choose the mean as the estimate at x, simulation chooses a random deviate from this normal distribution, selected according to a uniform random number in the interval [0, 1] representing the probability level (Zhang et al. 2009a). For example, if kriging provided a mean estimate of 12.0 with an estimated variance of 3.0^2, and if a uniform random number of $p = 0.84$ were generated for this grid node, then the realized Z value would be about $12.0 + 1 \times 3.0 = 15.0$, by referring to the quantile of the cumulative normal probability function (CDF) (as the quantile of $0.84 (= 0.16 + 0.68 = 1.00 - 0.16)$ is mean + 1 standard deviation approximately). Bohling (2005) provided a helpful illustration for stochastic simulation.

We use a random path to avoid artifacts induced by walking through the grid in a regular fashion. We include previously simulated grid nodes as "data" in order to preserve the proper covariance structure between the simulated values (Bohling 2005). When the data do not follow a Gaussian distribution, a normal score transform

(Equation 12.29a) is performed before simulation, followed by a back-transform (Equation 12.29b):

$$Z_G(x) = \phi\{Z(x)\} \tag{12.29a}$$

$$Z(x) = \phi^{-1}\{Z_G(x)\} \tag{12.29b}$$

This normal score transform ϕ needs to be monotonic or one-to-one and invertible so that the simulation results expressed as z_G values can be restored in the original z sample data units (Rossi et al. 1993).

To ensure that the correct spatial structure is reproduced, one needs to define the joint probability model of properties at all grid locations taken together, not one by one as done in kriging. A joint distribution is defined as

$$F(z_1, z_2, \ldots, z_N) = \Pr(Z(x_1) \leq z_1, \ldots, Z(x_N) \leq z_N) \tag{12.30a}$$

where N is the number of grid nodes. Since it is mathematically too difficult to define and draw from a joint distribution model, one relies on the fact that any joint distribution can be decomposed into a product of N conditional distributions:

$$\Pr\left(Z(x_1) \leq z_1, \ldots, Z(x_N) \leq z_N\right) = \Pr\left(Z(x_N) \leq z_N | Z(x_1) \leq z_1, \ldots, Z(x_{N-1}) \leq z_{N-1}\right) \times \cdots$$
$$\times \Pr\left(Z(x_2) \leq z_2 | Z(x_1) \leq z_1\right) \times \Pr\left(Z(x_1) \leq z_1\right)$$

$$\tag{12.30b}$$

which implies that drawing from a joint distribution is equivalent to drawing from N univariate conditional distributions, as described by Caers (2003) and Zhang et al. (2009a). This lays down the basis for sequential simulation whereby each node is simulated based on previously simulated nodes:

$$\Pr\left(Z(x_1) \leq z_1, \ldots, Z(x_N) \leq z_N | (n)\right) = \Pr\left(Z(x_N) \leq z_N | (n+N-1)\right) \times \cdots$$
$$\times \Pr\left(Z(x_2) \leq z_2 | (n+1)\right) \times \Pr\left(Z(x_1) \leq z_1 | (n)\right) \tag{12.31}$$

where conditional data at a grid node are denoted as (n) from the start or as $(n + N - 1)$ at the final grid node.

According to Caers (2003), the sequential simulation algorithms for simulating a single realization are outlined as follows:

1. Assign any conditioning data (n) to the grid.
2. Define a random path visiting all nodes $\{x_i\}$ in the grid.
3. Construct a conditional distribution $F_Z(x_i, z | (n + i - 1)) = \Pr(Z(x_i) \leq z | (n + i - 1))$, where i refers to a sequence of grid nodes being simulated (so $(n + i - 1)$

indicates the set of conditional data when simulation is being performed at node x_i).

4. Draw a simulated value $z(x_i)$ from the conditional distribution $F_Z(x_i, z|(n + i - 1))$.

5. Add the simulated value $z(x_i)$ to the dataset $(n + i - 1)$.

6. Continue to the next node along the random path until all nodes are exhausted.

Kyriakidis et al. (1999) proposed a geostatistical method for integrating elevation estimates derived from a portion of a USGS 1° DEM and elevation measurements of higher accuracy (e.g., elevation spot heights). Conditional simulation was performed for generating alternative, equiprobable surfaces regarding the unknown reference elevation values using both spot heights and DEM data. These numerical models reproduced the hard elevation data at their measurement locations, and a set of auto- and cross-covariance models quantified the spatial correlation between the two sources of information at various spatial scales. From this set of alternative representations of the reference elevation, various error statistics may be derived, such as joint uncertainty associated with spatial features observed in the DEM. Such metadata would be valuable for decision making concerning subareas where more accurate elevation measurements are required.

Now, we consider indicator stochastic simulation. From an original interval/ratio random variable Z, one can use the so-called indicator RFs as introduced in Chapter 4 to generate a set of indicator random variables (i.e., for various thresholds z_k ($k = 1, \ldots, K$), one defines a set of indicator variables $I(x, z_k) = 1$ (if $Z(x) \le z_k$); 0 otherwise). As described in Chapter 4, a spatial measure of connectivity between any two points in space separated by a lag h is the indicator variogram, $\gamma_1(h, z_k) = \text{Var}(I(x, z_k) - I(x + h, z_k))$, which can be modeled from the sample data by transforming each sample into a vector of size K of indicator data (zeros and ones).

Once the K indicator variograms are determined, one can perform indicator simulation. Indicator simulation relies on the simple rule in indicator kriging that $E = [I(x, z_k)] = \text{Pr}(I(x, z_k) = 1) = \text{Pr}(Z(x) \le z)$ (thus $E[I(x, z_k)|(n)] = \text{Pr}(Z(x) \le z|(n))$, as described in Chapter 6. This means that kriging an indicator variable is actually a quantification of the local uncertainty about the original Z variable. We might define thresholds, for example, at the 10th, 25th, 50th, 75th, and 90th percentiles of the data distribution, and perform kriging of the indicator values for a series of thresholds, building up a CDF at each grid node (Caers 2003).

If there are n sampled data at location x_α, we can code them as indicators $i(x_\alpha; z_k)$ for K thresholds or cutoffs: z_k, $k = 1, \ldots, K$. The conditional probability distribution at any unsampled location is found by computing the expected value of the indicators, z_k, across all cutoffs, $k = 1, \ldots, K$, using all data, $\alpha = 1, \ldots, n$:

$$\text{Prob}\left\{ Z(x) \le z_k \middle| Z(x_\alpha) = z_\alpha, k = 1, \ldots, K; \ \alpha = 1, \ldots, n \right\}$$
$$= E\left\{ I(x; z_k) \middle| I(x_\alpha; z_k) = i(x_\alpha; z_k), k = 1, \ldots, K; \ \alpha = 1, \ldots, n \right\} \quad (12.32a)$$

Rather than using all $K \cdot n$ conditioning data for each indicator simulation, we can adopt the following approximation by using only the indicators of the same threshold (Isaaks 1990; Caers 2003):

$$
\begin{aligned}
\text{Prob}&\left\{ Z(x) \le z_k \middle| Z(x_\alpha) = z_\alpha, \alpha = 1, \ldots, n \right\} \\
&\approx E\left\{ I(x; z_k) \middle| I(x_\alpha; z_k) = i(x_\alpha; z_k), \alpha = 1, \ldots, n \right\}
\end{aligned}
\tag{12.32b}
$$

Sequential indicator simulation (SIS) is very similar to sequential Gaussian simulation, except that indicator kriging is used to build up a discrete cumulative density function for the individual categories at each case and the node is assigned a category selected at random from this discrete CDF. Sequential indicator simulation works in the following steps (Caers 2003):

1. Assign the data (n) to the grid.
2. Define a random path visiting all nodes x.
3. Loop over all nodes x_i.
   - Construct a conditional distribution $F(x_i, z|(n + i - 1))$ by estimating $\Pr(Z(x) \le z_k|(n + i - 1))$ for various thresholds using indicator kriging.
   - Draw a simulated value $z(x_i)$ from the conditional distribution $F(x_i, z|(n + i - 1))$.
   - Add the simulated value to the dataset $(n + i - 1)$.
4. End simulation.

For an originally categorical RF $C(x)$, an indicator transform can be formulated as $I_k(x_\alpha) = 1$ if class k is present at x_α (i.e., $C(x_\alpha) = k$), or 0 if not, as shown in Chapter 4. There will be one indicator variable for each of the K different classes. We can then use kriging (based on indicator variograms) to produce a set of class membership probabilities at each grid point, build up a CDF from the probabilities (probabilities across all K classes will sum to 1.0), and select a class at random from the CDF (Goovaerts 1996, 1997; Goodchild et al. 2009).

Stochastic indicator simulation can be extended for assessing accuracy in thematic mapping based on remote sensing. As shown by Kyriakidis and Dungan (2001), it is possible to apply geostatistical simulation to generate multiple alternative realizations of the spatial distribution of the better-accuracy class labels over the study area. All simulated realizations are consistent with the available pieces of information, both hard (ground surveyed) and soft (image-based spectral class membership values) data. The simulated alternative class label representations can be used for assessing joint spatial accuracy (i.e., classification accuracy regarding entire spatial features read from the thematic map). Such realizations can also serve as input map layers to spatially explicit ecological models to drive the modeling of uncertainty regarding the ecological model's predictions.

12.4.2 Point-Support Simulation Conditional to Areal Data

Stochastic simulation conditional to areal data generates equiprobable, point-support realizations of Z, which should reproduce a point-support histogram, a point-support variogram model, and, when upscaled, the available areal data (Kyriakidis and Yoo 2005). Such point-support-simulated realizations can be very useful for uncertainty characterization in fine spatial resolution spatial models with data (model inputs) and model parameters inferred from them (both at coarser spatial resolution). Also, in parallel to change of support in kriging, as described in Chapter 6, it is of theoretical and practical interest to describe methods for geostatistical simulation with change of support. What follows concerns downscaling (i.e., area-to-point) simulation.

In principle, the stochastic simulation methods described in the previous section can be adapted for area-to-point simulation without much difficulty. The bottleneck may be the necessity of implementing kriging with both point and areal data along the random path during simulation runs. As a simulation run progresses along the fine grid nodes, kriging should be able to account for the existing areal data and previously simulated point data in the neighborhood (Kyriakidis and Yoo 2005). For insight into how we may overcome the hurdle of heterogeneous data supports in simulation, we consider a classical strategy for conditional simulation based on error simulation in which kriging plays the role of conditioning, following Kyriakidis and Yoo (2005).

Recall that area-to-point kriging has been described in Chapter 6. The area-to-point predictions $\hat{\mathbf{z}}(\mathbf{x})$ ($\mathbf{x}$ being the set of n points having been predicted) are coherent, as they reproduce the available areal-support data, $\mathbf{z}(\mathbf{v})$ ($\mathbf{v}$ denoting the set of areal data defined over N areas, $\{v_\beta, \beta = 1, ..., N\}$). Denote $\mathbf{B}$ as an $N \times n$ matrix of sampling kernels for linking the areal and point data, which consists of a convolution kernel vector $(1 \times n)$ $\mathbf{b}_\beta$ for each support v_β ($\beta = 1, ..., N$). The coherence means

$$\mathbf{B}\hat{\mathbf{z}}(\mathbf{x}) = \mathbf{z}(\mathbf{v}) \tag{12.33}$$

Thus, as proved in Chapter 6, we have $\mathbf{m}_{\hat{z}} = \mathbf{m}_z$, meaning that the mean of area-to-point kriging maintains the mean of the point-support RF Z. As shown by Kyriakidis and Yoo (2005), it can be proved that the point predictions are uncorrelated with the prediction errors so that the $(n \times n)$ matrix $\mathbf{C}_{\hat{z}, \hat{z}-z}$ of (cross-)covariance values between all pairs of point predictions and prediction errors is the zero matrix:

$$\mathbf{C}_{\hat{z}, z-\hat{z}} = \mathbf{C}_{\hat{z}z} - \mathbf{C}_{\hat{z}} = \mathbf{0} \tag{12.34}$$

For conditional simulation based on error simulation, consider the decomposition of the $(n \times 1)$ vector $\mathbf{z}$ of true point-support values:

$$\mathbf{z} = \hat{\mathbf{z}} + (\mathbf{z} - \hat{\mathbf{z}}) = \hat{\mathbf{z}} + \mathbf{e} \tag{12.35}$$

where $\mathbf{e}$ denotes an $(n \times 1)$ vector of prediction error. Since $\mathbf{m}_{\hat{z}} = \mathbf{m}_z$, the prediction error component $\mathbf{e}$ has zero mean. In addition, by Equation 12.34, error component $\mathbf{e}$ is uncorrelated with the kriging component $\hat{z}$. It follows that the $(n \times n)$ covariance matrix $\mathbf{C}_Z$ of the point-support values can be decomposed as (Kyriakidis and Yoo 2005):

$$\mathbf{C}_Z = \mathbf{C}_{\hat{z}} + \mathbf{C}_{Z-\hat{z}} \tag{12.36}$$

Recall the conventional error simulation technique in the stochastic simulation literature (Journel and Huijbregts 1978). It simulates the prediction error incurred by kriging, by repeating the kriging procedure using a simulated dataset with the same configuration as the original conditioning data. The simulated data (observation locations) are extracted from an unconditional simulation using the covariance model $C_Z(h)$. With both kriging (using the simulated dataset) and the unconditional simulation results available (simulation locations), one can derive a realization of the simulated prediction error (Kyriakidis and Yoo 2005).

To extend this paradigm into area-to-point simulation, we can generate a realization of point-support values (conditional to areal data $\mathbf{z}(\mathbf{v})$ and with covariance matrix $\mathbf{C}_Z$) and add a simulated error component (with covariance matrix $\mathbf{C}_{Z-\hat{z}}$) to the area-to-point predictions $\hat{\mathbf{z}}(\mathbf{x})$ (with covariance matrix $\mathbf{C}_{\hat{z}}$), as demonstrated by Kyriakidis and Yoo (2005). An overview is provided in the following paragraphs.

Let $\mathbf{z}^{(s)}$ denote an $(n \times 1)$ vector of unconditionally simulated point-support values with stationary mean vector $\mathbf{m}_z^{(s)}$ (set to equal to $\mathbf{m}_z$) and covariance matrix $\mathbf{C}_Z^{(s)} = \mathbf{C}_Z$. By convolving the simulated point values within each support v_β with the convolution kernel vector $(1 \times n)$ $\mathbf{k}_\beta$, one can generate an $(N \times 1)$ vector $\mathbf{z}(\mathbf{v})^{(s)}$ of simulated areal-support data with the same support configuration as the original data $(\mathbf{z}(\mathbf{v}))$:

$$\mathbf{z}(\mathbf{v})^{(s)} = \mathbf{B}\mathbf{z}^{(s)} \tag{12.37}$$

The mean vector of simulated areal data is given by $\mathbf{m}_{z(\mathbf{v})^{(s)}} = \mathbf{B}\mathbf{m}_{z^{(s)}}$, and the covariance matrix of simulated areal data is $\mathbf{C}_{z(\mathbf{v})^{(s)}} = \mathbf{B}\mathbf{C}_{z^{(s)}}\mathbf{B}^T = \mathbf{B}\mathbf{C}_Z\mathbf{B}^T = \mathbf{C}_{z(\mathbf{v})}$ (Kyriakidis and Yoo 2005).

By an error-simulation-based simulation strategy, the simulated values, conditioned (hence the subscript "c") on the areal-support data vector $\mathbf{z}(\mathbf{v})$, are

$$\mathbf{z}_c^{(s)} = \hat{\mathbf{z}} + \mathbf{e}^{(s)} = \hat{\mathbf{z}} + (\mathbf{z}^{(s)} - \hat{\mathbf{z}}^{(s)}) \tag{12.38}$$

where $\hat{\mathbf{z}}^{(s)}$ is an $(n \times 1)$ vector of area-to-point simple kriging predictions obtained from the simulated areal-support data $\mathbf{z}(\mathbf{v})^{(s)}$ of Equation 12.37, and $\mathbf{e}^{(s)} = \mathbf{z}^{(s)} - \hat{\mathbf{z}}^{(s)}$ is an $(n \times 1)$ vector of simulated prediction error values. Below, we verify that the conditionally simulated point-support Z surfaces reproduce the point-support covariance model $C_Z(h)$ and area-support conditional data.

The mean of the conditionally simulated point values is

$$E\left(\mathbf{z}_c^{(s)}\right) = E(\hat{\mathbf{z}}) + \left(E(\mathbf{z}^{(s)}) - E(\hat{\mathbf{z}}^{(s)})\right) = \mathbf{m}_Z \tag{12.39}$$

and the vector $\mathbf{z}^{(s)}$ of point-support simulated values has covariance matrix $\mathbf{C}_Z^{(s)} = \mathbf{C}_Z$. This entails that the simulated kriging component $\hat{\mathbf{z}}^{(s)}$ has the same covariance matrix as that of the point predictions $\hat{\mathbf{z}}(\mathbf{x})$, that is, $\mathbf{C}_{\hat{\mathbf{z}}^{(s)}} = \mathbf{C}_{\hat{\mathbf{z}}}$. Consequently, the resulting simulated prediction error component $\mathbf{e}^{(s)} = \mathbf{z}^{(s)} - \hat{\mathbf{z}}^{(s)}$ has the covariance matrix of prediction errors, that is, $\mathbf{C}_{\mathbf{z}^{(s)} - \hat{\mathbf{z}}^{(s)}} = \mathbf{C}_{Z-\hat{Z}}$, and is uncorrelated with the kriging component $\hat{\mathbf{z}}$. Therefore, the covariance matrix $\mathbf{C}_{\mathbf{z}_c^{(s)}}$ of the conditionally simulated vector $\mathbf{z}_c^{(s)}$ can be written as

$$\mathbf{C}_{\mathbf{z}_c^{(s)}} = \mathbf{C}_{\hat{Z}} + \mathbf{C}_{\mathbf{z}^{(s)} - \hat{\mathbf{z}}^{(s)}} = \mathbf{C}_{\hat{Z}} + \mathbf{C}_{Z-\hat{Z}} = \mathbf{C}_Z \tag{12.40}$$

meaning that the conditionally simulated point-support values reproduce the point-support covariance model $C_Z(h)$. Further detail is provided by Kyriakidis and Yoo (2005).

It is also straightforward to prove that the resulting area-to-point conditionally simulated realization $\mathbf{z}_c^{(s)}$ reproduces the areal-support data. Indeed, we can deduce from Equation 12.38:

$$\mathbf{B}\mathbf{z}_c^{(s)} = \mathbf{B}\hat{\mathbf{z}} + \mathbf{B}\mathbf{z}^{(s)} - \mathbf{B}\mathbf{C}_{zz(\mathbf{v})}\mathbf{C}_{z(\mathbf{v})}^{-1}\mathbf{z}(\mathbf{v})^{(s)} \tag{12.41a}$$

With the use of Equations 12.33 and 12.37, the above expression becomes

$$\mathbf{B}\mathbf{z}_c^{(s)} = \mathbf{z}(\mathbf{v}) + \mathbf{z}(\mathbf{v})^{(s)} - \mathbf{C}_{z(\mathbf{v})}\mathbf{C}_{z(\mathbf{v})}^{-1}\mathbf{z}(\mathbf{v})^{(s)} = \mathbf{z}(\mathbf{v}) \tag{12.41b}$$

Thus, any areal-support datum $z(v_\beta)$, defined as the convolution of point values within v_β with the sampling function $\mathbf{b}_\beta$, can be reproduced exactly by the resulting area-to-point simulated realizations (Equation 12.38) (Kyriakidis and Yoo 2005).

12.4.3 SIMULATION OVER BLOCKS

As opposed to the intent of the previous section, it is often necessary to be able to simulate RFs over block supports if available data are of point or quasi-point support. This requires modeling techniques that can account for the change of support. As another example, Deutsch (2006) described stochastic simulation of facies or geologic units, employing sequential indicator simulation, in which soft secondary data (on areal support) coming from geological interpretation or geophysical measurements can be incorporated to constrain indicator simulation.

As reviewed by Boucher and Dimitrakopoulos (2009), a simple approach to account for such a change of support consists of simulating dense grids of points that are subsequently averaged to estimate the simulated block support values needed. This process has two computational drawbacks: the need to process, store, and manage large sets of data and files, and time-consuming operations (typically averaging).

An alternative simulation method is termed direct block simulation, an extension of the block simulation described by Journel and Huijbregts (1978), in which the information stored in memory is minimized by retaining block values only in the physical memory. This procedure significantly speeds up the simulation process and reduces data storage requirements (Boucher and Dimitrakopoulos 2009).

Here, we first describe a variation of sequential Gaussian simulation developed by Emery and Ortiz (2011), in which each block value is simulated in turn, conditionally to the original data and to the previously simulated block values. In block simulation, there is no need to store point-support simulations. The algorithm proposed by Emery and Ortiz (2011) proceeds as follows:

1. Divide the problem domain D into nonoverlapping blocks.
2. Select a block v in the domain among the blocks not yet simulated, in a regular sequence or randomly.
3. Simulate the Gaussian random field Z_G at a set of nodes $\{x_i, i = 1, ..., n_v\}$, discretizing block v, conditionally to the data located in and around v.
4. Back-transform and regularize the values simulated in step 3:

$$z(v) \approx \sum_{i=1}^{n_v} w_i z(x_i) = \sum_{i=1}^{n_v} w_i \phi^{-1}(z_G(x_i)) \qquad (12.42)$$

where w_i's are the weights, which can be set uniformly, that is, $1/n_v$, or by kriging $z(v)$ from $z(x_i)$'s.
5. Summarize the point-support Gaussian values simulated in step 3 by a weighted sum:

$$z_G(v) = \sum_{i=1}^{n_v} \lambda_i z_G(x_i) \qquad (12.43)$$

where λ_i's are the weights attached to $z_G(x_i)$, with an equal weighting suggested.
6. Include the Gaussian variable $z_G(v)$ (on block support) into the set of conditioning data.
7. Go back to step 2 until all the blocks to be simulated have been exhausted.

This section has discussed univariate block simulation. As hinted previously, ground samples and remote-sensing measurements are often available. It is thus interesting to consider integrating multisource data via cokriging techniques and to assess spatial uncertainty in multivariate analysis. Co-simulation can play an important role therein.

Various joint simulation approaches have been researched for cross-correlated multivariable properties. Chilès and Delfiner (1999) describe a method based on the model of linear co-regionalization and conditioning of simulated correlated fields. Myers (1988) extends the conditional univariate LU decomposition method of Davis (1987) to joint simulation. The major drawbacks of these methods are that they require inference of cross-variograms and are often computationally expensive. An alternative is to factorize the variables involved to uncorrelated (orthogonal) factors, which can then be simulated independently. Simulated factors are subsequently back-transformed. However, conventional orthogonalization transformations, such as principal component analysis (PCA), de-correlate variables only at lag zero and require strong assumptions on the type of co-regionalization, as reviewed by Boucher and Dimitrakopoulos (2009).

Desbarats and Dimitrakopoulos (2000) proposed an improvement to the PCA approach by replacing PCA with minimum/maximum autocorrelation factors (MAFs), a factorization method developed for remote-sensing applications (Switzer and Green 1984). An important feature of MAFs is that it produces uncorrelated factors at all lags when the variogram model of the related variables follows a linear model of co-regionalization (Vargas-Guzmán and Dimitrakopoulos 2003). The efficiency of joint simulation with MAFs could be further enhanced if simulations are carried out directly on block support (Boucher and Dimitrakopoulos 2009).

In the remainder of this section, we review the MAF approach to the joint direct block simulation of multiple variables, as described by Boucher and Dimitrakopoulos (2009). Consider a stationary and ergodic non-Gaussian vector RF $\mathbf{Z}(x) = [Z_1(x), ..., Z_K(x)]^T$ representing K RFs on point support in the problem domain D. Suppose $\mathbf{Z}(x)$ has zero mean and covariance matrix $\mathbf{C}(h)$ or variogram matrix $\Gamma(h)$. Furthermore, assume that the K RFs are all measured at every sample location. For non-Gaussian $\mathbf{Z}(x)$, a normal transform (point-wise) is performed to obtain Gaussian vector RF $\mathbf{Z}_G(x)$

$$\mathbf{Z}_G(x) = \left[Z_{G_1}(x), ..., Z_{G_K}(x) \right]^T = \left[\phi_1\left(Z_1(x)\right), ..., \phi_K\left(Z_K(x)\right) \right]^T \quad (12.44)$$

Normal transformation of block-support RF $Z_{G_V}^k(v)$ ($v \in D$) is performed through averaging of the N point-support variables $Z_G^k(x_i)$ ($i = 1, ..., N$) discretizing block v

$$Z_{G_V}^k(v) = \frac{1}{N} \sum_{i=1}^{N} Z_G^k(x_i), \; x_i \in v, \; \forall i, k = 1, ..., K \quad (12.45)$$

As normal score transforms ϕ_k ($k = 1, ..., K$) are usually nonlinear, their inverse $Z^k(x) = \varphi_k^{-1}\left(Z_G^k(x)\right)$ ($k = 1, ..., K$) is known only at the point support. Thus, the block value $Z_V^k(v)$ is obtained by averaging the back-transformed point-support $\varphi_k^{-1}\left(Z_G^k(x)\right)$ discretizing block v:

$$Z_V^k(v) = \frac{1}{N} \sum_{i=1}^{N} \varphi_k^{-1}\left(Z_G^k(x_i)\right), \; x_i \in v, \; \forall i, k = 1, ..., K \quad (12.46)$$

(for details, see Boucher and Dimitrakopoulos 2009).

As described by Boucher and Dimitrakopoulos (2009), point-support multivariate simulation works via a vector of MAFs, $\mathbf{F}(x) = [F_1(x), \ldots, F_K(x)]^T$. They are defined as a linear combination of the multi-Gaussian vector RF $\mathbf{Z}_G(x)$ with a set of weights given by the orthogonalization coefficients matrix $\mathbf{O}$:

$$\mathbf{F}(x) = \mathbf{O}^T\mathbf{Z}_G(x) = \mathbf{O}^T\phi(\mathbf{Z}(x)) \tag{12.47}$$

The orthogonalization coefficients matrix $\mathbf{O}$ is composed of the eigenvectors of the following matrix (Switzer and Green 1984; Vargas-Guzmán and Dimitrakopoulos 2003)

$$2\Gamma_{Z_G}(h)\mathbf{C}_{Z_G}^{-1} = \mathbf{O}^T\Lambda\mathbf{O} \tag{12.48}$$

where $\mathbf{C}_{Z_G}$ is the variance/covariance matrix of $\mathbf{Z}_G(x)$ and $\Gamma_{Z_G}(h)$ is the variogram matrix at lag h, Λ being eigenvalues.

The steps in MAF-based point-support simulation are as follows (Boucher and Dimitrakopoulos 2009):

1. Transformation of the data to MAFs with Equations 12.47 and 12.48
2. Independent conditional simulation of each MAF with any point-support simulation algorithm
3. Back-transformation of the generated MAF realizations

For step 2 above, since the K RFs $\mathbf{F}(x) = [F_1(x), \ldots, F_K(x)]^T$ in Equation 12.47 are orthogonal by construction, each can be simulated independently with a univariate simulation method. The direct block simulation algorithm consists of visiting all the blocks in the domain randomly and simulating the N discretizing points for each block and for each factor $F_k^*(x_i)$ ($i = 1, \ldots, N$; $k = 1, \ldots, K$) conditional to both point and block data (Boucher and Dimitrakopoulos 2009). Once the MAFs on point support are simulated, the algorithm splits into two: (1) the simulated point-support values are averaged into the simulated block vector $\mathbf{F}_V^*(v)$ that is used for further conditioning along the simulation path, and (2) the simulated vector block value $\mathbf{Z}_V^*(v)$ at v, $\mathbf{Z}_v^*(v) = \left[Z_v^{1^*}(v), \ldots, Z_v^{K^*}(v) \right]^T$, is calculated with Equation 12.46 while only the K RFs $F_{V_k}^*(v)$ ($k = 1, \ldots, K$) are kept in memory for the sequential simulation process.

12.5 DISCUSSION

For error analysis, the classic probability theory is enormously important, as amply demonstrated in this chapter. Probability theory is the language of uncertainty here, because we have assumed implicitly that the spatial phenomena (both quantitative and qualitative properties) and objects are well defined and have footprints definable by Boolean logic (i.e., a location belongs to the object or not). Because of measurement errors in positioning, the locations defining the object become probabilistically defined. For continuous variables, we can derive probability distributions for the true values so that confidence intervals can be derived. For classification, we can assess accuracy through construction of error matrices and computation of accuracy measures, such as *PCC*, which are interpreted probabilistically.

Although not covered in this chapter, fuzzy sets and fuzzy logic are useful tools for describing and reasoning with spatial vagueness, a kind of spatial uncertainty. Fuzzy membership values convey an ordinal ranking of the memberships, so we can tell that a membership value of 0.2 is less than 0.5. Fuzziness may occur because of inexactness, inconsistency, ambiguity, and nonspecificity. For instance, there exists a great variation across soil classification schemes in the world. Even seemingly identical class names may actually have different definitions (Lagacherie et al. 1996). Further research on fuzziness, especially that in association with scale, would be a valuable extension to probabilistic analysis and interpretations of uncertainty.

Geostatistical simulation allows quantifying the uncertainty about unobserved spatial properties in terms of a set of alternative representations (i.e., realizations). The aim of stochastic simulations is to enforce various properties on these realizations, so that they should have a similar pattern of variability as the unknown truth and be constrained to any local sample data (Caers 2003). Once these realizations are simulated, they can be post-processed using any type of transfer function (GIS operations, terrain analysis, or environmental modeling) to determine the local distributions of the variable and their associated risks. This is further explained below.

For error modeling, we can define $\{Z(x) = m(x) + \varepsilon(x) | x \in D\}$, where $m(x)$ and $\varepsilon(x)$ refer to the mean and an error term, respectively. If an inquiry about uncertainty is about a set of locations $\{x\}$, the multipoint feature of uncertainty becomes apparent, and we attach great importance to the notion of spatial joint uncertainty. It is helpful to understand that the joint probability, $\text{Prob}\{Z(x_1) \leq z_c, ..., Z(x_{loc}) \leq z_c | (n)\}$ ((n) represents the set of conditional data at n sampled locations), will not be the same as the product of point-specific conditional probabilities $\text{Prob}\{Z(x_1) \leq z_c | (n)\}$ ($i = 1, ..., loc$). Instead, the joint probability can be evaluated by realizations of joint distributions at the *loc* locations, $\dfrac{1}{nsim} \sum_{l=1}^{nsim} \prod_{j=1}^{loc} i^{(l)}(x_j; z_c)$, where $i^{(l)}(x_j; z_c) = 1$, if $z(x) \leq z_c$; 0 otherwise. There are many other occasions where geostatistical simulation is indispensable.

For discrete objects *ID* $(x, attr)$, their uncertainty is usually described through positional accuracy with respect to the set $\{x\}$. Through this framework of uncertainty, $\{x\}$ will no longer be pure points or zero-width lines (or boundaries), but finite distributions $\{x + \delta x\}$; continuous fields will become $Z(x) + \delta_Z(x)$, which can be used to quantify various confidence intervals or joint probabilities; $C(x)$ will be described probabilistically through $(p_1(x), ..., p_K(x))$ (K being the total number of classes being considered), in which the components represent probabilities of individual classes occurring at location x. Similar remarks were made by Zhang (2010).

We have described cross-scale simulation techniques, either downscaling or upscaling, in correspondence to the fact that environmental modeling often implies cross-scale geo-processing. The need to operate at a range of scales, or to change between scales, introduces uncertainty because the dominant patterns and process controls may not be known at all scales or incorporated practically in models at the scales of interest (e.g., Hennings 2002).

Spatial modeling and applications often necessitate uncertainty analysis, which can be approached from the point of view of model inputs and models themselves. When model inputs are defined with a different support from that required by the model, these data must be aggregated or disaggregated, or the model must be redefined at an appropriate scale (Heuvelink and Pebesma 1999; Bierkens et al. 2000). The consequences of changes in support must be taken into account. Uncertainties in model structure and parameter values are also sensitive to changes in scale. However, unlike aggregation or disaggregation of data, model parameters cannot be transformed using the original quantities alone, but by (re)calibrating the model at another scale. This may require adaptation of the original model because functional relationships are typically nonlinear and process controls usually change with scale (Addiscott and Tuck 2001; Brown and Heuvelink 2005). Moreover, upscaling or downscaling of model parameters must be accompanied by simultaneous evaluation of model structures (process controls) for their relevance and sufficiency at other scales, as discussed by Brown and Heuvelink (2005). The fact that models representing spatial processes can vary considerably from scale to scale argues for using different models at different scales, and application of a model developed at a specific scale to another scale should be treated with great care. Uncertainty analysis will be helpful for quantifying the effects of scale mismatch between various components of a model, the relative contributions of input and model error to the output error in change of scale, and for achieving a balanced error budget (Heuvelink 1998; Jager and King 2004).

In summary, spatial patterns are complex. Thus, their modeling is limited in terms of predictive accuracy and scale transferability, as model predictions are fraught with uncertainty and modeling will rarely work in a linear mode across scales. The way ahead is to understand accuracy on the basis of scale-specific fitness for use. Both upscaling and downscaling are important techniques in uncertainty characterization and in translating information about error margins. Provision of spatial information and performance of informative spatial analysis should be backed up with techniques for analyzing scale effects and scaling of errors when they are propagated and cumulated in the resultant information and decision support.

Epilogue

SUMMARY AND CONCLUSION

This book has provided a broad coverage of the field of scale in spatial information and analysis, including the problems that must be tackled and a comprehensive set of methods for analyzing scale and for scaling data. Here, we take the opportunity to reflect on some of the fundamentals of scale and scaling before considering the likely *foci* of future research effort and potential solutions that need to be developed.

Building on Dutch innovation from Hans Lippershey and Zacharias Janssen in 1508, Galileo Galilei invented his first telescope in 1509 followed by improvements to the design of the compound microscope (Whitehouse 2009). Through technological innovation in the form of the ability to create arrangements of lenses that contracted or expanded narrow columns of light, science was transformed. Humans were for the first time able to view beyond the human scale, by "zooming in." This change of scale was a major catalyst for the Enlightenment and modern science. New fields were born such as astronomy, and science split into subfields leading eventually to modern disciplines such as microbiology and genetics. In short, the ability to view our world outside the human scale (i.e., that which is visible with the naked eye) has been crucial to modern thinking. In this context, it is interesting to reflect that humans evolved to make sense of the world at the human scale and that comprehending the information presented to us from beyond that scale has not been without its challenges.

In this book, we applied the concept of scale to three phenomena:

1. Data
2. Measurement
3. Reality (matter, energy, processes)

The view taken is that scales of variation exist in the relations between observations or data. We know this implicitly when we view a remotely sensed image. For example, in a panchromatic IKONOS image with a spatial resolution of 1 m, we can see large objects (e.g., agricultural fields, forest stands) and small objects (e.g., individual houses, individual trees). Loosely, these large objects can be considered to have a low frequency of variation and small objects a high frequency of variation given the spatial resolution of the image.

Much of geostatistics has been devoted to characterizing scales of spatial variation in data (e.g., Webster and Oliver 1990). The variogram and a range of other functions including the spatial covariance function and autocorrelation function have been used for this purpose as part of defining a random function (RF). Fundamentally, this fitting of the RF to data to characterize scales of spatial variation in data has been done to support a series of geostatistical operations such as spatial prediction, spatial simulation, regularization (change of measurement scale), and sample design (Journel and Huijbregts 1978; Goovaerts 1997). However, the characterization is interesting in itself, not least because it quantifies both the *magnitude* of spatial variation (through the semivariance or covariance) and the *scale(s)* of spatial variation through the form and parameters of the model fitted to the variogram or spatial covariance function (McBratney and Webster 1986).

The aforementioned approach is natural for datasets that are fixed in time, with greater utility for datasets that represent a reality that changes slowly through time. For example, in the traditional application areas of geostatistics such as mining and petroleum geology, a snapshot is a useful descriptor of the spatial variation now and tomorrow. This is not so for many other fields of interest where the subject of interest is dynamic at human scales. For example, if one's interest is the spatial distribution and attributes of a hurricane crossing the Eastern Seaboard of the United States or of shoppers in a supermarket, then a snapshot is likely to be of only limited value. Thus, for properties that vary quickly through time, it is necessary to consider time as well as space, leading to space–time characterizations. For properties that change slowly, such space–time characterizations are also increasingly useful where temporal data can be accumulated over long periods. Space–time geostatistics is the natural extension of geostatistics to the space–time domain, allowing the characterization of space–time cubes of data (Kyriakidis and Journel 1999).

As this book has demonstrated, all spatial and space–time data are arguably a function of two things: (i) the underlying reality of interest (which is what we usually think we are seeing) and (ii) the sampling framework (which we often forget). One can think of the underlying reality as having form (see Chapter 2 for examples of representations) in continuous space or space–time, being intersected by a sampling framework to create data. This concept is implicitly natural to humans; when we zoom in, we see more details in the data, leading us to accept *implicitly* that there exists an external reality and that zooming in and out are necessary to reveal its form fully across multiple scales of measurement. However, it takes some effort to make this view explicit as this book has demonstrated.

In addition to the data themselves, both the sampling framework and the underlying reality potentially have scale. The sampling framework has a clearly defined scale (or set of scales). The underlying reality is rather different, and we will come back to it. The sampling framework, as explained in this book, has three fundamental scale parameters: (spatial) extent (defined by the set of all observations), support (the space on which one measurement is made), and "grain" (Blöschl and Sivapalan 1995). The third scale parameter reflects the spatial resolution of the dataset, through the set of distances between neighboring observations (see Chapter 2). Strictly, the sample set of observations in combination defines the extent, support, and resolution, each of which might vary across the sample.

The human visual system also has a set of extents, supports, and resolutions that limit the amount of information that can be received by the brain. This fundamentally limits the snapshot utility of data that have larger extent-to-support ratios (e.g., very large images), meaning, for example, that large data images need to be scanned (i.e., sampled through time) by the human visual system to absorb more of the most important information (e.g., Massaro et al. 2012).

It is interesting to note that the human brain is set up to process data with the internally fixed sampling framework of the human visual system. Changing measurement scale for a human being means zooming in (either through moving closer to the object of interest or through aids such as telescopes and microscopes) or zooming out by moving further away. Thus, while the sampling framework remains internally constant, its scale parameters change in relation to reality (e.g., the extent of the field studied changes). Interestingly, remotely sensed images tend to have a fairly constant extent-to-support (-pixel size) ratio (Goodchild 2004). This is not determined by the human visual system, but by the trade-offs between the ability to record light signals and the cost of transmitting and processing information.

Returning to the third component, the underlying reality, we can potentially infer information about it by separating data and the sampling framework. Having formulated data as a function of the intersection of the sampling framework with the underlying reality, and being able to characterize scales in both, we can infer something of the underlying reality; for example, its state at a given location and time on a point support (i.e., no support). Note that measuring with zero support is a physical impossibility as data are always a function of both reality and sampling. In change-of-support (or change-of-scale) models in geostatistics, the variogram (a characterization of scales of spatial variation in data) is transformed to represent the same function that might be obtained on a point support. This punctual variogram amounts to attempting to characterize spatial form in reality independent of the sampling framework. This is an exciting possibility, but note that it is an ill-posed inverse problem for which no proper solution exists. Thus, it is common to use the pseudo-point variogram model as an intermediate between observations and prediction, both on a positive finite support.

A different approach involves the separation of process and form conceptually. In this common view of the real world, fundamental to classical Newtonian physics, forms are an outcome of a set of processes. We can measure forms (encoded in data as described above), but we can only infer that processes exist through change in observed forms through time. In classical physics, we think of processes as represented through forces acting on matter. From a modeling perspective, processes are represented as a set of rules, characterized in a variety of possible ways, from heuristics through to partial differential equations, some of which may embed physical laws. This dynamic modeling approach has grown hugely based on increased computing power and availability. An advantage of this characterization of processes through computer modeling is that one can create the universe of reality within a computer environment and observe that reality through a space–time sampling framework such as to create data, thus completing the hypothesized trio (reality, sampling, and data). This is attractive because it allows the possibility to explore the effect of the sampling framework when the underlying reality of both states and processes is known.

We should, of course, acknowledge that the above situation is potentially misleading, as it is *never* possible in practice to know reality perfectly. This may sound defeatist and points to a fundamental problem given that the first step in all principled scientific enquiry is to acknowledge that there exists an external reality. However, there is a difference between acknowledging that reality exists and being able to know it completely. If we acknowledge the necessity of an intervening sampling framework, then it follows that we always view reality through the blur of such a defined sampling framework.

This leads to another interesting consideration. If we can never know the underlying reality, then does it have to exist? Does it matter? The alternative to the model of data as a function of sampling and reality (what we will call the *forward model*) is to view reality and the sampling framework as an inferred function of data. This implies that we start with human experience and end with human experience. The sampling framework is driven in this view by the position and viewing angle of the observer, and that is all there is.

This alternative view is interesting because it relates to how humans are cognizant of changes in sampling scale (zooming in and out). It seems to us that ideas are scale invariant (or scale free). If we are comfortable representing an idea with a model (e.g., an analytical dynamic model), then it is clear that the idea or model does not depend on scale. Its parameters need to be estimated, of course, given the data that are available at a given location and time, on a given sampling framework, and these parameters may therefore be scale dependent, but the *concept* remains fixed, waiting to be tested against incoming data. In a similar way, many functions of the human brain are wired through evolution with specific abilities that are brought to bear when faced with an incoming signal, whether close to or far away from an object and whether using a visual aid or not. For example, the separate processes for analyzing form, color, and movement in the human visual system are all used whether viewing a subject through a telescope (large distant objects) or a microscope (micro-scale objects) and may therefore all be scale invariant. It is only by reminding ourselves of (i.e., adding information on) how close to or far away from the subject of interest we are that we can make sense of, and interact with, the world around us.

In psychology, there exists an important concept called "what you see is all there is" (WYSIATI) (Kahneman 2011). This concept links strongly to the discussion above; it has been supported through experiment and it has some interesting consequences (e.g., humans make judgments based on the average of what they see, not through summation). It implies a strongly brain-oriented conceptualization of reality, not a reality-oriented conceptualization. In short, what you experience is all there is. However, for now at least, scientists remain most comfortable with the conceptualization in which data are a function of reality and the sampling framework.

We have discussed above scales in measurements and in data and also introduced the idea that the underlying reality should be conceptualized as existing independently of measurement even though we can only observe it through measurement. We now consider the underlying processes. The idea that processes in reality have scale is important, but they do not vary spatially in the same way as for the sampling framework or data. Forces and the laws of nature are universal, and it is these forces that drive changes in the environment. The search for these fundamental laws is the root of scientific enquiry (Popper 1963). What does vary heterogeneously is the spatial and temporal arrangement of matter and energy. The application of universal forces to these arrangements leads to observable heterogeneity and dynamic changes that are, to a somewhat limited extent, predictable through time.

Of interest here is one of the most important laws governing the spatial arrangement of states: Newton's second law of thermodynamics published in Newton's *Principia* in 1687. In summary, this law states that the universe is tending to greater entropy, or greater disorder, through time. Indeed, the tendency to disorder is one way to diagnose the direction of time's arrow (see Flood and Lockwood (1986), for some interesting perspectives on the nature of time). However, there are some notable local exceptions to entropy, the most important being life. Life creates spatial structure through time. Plants grow into intricate structures from simpler seeds, reversing entropy locally. Humans, and their brains, the most complicated structure in the universe, grow from the division of a single cell. This structure is created by human beings traveling faster up an energy gradient than it is itself traveling downward, propelled by localizing an energy transform (Schrödinger 1955). More specifically, what travels through our bodies increases in entropy faster than our bodies decrease in entropy. This neat trick of life has led to some of the most beautiful structures in the known world, in the biosphere, and on the surface of planet Earth, and although by no means the only focus of this book, these structures are often our primary interest when sensing the environment from space.

FUTURE

SPACE–TIME

Scale is explored primarily through the spatial domain in this book, although the temporal and other (e.g., radiometric, spectral) dimensions were discussed, particularly in the context of remote sensing. Much attention has been devoted in recent years to the extension of spatial modeling to space–time problems. This has been driven by increasing availability of data. In particular, the early promise of remote sensing for monitoring applications is increasingly being realized through the availability of a long time series of remotely sensed images, especially from the Landsat archive, which is now freely available to download at http://landsat.usgs.gov/Landsat_Search_and_Download.php (Woodcock et al. 2008). Such a time series of images are transforming our understanding of Earth-surface processes, as explained above, and driving the development of new space–time modeling approaches.

Just as a transform of space to a scale that allows human comprehension and cognition has transformed our understanding of the physical world, so too does a transform of time through compression and dilation (speeding up and slowing down). Slowing down a time series of images of fast-moving phenomena such as flies flapping their wings can be highly informative and increase our understanding of the natural world (Srygley and Thomas 2002). Likewise, in environmental science, speeding up phenomena that change very slowly can make them interpretable to the human brain, which has evolved to make sense of the world at the human scale (McCoy et al. 2010). For example, animating a time series of remotely sensed images can bring Earth-surface processes to life, transforming our understanding of what is going on and the underlying processes (Gupta et al. 2013). Seeing the Earth oscillate through the seasons as the biosphere responds to the angle of the Earth relative to the Sun or seeing for the first time a mega-river meandering over a period of a few seconds can bring about a gestalt switch in understanding.

MODEL-BASED GEOSTATISTICS

Model-based geostatistics (MBG) was developed more than 15 years ago by Peter Diggle and others (Diggle et al. 1998). The term is a bit of a misnomer because classical geostatistics is also model based. However, the word model in MBG refers to the adoption of a fully Bayesian framework for model fitting and prediction, meaning that the parameters of the RF model are estimated as part of the prediction process. This is in contrast to classical geostatistics in which the RF model parameters are estimated first and then regarded as fixed for prediction. In MBG, geostatistics is generally used as a spatial random-effects term in a linear mixed model, that is, to model simultaneously the residuals from a linear model in which available covariates are used to explain a part of the variation of interest.

Space–time MBG can be achieved through a similar approach in which temporally varying covariates are first used to explain a part of the space–time variation of interest and then geostatistics is used to deal with the residuals. This may be important in situations where the temporal variation of interest has a seasonal profile, as for applications in vegetation phenology and tropical diseases, as geostatistics is not ideal for handling periodically varying phenomena.

Integrated nested Laplace approximation (e.g., implemented in the R-INLA package [http://www.r-inla.org/]) was introduced recently as a means of making geostatistics and autoregressive processes formally equivalent through the solution of partial differential equations (Rue et al. 2009). An interesting and important advantage of this integration is that processing for geostatistics can be speeded up significantly. R-INLA allows application of space–time geostatistics with explanatory covariates.

MULTIPLE-POINT GEOSTATISTICS

A recent advance in geostatistics has been the replacement of the traditional two-point approach with multiple-point statistics (MPS) and, specifically, multiple-point geostatistics (MPG) (Strebelle 2002). The variogram is effectively a two-point function because it represents the variance between *pairs* of points. However, a much richer spatial characterization can be achieved by extending to multiple points at a time, particularly through templates or local windows. The MPS formalism in geostatistics has led to some important benefits. One of the most important is the ability to characterize patterns that are complex, such as networks in continuous geographical space. In this way, it has been possible to capture connectivity and difficult properties such as permeability and porosity.

The shift toward MPG has brought some challenges. In particular, MPG is a data-based approach with no explicit RF model. A model exists, as it always must, but only in the loose sense of the definition of a template or local window, and of the rules for reconstructing predicted or simulated images from training images. If there is no RF model, then we must also lose the ability to model change of support. What is the equivalent of the model of regularization for multiple-point geostatistics? In the frame of MPS and MPG, the relations across scales are captured in data, not in the model. This means that we lose the power of the generic and must relearn the relations each time we encounter new data. An alternative is to assume scale invariance (in the fractal sense). This is appealing as it means that (as for the model of regularization) we do not require data at any support other than that of the original data, which may work remarkably well in some cases, but it must remain an assumption to be tested, not a modeling decision.

CROWDSOURCING AND GEOPORTALS

In recent years, the Internet has changed the way that data are recorded and made available, while at the same time, networks of sensors in the environment have expanded the ability to measure physical states (Neal et al. 2012). These two innovations have changed what is possible, and the current trend in spatial and space–time data analysis is increasingly to find ways of harnessing the power of these kinds of data. Crowdsourcing is the term used to describe the acquisition of data from distributed sources, often via the Internet (e.g., Goodchild and Glennon 2010). Approaches can be purposeful (e.g., volunteered geographic information) and unintended (e.g., harvesting Twitter feeds). In all cases, the data are from heterogeneous sources, often with large uncertainty, representing different views on the same underlying reality. The challenge then is to combine these error-prone and varying data sources so as to complete as precise as possible a picture of the underlying reality. This requires approaches for data integration that go beyond the multivariate geostatistical approaches described in this book.

A particularly interesting perspective, which relates to the tuple of (reality, sampling, and data), is to assume some underlying *scene model* (a term borrowed from remote sensing). Each data source can then be seen as a different view of the same underlying reality as observed through its specific sampling framework. This allows reconstruction of the underlying truth or scene through forward (initial guess at scene, specific sampling, generation of data, comparison to observation, error generation) and inverse modeling. This approach is particularly attractive in relation to change detection for major disasters where the before-and-after images may be acquired through different sensors, different times of day, etc.

Continuing with the theme of crowdsourcing, a current trend is for the analysis of Big Data obtained by "scraping" the web. The possibilities here, as for volunteered information, are great and lead to meta-analysis on a huge scale. For example, twitter feeds can be gathered to generate data that can be used to alert investigators to major events or outbreaks of disease. It is possible to search through academic publications automatically, even where these are in Adobe Portable Document Format (PDF). As an example, Professor Peter Murray-Rust at Cambridge has led the development of software to automatically extract crystallographic representations from academic publications in the field of chemistry and display them as three-dimensional images (Adams et al. 2008). We believe that the power of the Internet and the Web will be used increasingly by such tools to deliver transformational changes in the data available and, more importantly, the questions that can be asked of the data. In the context of the present book, we expect global mapping to emerge for properties for which this was previously unthinkable.

Geoportals are currently bringing together communities of researchers to share data and share software development in an open environment. This is changing the way that researchers and companies are thinking about what have previously been commercial propositions. For example, researchers are sharing information on disaster risk, which has previously been the domain of the lucrative reinsurance industry (Goodchild and Glennon 2010).

NETWORKS

Networks are attracting great interest. For example, complex anabranching and anastomosing river networks have drawn attention from geomorphologists (Carling et al. 2013). What causes rivers to change their planform from year to year? Although the specific pattern changes, its spatial characterization may not. Why is that? What does it tell us about the system? And what are the most appropriate space and space–time statistics to characterize these patterns?

Networks are important in disease transmission. Interestingly, from a geographer's perspective, one might expect space to be important in the transmission process. This is certainly true. However, it is only important in defining the individual-to-individual contact network. Once that is defined, possibly changing through time, one has all that is needed to model the transmission process through a population (Bansal et al. 2010). For a geographer, this can be quite disconcerting. It begs several questions: What is the role for spatial analysis? What is the appropriate transform between spatial representation and network representation? What parameters are needed to create this transform? It is possible, of course, to build a detailed spatially distributed dynamic model (e.g., agent-based model) with which to simulate reality in space–time and, thereby, to build a realistic contact network. However, are there are other, less intensive ways of achieving the same outcome?

In recent years, a large number of data, and in particular images, have been produced about the human brain. These data range from geodesic sensor nets (Tucker et al. 2004) that provide detailed temporal data, to computed tomography scans and functional magnetic resonance (fMR) images that produce fine-resolution spatial representations (Woolrich et al. 2001). These data are transforming our knowledge of the human brain. However, ultimately, the brain is a complex *network* for processing information. Thus, although such raster representations can identify functional areas, they are unlikely to be adequate to *characterize* the functioning of the brain. However, parallels to the above discussion should be obvious: an underlying process, represented on a network, observed through a sampling framework defined with a given extent, support, and grain. The above questions come immediately into play: what can be inferred about the underlying reality given the observations made? What assumptions do we need to make about the nature of the underlying reality in order to make useful inferences? Given the complex processing of the brain, what does a complexity-science view bring to our understanding? And to what extent can the observations be viewed as emergent outcomes of the underlying complex process? Can we create a complex network in a computer environment and, with forward modeling through a sampling framework, learn anything about the observations that are made through fMR imaging? Whatever the questions, how strange to be able to ask questions of the thing that is doing the asking!

COMPLEXITY SCIENCE

Complex systems and complexity science have been foregrounded in much environmental research in recent years (e.g., Paola and Leeder 2011). The concept of chaos theory is at its heart. Chaos has some central properties that are worth discussing here in relation to spatial processes. Firstly, even very simple dynamic systems can exhibit nonlinear behavior because future states are a function of present states. Even a very simple three-parameter system can exhibit chaotic behavior (Gleick 1987). Secondly, chaotic systems are unpredictable and this arises because of the stretching and folding action of the system. The stretching aspect means that the numerical precision of starting conditions quickly evaporates as the system iterates through time, meaning that one can never measure precisely enough to determine future states.

Complex systems and complexity science build on these fundamental concepts of chaos theory, integrating them with a systems view of real-world processes. Looking at, say, environmental processes from a complexity-science perspective, one would admit that future states are highly unpredictable and would see measured forms (observed through a sampling framework) as emergent given the underlying process (which may be very simple). Given this view, the classical complexity-science approach to a system is to observe states or forms through time and, rather than trying to model the system in detail, use analog models or knowledge of classes of previously modeled systems to characterize the behavior of the observed system and predict likely future shifts in behavior and state. One might also concentrate on the ability to replicate emergent spatial and space–time patterns rather than on the ability to predict at a point, as a means of assessing the representativeness of any model (Paola and Leeder 2011).

The two aspects of the above that are not common currently in complexity science are (i) the intervening effect of the sampling framework, which is often neglected, and, to a lesser extent, (ii) the representation of realistic spatial patterns (often, systems are represented at a coarse resolution or as a spatial average state through time). In the context of this book, it should be obvious why the sampling framework is an important consideration for complexity science. Specifically, complexity thinking is currently applied to temporal or space–time data and not to the underlying reality. What happens when a system is convolved spatially? What happens to the emergent properties and patterns? Can the system be represented without a sampling framework? And if not, what of reality? These are open questions in science.

CONCLUSION

In this book, we have introduced scale in terms of data, sampling, and reality. We have shown how various spatially explicit models can be used to characterize the scales of spatial and space–time variation in data and how an understanding of the effect of sampling on data is necessary to handle data appropriately in spatial and space–time analysis. Moreover, given appropriate spatial and space–time tools, we have shown that it is possible to change the support of predictions in order to "zoom out" or "zoom in" *without the need for further measurement.* Such spatial scaling has important consequences where the objective is to use the predictions in further spatial analysis. The utility of such scaling operations depends on a reasonable model of the effect of spatial and space–time sampling on data. In turn, the demonstrated high information-retention and -prediction accuracy of current methods gives confidence that we are on the right track.

In this book, scale has been shown to be well complemented by information content and uncertainty, the other important metrics for spatial data and analysis. Their links can be elucidated through detailed study of their interactions in various settings (e.g., between scale and information density, and between scale and accuracy). Furthermore, we have shown that our understanding of scale needs to be enriched through concepts such as semantic scale and scale-incorporated ontologies in the domains of area classes and discrete entities. In this epilog, we have discussed additional topics of far-reaching significance, such as space–time analysis, multi-point spatial statistics, crowdsourced geo-information, networks (as forms and processes), and complexity. Concerted research on these and related issues will help advance geographic information and analysis, in support of the information society and consumers.

This epilog has provided a summary of the main foundations of this book and, hopefully, reminded the reader of the centrality of scale and scaling, not only to spatial analysis, but to all human experience. It has also provided a forward look that points the reader to a diverse world of possible avenues for spatial and space–time analysis, in which scale and scaling are likely to be increasingly central.

References

Aarts, G., J. Fieberg, and J. Matthiopoulos. 2012. Comparative interpretation of count, presence–absence and point methods for species distribution models. *Methods Ecol. Evol.* 3: 177–187.

Abdelguerfi, M., C. Wynne, E. Cooper, L. Roy, and K. Shaw. 1998. Representation of 3-D elevation in terrain database using hierarchical triangulated irregular networks: A comparative analysis. *Int. J. Geogr. Inf. Sci.* 12(8): 853–873.

Adams, N., J. Winter, P. Murray-Rust, and H. Rzepa. 2008. Chemical Markup, XML and the World-Wide Web. 8. Polymer Markup Language. *J. Chem. Inf. Model.* 48: 2118–2128.

Addiscott, T. M., and G. Tuck. 2001. Non-linearity and error in modelling soil processes. *Eur. J. Soil Sci.* 52: 129–138.

Ahmed, S., and G. De Marsily. 1987. Comparison of geostatistical methods for estimating transmissivity using data on transmissivity and specific capacity. *Water Resour. Res.* 23: 1717–1737.

Aiazzi, B., L. Alparone, and S. Baronti. 1999. Estimation based on entropy matching for generalized Gaussian PDF modeling. *IEEE Signal Proc. Lett.* 6(6): 138–140.

Aiazzi, B., L. Alparone, S. Baronti, and A. Garzelli. 2002. Context-driven fusion of high spatial and spectral resolution images based on oversampled multiresolution analysis. *IEEE Trans. Geosci. Remote Sens.* 10(40): 2300–2312.

Aiazzi, B., L. Alparone, and S. Baronti. 2004. An information-theoretic SAR heterogeneity feature. *Proceedings of ESA-EUSC 2004—Theory and Applications of Knowledge-Driven Image Information Mining with Focus on Earth Observation (ESA SP-553)*, pp. 17–18.

Aiazzi, B., S. Baronti, and L. Alparone. 2009. Lossless compression of hyperspectral images using multiband lookup tables. *IEEE Signal Proc. Lett.* 16(6): 481–484.

Akima, H. 1978. A method of bivariate interpolation and smooth surface fitting for irregularly distributed data points. *ACM Trans. Math. Soft.* 4(2): 148–159.

Alexander, R. B., A. H. Elliott, U. Shankar, and G. B. McBride. 2002. Estimating the sources and transport of nutrients in the Waikato River basin, New Zealand. *Water Resour. Res.* 38(12): 1268.

Allain, C., and M. Cloitre. 1991. Characterizing the lacunarity of random and deterministic fractal sets. *Phys. Rev. A* 44(6): 3552–3558.

Allen, J. C., and D. F. Barnes. 1985. The causes of deforestation in developing countries. *Ann. Assoc. Am. Geogr.* 75(2): 163–184.

Alparone, L., S. Baronti, A. Garzelli, and F. Nencin. 2004. Landsat ETM+ and SAR image fusion based on generalized intensity Modulation. *IEEE Trans. Geosci. Remote Sens.* 42(12): 2832–2839.

American Society for Photogrammetry and Remote Sensing (ASPRS) Specifications and Standards Committee. 1990. ASPRS accuracy standards for large-scale maps. *Photogramm. Eng. Remote Sens.* 56(7): 1068–1070.

Amrhein, C., and D. Wong. 1996. Research on the MAUP: Old wine in a new bottle or real breakthrough? *Geogr. Sys.* 3: 73–76.

Amolins, K., Y. Zhang, and P. Dare. 2007. Wavelet based image fusion techniques—An introduction, review and comparison. *ISPRS J. Photogramm. Remote Sens.* 62(4): 249–263.

Anastassiou, D., and D. J. Sakrison. 1982. Some results regarding the entropy rates of random fields. *IEEE Trans. Inf. Theory* 28(2): 340–343.

Anderson, J. R., E. E. Hardy, J. T. Roach, and R. E. Witmer. 1976. *A Land Use and Land Cover Classification System for Use with Remote Sensor Data*. Washington, DC: United States Government Printing Office for the U.S. Department of the Interior, Geological Survey, 2001: Professional Paper 964.

Anselin, L. 1995. Local indicators of spatial autocorrelation—LISA. *Geogr. Anal.* 27: 93–115.

Anuta, P. E., L. A. Bartolucci, M. E. Dean, D. F. Lozano, E. Malaret, C. D. McGillem, J. A. Valdes, and C. R.Valenzuela. 1984. Landsat-4 MSS and Thematic Mapper quality and information content analysis. *IEEE Trans. Geosci. Remote Sens.* GE-22(3): 222–236.

Araújo, M. B., W. Thuiller, P. H. Williams, and I. Reginster. 2005. Downscaling European species atlas distributions to a finer resolution: Implications for conservation planning. *Global Ecol. Biogeogr.* 14(1): 17–30.

Arbia, G., D. Griffith, and R. Haining. 1998. Error propagation modelling in raster GIS: Overlay operations. *Int. J. Geogr. Inf. Sci.* 12(2): 145–167.

Atkinson, P. M. 2001. Regularization in remote sensing. In *Modeling Scale in Geographical Information Science*, eds. N. J. Tate and P. M. Atkinson, 237–260. Chichester: Wiley.

Atkinson, P. M. 2006. Resolution manipulation and sub-pixel mapping. In *Remote Sensing Image Analysis: Including the Spatial Domain*, eds. S. M. de Jong and F. D. van der Meer, 51–70. Dordrecht, the Netherlands: Springer.

Atkinson, P. M., and P. Aplin. 2004. Spatial variation in land cover and choice of spatial resolution for remote sensing. *Int. J. Remote Sens.* 25(18): 3687–3702.

Atkinson, P. M., and P. J. Curran. 1997. Choosing an appropriate spatial resolution for remote sensing investigation. *Photogramm. Eng. Remote Sens.* 63: 1345–1351.

Atkinson, P. M., and C. D. Lloyd. 2007. Non-stationary variogram models for geostatistical sampling optimisation: An empirical investigation using elevation data. *Comput. Geosci.* 33(10): 1285–1300.

Atkinson, P. M., and R. Massari. 1998. Generalised linear modelling of susceptibility to landsliding in the central Apennines, Italy. *Comput. Geosci.* 24: 373–385.

Atkinson, P. M., E. Pardo-Iguzquiza, and M. Chica-Olmo. 2008. Downscaling cokriging for super-resolution mapping of continua in remotely sensed images. *IEEE Trans. Geosci. Remote Sens.* 46(2): 573–580.

Atkinson, P. M., and N. J. Tate. 2000. Spatial scale problems and geostatistical solutions: A review. *Prof. Geogr.* 52(4): 607–623.

Bailey, R. G. 1983. Delineation of ecosystem regions. *Environ. Manage.* 7(4): 365–373.

Bailey, R. G. 1996. *Ecosystem Geography*. New York: Springer.

Bailey, T. C., and A. C. Gatrell. 1995. *Interactive Spatial Data Analysis*. Harlow: Longman.

Bailey, T. C., and W. J. Krzanowski. 2012. An overview of approaches to the analysis and modelling of multivariate geostatistical data. *Math. Geosci.* 44(4): 381–393.

Baker, S., and T. Kanade. 2002. Limits on super-resolution and how to break them. *IEEE Trans. Pattern Anal. Mach. Intell.* 24(9): 1167–1183.

Band, L. E. 1986. Topographic partitioning of watersheds with digital elevation models. *Water Resour. Res.* 22(1): 15–24.

Band, L. E. 1989. Spatial aggregation of complex terrain. *Geogr. Anal.* 21: 279–293.

Bannari, A., D. Morin, G. B. Bénié, and F. J. Bonn. 1995. A theoretical review of different mathematical models of geometric corrections applied to remote sensing images. *Remote Sens. Rev.* 13(1–2): 27–47.

Bansal, S., J. Read, B. Pourbohloul, and L. Ancel Meyers. 2010. The dynamic nature of contact networks in infectious disease epidemiology, *J. Biol. Dyn.* 4: 478–489.

Baraniuk, R. G. 2007. Compressive sensing. *IEEE Signal Proc. Mag.* 24(4): 118–121.

Barnsley, M. J., S. L. Barr, and T. Tsang. 1997. Scaling and generalisation in land cover mapping from satellite sensors. In *Scaling-up from Cell to Landscape*, eds. P. R. van Gardingen, G. M. Foody, and P. J. Curran, 173–199. New York: Cambridge University Press.

Baskent, E. Z., and G. A. Jordan. 1995. Characterizing spatial structure of forest landscapes. *Can. J. Forest Res.* 25(11): 1830–1849.

Basseville, M., A. Benveniste, K. C. Chou, S. A. Golden, R. Nikoukhah, and A. S. Willsky. 1992. Modeling and estimation of multiresolution stochastic processes. *IEEE Trans. Inf. Theory* 38(2): 766–784.

Battersby, S. E., and Clarke, K. C. 2003. Quantifying Information content in line generalization. *Proceedings of the 21st International Cartographic Conference (ICC)*, Durban, South Africa, pp. 118–126.

Batty, M. 1974. Spatial entropy. *Geogr. Anal.* 6: 1–31.

Beasom, S. L., E. P. Wiggers, and J. R. Giardino. 1983. A technique for assessing land surface ruggedness. *J. Wildl. Manage.* 47: 1163–1166.

Bell, M. R. 1988. Information theory and radar: Mutual information and the design and analysis of radar waveforms and systems. PhD Thesis. California Institute of Technology.

Benveniste, A., R. Nikoukhah, and A. S. Willsky. 1994. Multiscale system theory. *IEEE Trans. Circ. Syst. I: Fundam. Theory Appl.* 41: 2–15.

Besag, J. E. 1974. Spatial interaction and the statistical analysis of lattice systems (with discussion). *J. R. Stat. Soc.: Ser. B* 36: 192–236.

Beven, K., and M. J. Kirkby. 1979. A physically based variable contributing area model of basin hydrology. *Hydrol. Sci. Bull.* 24: 43–69.

Bian, L., and R. Butler. 1999. Comparing effects of aggregation methods on statistical and spatial properties of simulated spatial data. *Photogramm. Eng. Remote Sens.* 65(1): 73–84.

Bian, L., and S. J. Walsh. 1993. Scale dependencies of vegetation and topography in a mountainous environment of Montana. *Prof. Geogr.* 45: 1–11.

Biba, S., K. M. Curtin, and G. Manca. 2010. A new method for determining the population with walking access to transit. *Int. J. Geogr. Inf. Sci.* 24: 347–364.

Bierkens, M. F. P., P. A. Finke, and P. D. Willigen (eds.). 2000. *Upscaling and Downscaling Methods for Environmental Research*. Dordrecht: Kluwer Academic Publishers.

Bindlish, R., and A. P. Barros. 1996. Aggregation of digital terrain data using a modified fractal interpolation scheme. *Comput. Geosci.* 22(8): 907–917.

Birch, C. P. D., S. P. Oom, and J. A. Beechman. 2007. Rectangular and hexagonal grids used for observation, experiment and simulation in ecology. *Ecol. Model.* 206: 347–359.

Bishop, T. F. A., A. B. McBratney, and B. M. Whelan. 2001. Measuring the quality of digital soil maps using information criteria. *Geoderma* 103(1–2): 95–111.

Bjørke, J. T. 1996. Framework for entropy-based map evaluation. *Cartogr. Geogr. Inf. Syst.* 23(2): 78–95.

Blacknell, D., and C. J. Oliver. 1993. Information content of coherent images. *J Phys. D: Appl. Phys.* 26: 1364–1370.

Blaschke, T. 2010. Object based image analysis for remote sensing. *ISPRS J. Photogramm. Remote Sens.* 65(1): 2–16.

Blöschl, G., R. B. Grayson, and M. Sivapalan. 1995. On the representative elementary area (REA) concept and its utility for distributed rainfall-runoff modelling. *Hydrol. Proc.* 9(3–4): 313–330.

Blöschl, G., and M. Sivapalan. 1995. Scale issues in hydrological modelling: A review. *Hydrol. Proc.* 9: 251–290.

Bohling, G. 2005. Stochastic simulation and reservoir modeling workflow. Kansas University and Kansas Geological Survey: C&PE 940.

Bolstad, P. V., and T. Stowe. 1994. An evaluation of DEM accuracy: Elevation, slope, and aspect. *Photogramm. Eng. Remote Sens.* 60: 1327–1332.

Bóo, M., and M. Amor. 2009. Dynamic hybrid terrain representation based on convexity limits identification. *Int. J. Geogr. Inf. Sci.* 23(4): 417–439.

Boots, B. 2002. Local measures of spatial association. *Ecoscience* 9: 168–176.

Boots, B. 2003. Developing local measures of spatial association for categorical data. *J. Geogr. Sys.* 5: 139–160.

Boots, B., and A. Okabe. 2007. Local statistical spatial analysis: Inventory and prospect. *Int. J. Geogr. Inf. Sci.* 21: 355–375.

Borer, E. T., E. W. Seabloom, and D. Tilman. 2012. Plant diversity controls arthropod biomass and temporal stability. *Ecol. Lett.* 15(12): 1457–1464.

Boucher, A. 2007. Downscaling of satellite remote sensing data: Application to land cover mapping. PhD Dissertation. Stanford: Stanford University.

Boucher, A., and R. Dimitrakopoulos. 2009. Block Simulation of multiple correlated variables. *Math. Geosci.* 41: 215–237.

Bourennane, H., and D. King. 2003. Using multiple external drifts to estimate a soil variable. *Geoderma* 114(1–2): 1–18.

Bourgault, G., and D. Marcotte. 1991. Multivariable variogram and its application to the linear model of coregionalization. *Math. Geol.* 23(7): 899–928.

Bowyer, A. 1981. Computing Dirichlet tessellations. *Comput. J.* 24(2): 162–166.

Bracken, I., and D. Martin. 1989. The generation of spatial population distributions from census centroid data. *Environ. Plan. A* 21(4): 537–543.

Brasington, J., and K. Richards. 1998. Interactions between model predictions, parameters and DTM scales for topmodel. *Comput. Geosci.* 24(4): 299–314.

Brown, D., and T. Bara. 1994. Recognition and reduction of systematic error in elevation and derivative surfaces from 7.5-Minute DEMs. *Photogramm Eng. Remote Sens.* 60(2): 189–194.

Brown, D., P. Goovaerts, A. Burnicki, and M. Y. Li. 2002. Stochastic simulation of land-cover change using geostatistics and generalized additive models. *Photogramm. Eng. Remote Sens.* 68: 1051–1061.

Brown, J. D., and G. B. M. Heuvelink. 2005. Assessing uncertainty propagation through physically based models of soil water flow and solute transport. In *Encyclopedia of Hydrological Sciences*, ed. M. G. Anderson, 1181–1195. Chichester, UK: Wiley.

Brunsell, N. A., and M. C. Anderson. 2011. Characterizing the multi-scale spatial structure of remotely sensed evapotranspiration with information theory. *Biogeosciences* 8: 2269–2280.

Brus, D. J., and J. J. de Gruijter. 1997. Random sampling or geostatistical modelling? Choosing between design-based and model-based sampling strategies for soil (with discussion). *Geoderma* 80(1–2): 1–44.

Bruzzone, L., D. F. Prieto, and S. B. Serpico. 1999. A neural-statistical approach to multitemporal and multisource remote-sensing image classification. *IEEE Trans. Geosci. Remote Sens.* 37(3): 1350–1359.

Burrough, P. A., and A. U. Frank (eds.). 1996. *Geographic Objects with Indeterminate Boundaries*. London: Taylor & Francis.

Burrough, P. A., and R. A. McDonnell. 1998. *Principles of Geographical Information Systems*. Oxford, UK: Oxford University Press.

Busby, J. R. 2002. Biodiversity mapping and monitoring. In *Environmental Modelling with GIS and Remote Sensing*, ed. A. Skidmore, 145–165. London: Taylor & Francis.

Butler, J. B., S. N. Lane, and J. H. Chandler. 1998. Assessment of DEM quality for characterizing surface roughness using close range digital photogrammetry. *Photogr. Rec.* 16(92): 271–291.

Buyantuyev, A., and J. Wu. 2007. Effects of thematic resolution on landscape pattern analysis. *Land. Ecol.* 22(1): 7–13.

Candès, E., J. Romberg, and T. Tao. 2006. Robust uncertainty principles: Exact signal reconstruction from highly incomplete frequency information. *IEEE Trans. Inf. Theory* 52(2): 489–509.

Caers, J. 2003. Geostatistics: From pattern recognition to pattern reproduction. *Dev. Petrol. Sci.* 51: 97–115.

Cairns, D. M. 2001. A comparison of methods for predicting vegetation type. *Plant Ecol.* 156(1): 3–18.

Câmara, G., A. M. V. Monteiro, J. A. Paiva, J. Gomes, and L. Velho. 2000. Towards a unified framework for geographical data models. *J. Braz. Comput. Soc.* 7: 17–25.

Canada Centre for Remote Sensing (CCRS). 2007. Tutorial: Fundamentals of Remote Sensing. http://www.nrcan.gc.ca/earth-sciences/geography-boundary/remote-sensing/fundamentals/1430 (accessed February 28, 2013).

Carasso, A. 2001. Direct blind deconvolution. *SIAM J. Appl. Math.* 61(6): 1980–2007.

Carle, S. F., and G. E. Fogg. 1996. Transition probability-based indicator geostatistics. *Math. Geol.* 28: 453–476.

Carling, P. A., J. Jansen, and L. Meshkova. 2013. Multichannel rivers: Their definition and classification. *Earth Surf. Proc. Landforms*, doi: 10.1002/esp.3419.

Carper, W. J., T. M. Lillesand, and R. W. Kiefer. 1990. The use of intensity-hue-saturation transformations for merging SPOT panchromatic and multispectral image data. *Photogramm. Eng. Remote Sens.* 56(4): 459–467.

Carter, J. 1992. The effect of data precision on the calculation of slope and aspect using gridded DEMs. *Cartographica* 29(1): 22–34.

Casas, A., G. Benito, V. R. Thorndycraft, and M. Rico. 2006. The topographic data source of digital terrain models as a key element in the accuracy of hydraulic flood modeling. *Earth Surf. Proc. Landforms* 31: 444–456.

Castleman, J. R. 1995. *Digital Image Processing.* Upper Saddle River: Prentice Hall, Inc.

Central Intelligence Agency. 2009. *CIA World Factbook.* Washington, DC: Central Intelligence Agency.

Chalmond, B. 1991. PSF estimation for image deblurring. *CVGIP: Graph. Models Image Proc.* 53(4): 364–372.

Chan, Y. K., and V. C. Koo. 2008. An introduction to synthetic aperture radar (SAR). *Prog. Electromagnet. Res. B* 2: 27–60.

Chandler, D., and D. Field. 2007. Estimates of the information content and dimensionality of natural scenes from proximity distributions. *J. Opt. Soc. Am. A* 24: 922–941.

Chatterjee, A., A. M. Michalak, R. A. Kahn, S. R. Paraside, A. J. Braverman, and C. E. Miller. 2010. A geostatistical data fusion technique for merging remote sensing and ground-based observations of aerosol optical thickness. *J. Geophys. Res.* 115(D20207).

Chavez, P. S. Jr., and J. A. Bowell. 1988. Comparison of the spectral information content of Landsat Thematic Mapper and SPOT for three different sites in Phoenix, Arizona region. *Photogramm. Eng. Remote Sens.* 54(12): 1699–1708.

Chen, C., and Y. Li. 2012. An adaptive method of non-stationary variogram modeling for DEM error surface simulation. *Trans. GIS* 16(6): 885–899.

Chen, C.-C., C. A. Knoblock, and C. Shahabi. 2008. Automatically and accurately conflating raster maps with orthoimagery. *Geoinformatica* 12(3): 377–410.

Chen, D. M. 1998. *Multiresolution Models on Image Analysis and Classification.* Paper submitted to UCGIS Summer Assembly and Retreat.

Chen, M., and H. Jäenicke. 2010. An information-theoretic framework for visualization. *IEEE Trans Visual. Comput. Graph.* 16(6): 1206–1215.

Chen, T. M., D. H. Staelin, and R. B. Arps. 1987. Information content analysis of Landsat image data for compression. *IEEE Trans. Geosci. Remote Sens.* 25: 499–501.

Chen, Z. T., and J. A. Guevara. 1987. Systematic selection of very important points (VIP) from digital terrain model for constructing triangular irregular networks. In *Proceedings of AUTO-CARTA 8,* ed., N. Chrisman, pp. 50–56. Falls Church, VA: American Congress of Surveying and Mapping.

Cheung, S., K. C. Slatton, and H. Cho. 2012. Fusing landscape accuracy-dependent SRTM elevation data with NGDC and LiDAR data for the Florida coastline. *Remote Sens. Lett.* 3(8): 687–696.

Chilès, J. P., and P. Delfiner. 1999. *Geostatistics: Modeling Spatial Uncertainty.* New York: Wiley.

Chokmani, K., and T. B. M. J. Ouarda. 2004. Physiographical space-based kriging for regional flood frequency estimation at ungauged sites. *Water Resour. Res.* 40(12): 1–13.

Choo, L., and S. G. Walker. 2008. A new approach to investigating spatial variations of disease. *J. R. Stat. Soc.: Ser. A* 171: 395–405.

Chou, K. C., A. S. Willsky, and A. Benveniste. 1994. Multiscale recursive estimation, data fusion, and regularization. *IEEE Trans. Autom. Control* 39(3): 464–477.

Chou, Y. H., P. S. Liu, and R. J. Dezzani. 1999. Terrain complexity and reduction of topographic data. *Geogr. Sys.* 1(2): 179–197.

Chow, T. E., and M. E. Hodgson. 2009. Effects of lidar post-spacing and DEM resolution on mean slope estimation. *Int. J. Geogr. Inf. Sci.* 23(10): 1277–1295.

Chrisman, N. R. 1989. Modelling error in overlaid *categorical maps*. In *Accuracy of Spatial Databases*, eds. M. F. Goodchild and S. Gopal, 21–34. London: Taylor & Francis.

Cliff, A. D., and J. K. Ord. 1969. The problem of spatial autocorrelation. In *Studies in Regional Science*, ed. A. J. Scott, 25–55. London: Pion Press.

Cliff, A. D., and J. K. Ord. 1970. Spatial autocorrelation: A review of existing and new measures with applications. *Econ. Geogr.* 46: 269–292.

Cliff, A. D., and J. K. Ord. 1973. *Spatial Autocorrelation*. London: Pion Press.

Cliff, A. D., and J. K. Ord. 1981. *Spatial Processes: Models and Applications*. London: Pion Press.

Cochran, W. G. 1977. *Sampling Techniques*, 3rd ed. New York: Wiley.

Codd, E. F. 1970. A relational model of data for large shared data banks. *Commun. ACM* 13: 377–387.

Coe, M. T. 2000. Modeling terrestrial hydrological systems at the continental scale: Testing the accuracy of an atmospheric GCM. *J. Clim.* 13(4): 686–704.

Collingham, Y. C., R. A. Wadsworth, B. Huntley, and P. E. Hulme. 2000. Predicting the spatial distribution of non-indigenous riparian weeds: Issues of spatial scale and extent. *J. Appl. Ecol.* 37(s1): 13–27.

Collins, J., and C. Woodcock. 1999. Geostatistical estimation of resolution-dependent variance in remotely sensed images. *Photogramm. Eng. Remote Sens.* 65: 41–50.

Colvocoresses, A. P. 1979. Effective resolution element (ERE) of remote sensor. Memorandum for the Record, U.S. Department of the Interior.

Congalton, R. G. 1991. A review of assessing the accuracy of classifications of remotely sensed data. *Remote Sens. Environ.* 37: 35–46.

Congalton, R. G., and K. Green. 2009. *Assessing the Accuracy of Remotely Sensed Data: Principles and Practices*, 2nd ed. Boca Raton, FL: CRC Press.

Cooper, M. A. R., and P. A. Cross. 1988. Statistical concepts and their application in photogrammetry and surveying. *Photogr. Rec.* 12(71): 637–663.

Coppin, P., I. Jonckheere, K. Nackaerts, B. Muys, and E. Lambin. 2004. Digital change detection methods in ecosystem monitoring: A review. *Int. J. Remote Sens.* 25(9): 1565–1596.

Corner, B. R., R. M. Narayanan, and S. E. Reichenbach. 2003. Noise estimation in remote sensing imagery using data masking. *Int. J. Remote Sens.* 24(4): 689–702.

Corripio, J. G. 2003. Vectorial algebra algorithms for calculating terrain parameters from DEMs and solar radiation modeling in mountainous terrain. *Int. J. Geogr. Inf. Sci.* 17: 1–23.

Corstanje, R., S. Grunwald, and R. M. Lark. 2008. Inferences from fluctuations in the local variogram about the assumption of stationarity in the variance. *Geoderma* 143: 123–132.

Costanza, R., and T. Maxwell. 1994. Resolution and predictability: An approach to the scaling problem. *Land. Ecol.* 9(1): 47–57.

Cova, T. J., and M. F. Goodchild. 2002. Extending geographical representation to include fields of spatial objects. *Int. J. Geogr. Inf. Sci.* 16: 509–532.

Cover, T. M., and P. E. Hart. 1967. Nearest neighbor pattern classification. *IEEE Trans. Inf. Theory* 13(1): 21–27.

Cover, T. M., and J. A. Thomas. 2006. *Elements of Information Theory*, 2nd ed. New York: Wiley.

Cox, I. J., and C. J. R. Sheppard. 1986. Information capacity and resolution in an optical system. *J. Opt. Soc. Am.* 3(8): 1152–1158.

Cressie, N. A. C. 1993. *Statistics for Spatial Data*, revised edition. New York: Wiley.

Cressie, N., and N. H. Chan. 1989. Spatial modeling of regional variables. *J. Am. Stat. Assoc.* 393–401.

Cressie, N., and J. Kornak. 2003. Spatial statistics in the presence of location error with an application to remote sensing of the environment. *Statist. Sci* 18(4): 436–456.

Crist, E. P., and R. C. Cicone. 1984. A physically-based transformation of Thematic-Mapper data-the TM Tasselled Cap. *IEEE Trans. Geosci. Remote Sens.* 22(3): 256–263.

Cross, A. M., J. J. Settle, N. A. Drake, and R. T. M. Paivinen. 1991. Subpixel measurement of tropical forest cover using AVHRR data. *Int. J. Remote Sens.* 12(5): 1119–1129.

Crow, W. T., D. Ryu, and J. S. Famiglietti. 2005. Upscaling of field-scale soil moisture measurements using distributed land surface modeling. *Adv. Water Resour.* 28(1): 1–14.

Crowley, J. M. 1967. Biogeography. *Can. Geogr.* 11(4): 312–326.

Cuba, M. A., O. Leuangthong, and J. M. Ortiz. 2012. Detecting and quantifying sources of non-stationarity via experimental semivariogram modeling. *Stoch. Environ. Res. Risk Assess.* 26(2): 247–260.

Cumming, I. G., and F. H. Wong. 2005. *Digital Processing of Synthetic Aperture Radar Data: Algorithms and Implementation*. Norwood, MA: Artech House, Inc.

Curlander, J. C. 1982. Location of spaceborne SAR imagery. *IEEE Trans. Geosci. Remote Sens.* GE-20(3): 359–364.

Curlander, J., R. Kwok, and S. S. Pang. 1987. A post-processing system for automated rectification and registration of spaceborne SAR imagery. *Int. J. Remote Sens.* 8(4): 621–638.

Curlander, J. C., and R. N. McDonough. 1991. *Synthetic Aperture Radar: Systems and Signal Processing*. New York: Wiley.

Curran, P. J., and P. M. Atkinson. 1998. Geostatistics and remote sensing. *Prog. Phys. Geogr.* 22(1): 61–78.

Curran, P. J., and P. M. Atkinson. 1999. Issues of scale and optimal pixel size. In *Spatial Statistics for Remote Sensing*, eds. A. Stein, F. D. van der Meer, and B. Gorte, 115–133. Dordrecht: Kluwer Academic Publishers.

Currie, A. 1991. Synthetic aperture radar. *Electron. Commun. Eng. J.* 3(4): 159–170.

Czaplewski, R. L. 2003. Can a sample of Landsat sensor scenes reliably estimate the global extent of tropical deforestation? *Int. J. Remote Sens.* 24(6): 1409–1412.

Czaplewski, R. L., and G. P. Catts. 1992. Calibration of remotely sensed proportion of area estimates for misclassification error. *Remote Sens. Environ.* 39: 29–43.

Dai, X., and E. Maarel. 1997. Transect-based patch size frequency analysis. *J. Veg. Sci.* 8(6): 865–872.

Dale, M. R. T., P. Dixon, M.-J. Fortin, P. Legendre, D. E. Myers, and M. S. Rosenberg. 2002. Conceptual and mathematical relationships among methods for spatial analysis. *Ecography* 25: 558–577.

Dale, M. R. T., and M. Mah. 1998. The use of wavelets for spatial pattern analysis in ecology. *J. Veg. Sci.* 9: 805–814.

Davis, F. W., and J. Dozier. 1990. Information analysis of a spatial database for ecological land classification. *Photogramm. Eng. Remote Sens.* 56: 605–613.

Davis, M. 1987. Production of conditional simulations via the LU decomposition of the covariance matrix. *Math. Geol.* 19(2): 91–98.

de Berg, M., O. Cheong, M. van Kreveld, and M. Overmars. 2008. *Computational Geometry: Algorithms and Applications*, 3rd ed. Berlin: Springer.

De Floriani, L., and P. Magillo. 1997. Visibility computations on hierarchical triangulated terrain models. *GeoInformatica* 1: 219–250.

De Floriani, L., B. Falcidieno, C. Pienovi, and G. Nagy. 1984. A hierarchical data structure for surface approximation. *Comput. Graph.* 8: 475–484.

Desbarats, A. J., and R. Dimitrakopoulos. 2000. Geostatistical simulation of regionalized pore-size distributions using min/max autocorrelation factors. *Math. Geol.* 32(8): 919–941.

de Smith, M. J., M. F. Goodchild, and P. A. Longley. 2009. *Geospatial Analysis: A Comprehensive Guide to Principles, Techniques, and Software Tools.* Leicester, UK: Troubador Publishing, Ltd. (on behalf of The Winchelsea Press) (a PDF e-book at http://www.spatialanalysisonline.com/).

Defense Mapping Agency. 1990. *Digitizing the Future*, 3rd ed., 105. Washington, DC: Defense Mapping Agency.

DeFries, R., M. Hansen, M. Steininger, R. Dubayah, R. Sohiberg, and J. Townshend. 1997. Subpixel forest cover in central Africa from multisensor, multitemporal data. *Remote Sens. Environ.* 60(3): 228–246.

DeFries, R. S., and J. R. G. Townshend. 1994. NDVI-derived land cover classification at a global scale. *Int. J. Remote Sens.* 15(17): 3567–3586.

den Dekker, A. J., and A. van den Bos. 1997. Resolution: A survey. *J. Opt. Soc. Am.* 14(3): 547–557.

Deng, Y., J. P. Wilson, and B. O. Bauer. 2007. DEM resolution dependencies of terrain attributes across a landscape. *Int. J. Geogr. Inf. Sci.* 21: 187–213.

Dereniak, E. L., and G. D. Boreman. 1996. *Infrared Detectors and Systems.* New York: Wiley.

Derksen, C., P. Toose, J. Lemmetyinen, J. Pulliainen, A. Langlois, N. Rutter, and M. C. Fuller. 2012. Evaluation of passive microwave brightness temperature simulations and snow water equivalent retrievals through a winter season. *Remote Sens. Environ.* 117: 236–248.

Deutsch, C. V. 2006. A sequential indicator simulation program for categorical variables with point and block data: BlockSIS. *Comput. Geosci.* 32(10): 1669–1681.

Deutsch, C. V., and A. G. Journel. 1998. *GSLIB: Geostatistical Software Library and User's Guide*, 2nd ed. New York: Oxford University Press.

Diggle, P. J., R. A. Moyeed, and J. A. Tawn. 1998. Model-based geostatistics (with discussion). *Appl. Stat.* 47: 299–350.

Diodato, N. 2005. The influence of topographic co-variables on the spatial variability of precipitation over small regions of complex terrain. *Int. J. Climatol.* 25(3): 351–363.

Dobson, M. C., F. T. Ulaby, and L. E. Pierce. 1995. Land-cover classification and estimation of terrain attributes using synthetic aperture radar. *Remote Sens. Environ.* 51(1): 199–214.

Dodgson, N. A. 1997. Quadratic interpolation for image resampling. *IEEE Trans. Image Process.* 6(9): 1322–1326.

Donoho, D. 2006. Compressed sensing. *IEEE Trans. Inf. Theory* 52(4): 1289–1306.

Douglas, D. H. 1986. Experiments to locate ridges and channels to create a new type of digital elevation model. *Cartographica* 23(4): 29–61.

Dowman, I., and G. Peacegood. 1989. Information content of high resolution satellite imagery. *Photogrammetria* 43(6): 295–310.

Doytsher, Y., S. Filin, and E. Ezra. 2001. Transformation of datasets in a linear-based map conflation. *Surv. Land Inf. Syst.* 61: 159–169.

Drăguţ, L., and C. Eisank. 2012. Automated object-based classification of topography from SRTM data. *Geomorphology* 141–142: 21–33.

Drăguţ, L., Eisank, C., and T. Strasser. 2011. Local variance for multi-scale analysis in geomorphometry. *Geomorphology* 130(3–4): 162–172.

Drummond, M. A., R. F. Auch, K. A. Karstensen, K. L. Sayler, J. L. Taylor, and T. R. Loveland. 2012. Land change variability and human environment dynamics in the United States Great Plains. *Land Use Pol.* 29(3): 710–723.

Dubayah, R., and P. M. Rich. 1995. Topographic solar radiation models for GIS. *Int. J. Geogr. Inf. Syst.* 9: 405–413.

Dubayah, R., E. F. Wood, and D. Lavallee. 1997. Multiscaling analysis in distributed modeling and remote sensing. In *Scale in Remote Sensing and GIS*, eds. D. A. Quattrochi and M. F. Goodchild, 93–112. Boca Raton, FL: Lewis Publishers.

Dubes, R. C., and A. K. Jain. 1989. Random field models in image analysis. *J. Appl. Stat.* 16(2): 131–164.

Duda, R. O., P. E. Hart, and D. G. Stork. 2001. *Pattern Classification.* New York: Wiley.

Duranleau, D. L. 1999. Random sampling of regional populations: A field test. *Field Meth.* 11(1): 61–67.

Dutton, G. 1989. Modelling locational uncertainty via hierarchical tessellation. In *Accuracy of Spatial Databases,* eds. M. Goodchild and S. Gopal, 125–140. London: Taylor & Francis.

Dutton, G. 1991. Polyhedral hierarchical tessellations: The shape of GIS to come. *Geo Info Syst.* 1: 49–55.

Dymond, J. R., J. D. Shepherd, G. C. Arnold, and C. M. Trotter. 2008. Estimating area of forest change by random sampling of change strata mapped using satellite imagery. *Forest Sci.* 54(5): 475–480.

Edem, S., and K. Duadze. 2004. *Land Use and Land Cover Study of the Savannah Ecosystem in the Upper West Region (Ghana) Using Remote Sensing.* Göttingen: Cuvillier Verlag.

Edwards, T. C. Jr., D. R. Cutler, N. E. Zimmermann, L. Geiser, and G. G. Moisen. 2006. Effects of sample survey design on the accuracy of classification tree models in species distribution models. *Ecol. Model.* 199(2): 132–141.

Edwards, T. C. Jr., G. G. Moisen, and D. R. Cutler. 1998. Assessing map accuracy in a remotely sensed, ecoregion-scale cover map. *Remote Sens. Environ.* 63(1): 73–83.

Elachi, C. 1980. Spaceborne imaging radar: Geologic and oceanographic applications. *Science* 209(4461): 1073–1082.

Elith, J., and J. R. Leathwick. 2009. Species distribution models: Ecological explanation and prediction across space and time. *Ann. Rev. Ecol. Evol. Syst.* 40: 677–697.

Emerson, C. W., N. S. N. Lam, and D. A. Quattrochi. 1999. Multi-scale fractal analysis of image texture and pattern. *Photogramm. Eng. Remote Sens.* 65(1): 51–61.

Emery, X. 2010. Iterative algorithms for fitting a linear model of coregionalization. *Comput. Geosci.* 36: 1150–1160.

Emery, X., and J. M. Ortiz. 2011. Two approaches to direct block-support conditional co-simulation. *Comput. Geosci.* 37(8): 1015–1025.

Endreny, T. A., and E. F. Wood. 2001. Representing elevation uncertainty in runoff modelling and flowpath mapping. *Hydrol. Proc.* 15(12): 2223–2236.

Engman, E. T. 1991. Applications of microwave remote sensing of soil moisture for water resources and agriculture. *Remote Sens. Environ.* 35(2–3): 213–226.

Environmental Systems Research Institute, National Center for Geographic Information and Analysis, and The Nature Conservancy. 1994. Accuracy Assessment Procedures: NBS/NPS Vegetation Mapping Program. Report prepared for the National Biological Survey and National Park Service, Redlands, CA; Santa Barbara, CA; and Arlington, VA, USA.

Evans, I. S. 1980. An integrated system of terrain analysis and slope mapping. *Zeitsch. Geomorphol., Suppl.-Bd.* 36: 274–295.

Evans, M., and J. Lindsay. 2010. Quantifying gully erosion in upland peatlands at the landscape scale: Gully mapping from a high resolution LiDAR DEM. *Earth Surf. Proc. Land.* 35: 876–886.

Evans, T. P., and H. Kelley. 2004. Multi-scale analysis of a household level agent-based model of landcover change. *J. Environ. Manage.* 72: 57–72.

Fales, C. L., and F. O. Huck. 1991. An information theory of image gathering. *Inf. Sci.* 57: 245–285.

Falorni, G., V. Teles, E. R. Vivoni, and R. L. Bras. 2005. Analysis and characterization of the vertical accuracy of digital elevation models from the Shuttle Radar Topography Mission. *J. Geophys. Res.* 110, doi: 10.1029/2003JF000113.

Farina, A. 2006. *Principles and Methods in Landscapoe Ecology: Towards a Science of Landscape.* Dordrecht, the Netherlands: Springer.

Farr, T. G., P. A. Rosen, E. Caro, R. Crippen, R. Duren, S. Hensley, M. Kobrick et al. 2007. The shuttle radar topography mission. *Rev. Geophys.* 45, RG2004, doi: 10.1029/2005RG000183.

Farsiu, S., D. Robinson, M. Elad, and P. Milanfar. 2004. Advances and challenges in super-resolution. *Int. J. Imag. Syst. Technol.* 14(2): 47–57.

Fellgett, P. B., and E. H. Linfoot. 1955. On the assessment of optical images. *Philos. Trans. R. Soc.* 247: 369–407.

Fenneman, N. M. 1928. Physiographic divisions of the United States. *Ann. Assoc. Am. Geogr.* 18(4): 261–353.

Fernandes, R. A., J. R. Miller, J. M. Chen, and I. G. Rubinstein. 2004. Evaluating image-based estimates of leaf area index in boreal conifer stands over a range of scales using high-resolution CASI imagery. *Remote Sens. Environ.* 89: 200–216.

Fieguth, P. W., W. C. Karl, A. S. Willsky, and C. Wunsch. 1995. Multiresolution optimal inter-polation and statistical analysis of TOPEX/POSEIDON satellite altimetry. *IEEE Trans. Geosci. Remote Sens.* 33(2): 280–292.

Figueira, R., P. C. Tavares, L. Palma, P. Beja, and C. Sergio. 2009. Application of IK to the complementary use of bioindicators at three trophic levels. *Environ. Pollut.* 157: 2689–2696.

Finn, J. T. 1993. Use of the average mutual information index in evaluating classification error and consistency. *Int. J. Geogr. Inf. Syst.* 7(4): 349–366.

Fiorentino, M., P. Claps, and V. P. Singh. 1993. An entropy-based morphological analysis of river basin networks. *Water Resour. Res.* 29(4): 1215–1224.

Flood, R., and M. Lockwood. 1986. *The Nature of Time.* Oxford: Blackwell Publishing.

Florinsky, I. V. 1998. Accuracy of local topographic variables derived from digital elevation models. *Int. J. Geogr. Inf. Sci.* 12(1): 47–61.

Florinsky, I. V., and G. A. Kuryakova. 2000. Determination of grid size for digital terrain mod-elling in landscape investigations exemplified by soil moisture distribution at a micro-scale. *Int. J. Geogr. Inf. Sci.* 14: 815–832.

Foley J. A., R. Defries, G. P. Asner, C. Barford, G. Bonan et al. 2005. Global consequences of land use. *Science* 309(5734): 570–574.

Fonseca, T. C. O., and M. F. J. Steel. 2011. A general class of nonseparable space–time covariance models. *Environmetrics* 22: 224–242.

Foody, G. M. 2002. Status of land cover classification accuracy assessment. *Remote Sens. Environ.* 80: 185–201.

Foody, G. M. 2010. Assessing the accuracy of land cover change with imperfect ground refer-ence data. *Remote Sens. Environ.* 114: 2271–2285.

Forman, R. T. T., and M. Godron. 1986. *Land. Ecol.* New York: Wiley.

Forshaw, M. R. B., A. Haskell, P. F. Miller, D. J. Stanley, and J. R. G. Townshend. 1983. Spatial resolution of remotely sensed imagery: A review paper. *Int. J. Remote Sens.* 4(3): 497–520.

Fortin, M. J., M. R. T. Dale, and J. Hoef. 2002. Spatial analysis in ecology. In *Encyclopedia of Environmetrics*, eds. A. H. El-Shaarawi and W. W. Piegorsch, 4: 2051–2058. Chichester: Wiley.

Fortin, M. J., P. James, A. MacKenzie, S. J. Melles, and B. Rayfield. 2012. Spatial statistics, spatial regression, and graph theory in ecology. *Spatial Stat.* 1: 100–109.

Fotheringham, A. S. 1997. Trends in quantitative methods I: Stressing the local. *Prog. Hum. Geogr.* 21: 88–96.

Fotheringham, A. S. 2009. "The problem of spatial autocorrelation" and local spatial statistics. *Geogr. Anal.* 41: 398–403.

Fotheringham, A. S., and M. E. O'Kelly. 1989. *Spatial Interaction Models: Formulations and Applications.* Boston: Kluwer Academic Publishers.

Fotheringham, A. S., C. Brunsdon, and M. Charlton. 2002. *Geographically Weighted Regression: The Analysis of Spatially Varying Relationships.* Chichester, UK: Wiley.

Franklin, J., F. W. Davis, M. Ikegami, A. D. Syphard, L. E. Flint, A. L. Flint, and L. Hannah. 2012. Modeling plant species distributions under future climates: How fine scale do climate projections need to be? *Glob. Change Biol.* doi: 10.1111/gcb.12051.

Frieden, B. R. 1983. *Probability, Statistical Optics, and Data Testing.* Berlin: Springer.

Frost, V. S., and K. S. Shanmugam. 1983. The information content of synthetic apertaure radar images of terrain. *IEEE Trans. Aerosp. Elect. Syst.* 19: 768–774.

Friedl, M. A., D. Sulla-Menashe, B. Tan, A. Schneider, N. Ramankutty, A. Sibley, and X. Huang. 2010. MODIS Collection 5 global land cover: Algorithm refinements and characterization of new datasets. *Remote Sens. Environ.* 114(1): 168–182.

Fuderer, M. 1988. The information content of MR images. *IEEE Trans. Med. Imag.* 7(4): 368–380.

Fuentes, M. 2002. Interpolation of nonstationary air pollution process: A spatial spectral approach. *Stat. Model.* 2: 281–298.

Furrer, R., and M. G. Genton. 2011. Aggregation-cokriging for highly-multivariate spatial data. *Biometrika* 98: 615–631.

Gabriel, A. K., R. M. Goldstein, and H. A. Zebker. 1989. Mapping small elevation changes over large areas: Differential radar interferometry. *J. Geophys. Res.* 94(B7): 9183–9191.

Gad-el-Hak, M. 2008. *Large-Scale Disasters: Prediction, Control, and Mitigation.* Cambridge: Cambridge University Press.

Gallant, A. L., T. R. Loveland, T. L. Sohl, and D. E. Napton. 2004. Using an ecoregion framework to analyze land-cover and land-use dynamics. *Environ. Manage.* 34 Suppl 1(Hart 1996): S89–S110.

Gallant, J., and M. F. Hutchinson. 1996. Towards an understanding of landscape scale and structure. In *Proceedings Third International Workshop/Conference on Integrating GIS and Environmental Modelling*, Santa Fe, NM, January 21–26. Santa Barbara, CA: National Centre for Geographic Information and Analysis.

Garbrecht, J., and L. Martz. 1994. Grid size dependency of parameters extracted from digital elevation models. *Comput. Geosci.* 20(1): 85–87.

Garcia-Gigorro, S., and S. Saura. 2005. Forest fragmentation estimated from remotely sensed data: Is comparison across scales possible? *Forest Sci.* 51: 51–63.

Gardner, R. H., T. R. Lookingbill, P. A. Townsend, and J. Ferrari. 2008. A new approach for rescaling land cover data. *Land. Ecol.* 23(5): 513–526.

Gargantini, I. 1982. An effective way to represent quadtrees. *Commun. ACM* 25: 905–910.

Garrigues, S., D. Allard, F. Baret, and J. Morisette. 2008. Multivariate quantification of landscape spatial heterogeneity using variogram models. *Remote Sens. Environ.* 112(1): 216–230.

Garrigues, S., D. Allard, F. Baret, and M. Weiss. 2006. Quantifying spatial heterogeneity at the landscape scale using variogram models. *Remote Sens. Environ.* 103(1): 81–96.

Gatrell, A. C. 1977. Complexity and redundancy in binary maps. *Geogr. Anal.* 9: 29–41.

Gatrell, A. C. 1979. Autocorrelation in spaces. *Environ. Plan. A* 11: 507–516.

Gauch, H. G., Jr. 1982. *Multivariate Analysis in Community Structure.* Cambridge: Cambridge University Press.

Geary, R. C. 1954. The contiguity ratio and statistical mapping. *Inc. Stat.* 5: 115–141.

Gelfand, A. E., L. Zhu, and B. P. Carlin. 2001. On the change of support problem for spatio-temporal data. *Biostatistics* 2(1): 31–45.

Geman, S., and D. Geman. 1984. Stochastic relaxation, Gibbs distributions, and the Bayesian restoration of images. *IEEE Trans. Pattern Anal. Mach. Intell.* 6: 721–741.

Gertner, G., G. Wang, A. B. Anderson, and H. Howard. 2007. Combining stratification and up-scaling method-block cokriging with remote sensing imagery for sampling and mapping an erosion cover factor. *Ecol. Inf.* 2(4): 373–386.

Getis, A. 2009. Spatial weights matrices. *Geogr. Anal.* 41: 404–410.

Getis, A., and J. K. Ord. 1992. The analysis of spatial association by use of distance statistics. *Geogr. Anal.* 24: 189–206.

Getis, A., and J. K. Ord. 1996. Local spatial statistics: An overview. In *Spatial Analysis: Modelling in a GIS Environment*, eds. P. Longley and M. Batty, 261–277. Cambridge: GeoInformation International.

Gleick, J. 1987. *Chaos: Making a New Science*. New York: Penguin.

Göckede, M., A. M. Michalak, D. Vickers, D. P. Turner, and B. E. Law. 2010. Atmospheric inverse modeling to constrain regional-scale CO_2 budgets at high spatial and temporal resolution. *J. Geophys. Res.* 115: D15113.

Gonga-Saholiariliva, N., Y. Gunnell, C. Petit, and C. Mering. 2011. Techniques for quantifying the accuracy of gridded elevation models and for mapping uncertainty in digital terrain analysis. *Prog. Phys. Geogr.* 35(6): 739–764.

Gonzalez, R. C., R. E. Woods, and S. L. Eddins. 2003. *Digital Image Processing Using Matlab*. Prentice Hall, Upper Saddle River, NJ, USA.

Goodall, J., P. Zacharias, T. Olckers, and T. Edwards. 2006. A contiguous-quadrat sampling exercise in a shrub-invaded grassland patch: Size matters but biggest is not best. *Afr. J. Range Forage Sci.* 23(2): 123–130.

Goodchild, M. F. 1980a. Fractals and the accuracy of geographical measures. *Math. Geol.* 12: 85–98.

Goodchild, M. F. 1980b. Simulation of autocorrelation for aggregate data. *Environ. Plan. A* 12: 1073–1081.

Goodchild, M. F. 1982. The fractional Brownian process as a terrain simulation model. *Model. Simul.* 13: 1133–1137.

Goodchild, M. F. 1986. *Spatial Autocorrelation*. CATMOG 47. Norwich: Geo Books.

Goodchild, M. F. 1991. Geographical data modeling. *Comput. Geosci.* 18: 400–408.

Goodchild, M. F. 1992. Geographical information science. *Int. J. Geogr. Inf. Syst.* 6: 31–46.

Goodchild, M. F. 1993. Data models and data quality: Problems and prospects. In *Environmental Modeling with GIS*, eds. M. F. Goodchild, B. O. Parks, and L. T. Steyaert, 94–103. New York: Oxford University Press.

Goodchild, M. F. 1994. Integrating GIS and remote sensing for vegetation analysis and modelling: Methodological issues. *J. Veg. Sci.* 5(5): 615–626.

Goodchild, M. F. 1995. Attribute accuracy. In *Elements of Spatial Data Quality*, eds. S. C. Guptill and J. L. Morrison, 59–80. New York: Elsevier.

Goodchild, M. F. 2001. Metrics of scale in remote sensing and GIS. *Int. J. Appl. Earth Obs. Geoinf.* 3(2): 114–120.

Goodchild, M. F. 2003. The nature and value of geographic information. In *Foundations of Geographic Information Science*, eds. M. Duckham, M. F. Goodchild, and M. F. Worboys, 19–32. New York: Taylor & Francis.

Goodchild, M. F. 2004. Scales of cybergeography. In *Scale and Geographic Inquiry: Nature, Society, and Method*, eds. E. Sheppard and R. B. McMaster, 154–169. Malden, MA: Blackwell.

Goodchild, M. F. 2011. Scale in GIS: An overview. *Geomorphology* 130(1–2): 5–9.

Goodchild, M. F., and O. Dubuc. 1987. A model of error for choropleth maps with applications to geographic information systems. *Proc. Auto-Carto.* 8: 165–174.

Goodchild, M. F., L. Anselin, and U. Deichmann. 1993. A framework for the areal interpolation of socioeconomic data. *Environ. Plan. A* 25: 383–397.

Goodchild, M. F., G. S. Biging, R. G. Congalton, P. G. Langley, N. R. Chrisman, and F. W. Davis. 1994. Final Report of the Accuracy Assessment Task Force. California Assembly Bill AB1580. Santa Barbara, CA: National Center for Geographic Information and Analysis (NCGIA), University of California.

Goodchild, M. F., F. W. Davis, M. Painho, and D. M. Stoms. 1991. The Use of Vegetation Maps and Geographic Information Systems for Assessing Conifer Lands in California, NCGIA Technical report 91–23. Santa Barbara, CA: NCGIA, 75 pp.

Goodchild, M. F., M. J. Egenhofer, K. K. Kemp, D. M. Mark, and E. Sheppard. 1999. Introduction to the Varenius project. *Int. J. Geogr. Inf. Sci.* 13: 731–745.

Goodchild, M. J., and A. Glennon. 2010. Crowdsourcing geographic information for disaster response: A research frontier. *Int. J. Digit. Earth* 3: 231–241.

Goodchild, M. F., and G. J. Hunter. 1997. A simple positional accuracy measure for linear features, *Int. J. Geogr. Inf. Sci.* 11(3): 299–306.

Goodchild, M. F., and K. K. Kemp (eds.). 1990. *NCGIA Core Curriculum in GIS.* Santa Barbara, CA: National Center for Geographic Information and Analysis (NCGIA), University of California.

Goodchild, M. F., and D. M. Mark. 1987. The fractal nature of geographic phenomena. *Ann. Assoc. Am. Geogr.* 77: 265–278.

Goodchild, M. F., G. Sun, and S. Yang. 1992. Development and test of an error model for categorical data. *Int. J. Geogr. Inf. Syst.* 6(2): 87–104.

Goodchild, M. F., and M. H. Wang. 1989. Modeling error for remotely sensed data input to GIS. *Proc. Auto-Carto.* 9: 530–537.

Goodchild, M. F., and S. Yang. 1992. A hierarchical spatial data structure for global geographic information systems. *Comput. Vis. Graph. Image Process.* 54: 31–44.

Goodchild, M. F., M. Yuan, and T. J. Cova. 2007. Towards a general theory of geographic representation in GIS. *Int. J. Geogr. Inf. Sci.* 21(3): 239–260.

Goodchild, M. F., J. X. Zhang, and P. Kyriakidis. 2009. Discriminant models of uncertainty in nominal fields. *Trans. GIS* 13: 7–23.

Goodman, J. W. 2005. *Introduction to Fourier Optics*, 3rd ed. Greenwood Village, CO: Roberts and Company Publishers.

Goovaerts, P. 1996. Stochastic simulation of categorical variables using a classification algorithm and simulated annealing. *Math. Geol.* 28(7): 909–921.

Goovaerts, P. 1997. *Geostatistics for Natural Resources Evaluation.* New York: Oxford University Press.

Goovaerts, P. 1999. Geostatistics in soil science: State-of-the-art and perspectives. *Geoderma* 89: 1–45.

Goovaerts, P. 2008. Kriging and semivariogram deconvolution in the presence of irregular geographical units. *Math. Geol.* 40(1): 101–128.

Goshtasby, A. A., and S. Nikolov. 2007. Image fusion: Advances in the state of the art. *Inf. Fusion* 8: 114–118.

Gotway Crawford, C. A. 2006. Sample support. *Encycl. Environmetrics.* doi: 10.1002/9780470057339.vas004.

Gotway, C. A., and W. W. Stroup. 1997. A generalized linear model approach to spatial data analysis and prediction. *J. Agric. Biol. Environ. Stat.* 2(2): 157–178.

Gotway, C. A., and L. J. Young. 2002. Combining incompatible spatial data. *J. Am. Stat. Assoc.* 97(458): 632–648.

Gotway, C. A., and L. J. Young. 2007. A geostatistical approach to linking geographically-aggregated data from different sources. *J. Comput. Graph. Stat.* 16(1): 1–21.

Gourdji, S. M., A. I. Hirsch, K. L. Mueller, A. E. Andrews, and A. M. Michalak. 2009. Regional-scale geostatistical inverse modeling of North American CO_2 fluxes: A synthetic data study. *Atm. Chem. Phys.* 9: 22407–22458.

Graps, A. 1995. An introduction to wavelets. *IEEE Comput. Sci. Eng.* 2(2): 50–61.

Griffith, D. A. 1996. Some guidelines for specifying the geographic weights matrix contained in spatial statistical models. In *Practical Handbook of Spatial Statistics*, ed. S. L. Arlinghaus, 65–82. Boca Raton, FL: CRC Press.

Griffith, D. A., and G. Arbia. 2010. Detecting negative spatial autocorrelation in georeferenced random variables. *Int. J. Geogr. Inf. Sci.* 24(3): 417–437.

Grossman, D., K. L. Goodin, X. Li, C. Wisnewski, D. Faber-Langendoen, M. Anderson, L. Sneddon, D. Allard, M. Gallyoun, and A. Weakley. 1994. Standardized National Vegetation Classification System. Report by The Nature Conservancy and Environmental Systems Research Institute for the NBS/NPS Vegetation Mapping Program. Denver: National Biological Service.

Grubesic, T. H. 2008. Zip codes and spatial analysis: Problems and prospects. *Soc.-Econ. Plan. Sci.* 42(2): 129–149.

Gruhier, C., P. de Rosnay, S. Hasenauer, T. Holmes, R. de Jeu, Y. Kerr, E. Mougin, E. Njoku, F. Timouk, W. Wagner, and M. Zribi. 2010. Soil moisture active and passive microwave products: Intercomparison and evaluation over a Sahelian site. *Hydrol. Earth Syst. Sci.* 14: 141–156.

Guisan, A., T. C. Edwards, and T. Hastie. 2002. Generalized linear and generalized additive models in studies of species distributions: Setting the scene. *Ecol. Model.* 157(2–3): 89–100.

Guo, D. 2010. Local entropy map: A nonparametric approach to detecting spatially varying multivariate relationships. *Int. J. Geogr. Inf. Sci.* 24(9): 1367–1389.

Gupta, N., P. M. Atkinson, and P. A. Carling. 2013. Decadal length changes in the fluvial planform of the River Ganga: Bringing a mega-river to life with Landsat archives. *Remote Sens. Lett.* 4(1): 1–9.

Gupta, S., and S. P. S. Virdi. 1989. Information content in astronomical images (Letter to the Editor). *Astrophys. Space Sci.* 162(1):159–161.

Gupta, S. K. 2011. *Modern Hydrology and Sustainable Water Development.* Oxford: Wiley-Blackwell.

Gustafson, E. J. 1998. Quantifying landscape spatial pattern: What is the state of the art? *Ecosystems* 1(2): 143–156.

Guth, P. 1999. Contour line "ghosts" in USGS Level 2 DEMs. *Photogramm. Eng. Remote Sens.* 65(3): 289–296.

Gyasi-Agyei, Y., G. Willgoose, and F. De Troch. 1995. Effects of vertical resolution and map scale of digital elevation models on geomorphological parameters used in hydrology. *Hydrol. Proc.* 9: 363–382.

Haining, R. 2003. *Spatial Data Analysis: Theory and Practice.* Cambridge, UK: Cambridge University Press.

Haining, R. 2009. Spatial autocorrelation and the quantitative revolution. *Geogr. Anal.* 41: 364–374.

Haining, R. P., and G. Arbia. 1993. Error propagation through map operations. *Technometrics* 35(3): 293–305.

Hallikainen, M. T., F. T. Ulaby, M. C. Dobson, M. A. El-Rayes, and W. Lil-Kun. 1985. Microwave dielectric behavior of wet soil—Part 1: Empirical models and experimental observations. *IEEE Trans. Geosci. Remote Sens.* GE-23(1): 25–34.

Hammond, E. H. 1954. Small-scale continental landform maps. *Ann. Assoc. Am. Geogr.* 44: 33–42.

Hammond, E. H. 1964. Analysis of properties in landform geography: An application to broad scale landform mapping. *Ann. Assoc. Am. Geogr.* 54: 11–19.

Hansen, M. C., R. S. DeFries, J. R. G. Townshend, M. Carroll, C. Dimiceli, and R. A. Sohlberg. 2003. Global percent tree cover at a spatial resolution of 500 meters: First results of the MODIS vegetation continuous fields algorithm. *Earth Interact.* 7(10): 1–15.

Hansen, M. C., J. R. G. Townshend, R. S. DeFries, and M. Carroll. 2005. Estimation of tree cover using MODIS data at global, continental and regional/local scales. *Int. J. Remote Sens.* 26(19): 4359–4380.

Hao, Y. 2011. Generalizing sampling theory for time-varying Nyquist rates using self-adjoint extensions of symmetric operators with deficiency indices (1, 1) in Hilbert spaces. PhD Thesis. University of Waterloo.

Harrie, L., and H. Stigmar. 2007. An evaluation of measures for quantifying complexity of a map. In *International Archives of Photogrammetry, Remote Sensing and Spatial Information Sciences* Volume XXXVI-4/W45 covering contributions of the Joint Workshop "Visualization and Exploration of Geospatial Data," June 27–29, 2007, Stuttgart, Germany.

Harrie, L., and H. Stigmar. 2010. An evaluation of measures for quantifying map information. *ISPRS J. Photogramm. Remote Sens.* 65(3): 266–274.

Harris, J. L. 1964. Diffraction and resolving power. *J. Opt. Soc. Am.* 54(7): 931–936.

Hastings, J. T. 2008. Automated conflation of digital gazetteer data. *Int. J. Geogr. Inf. Sci.* 22(10): 1109–1127.

Hawley, K., and H. Moellering. 2005. A comparative analysis of areal interpolation methods. *Cartogr. Geogr. Inf. Sci.* 32: 411–423.

Hay, S. I., A. M. Noor, A. Nelson, and A. J. Tatem. 2005. The accuracy of human population maps for public health application. *Trop. Med. Int. Health* 10: 1073–1086.

He, H. S., S. J. Ventura, and D. J. Mladenoff. 2002. Effects of spatial aggregation approaches on classified satellite imagery. *Int. J. Geogr. Inf. Sci.* 16(1): 93–109.

Heikkila, E. J., and L. Hu. 2006. Adjusting spatial-entropy measures for scale and resolution effects. *Environ. Plan. B: Plan. Des.* 33: 845–861.

Hennings, V. 2002. Accuracy of coarse-scale land quality maps as a function of the upscaling procedure used for soil data. *Geoderma* 107: 177–196.

Herman, M. A., and T. Strohmer. 2009. High-resolution radar via compressed sensing. *IEEE Trans. Signal Process.* 57(6): 2275–2284.

Hess, L., J. M. Melack, and D. S. Simonett. 1990. Radar detection of flooding beneath the forest canopy: A review. *Int. J. Remote Sens.* 11(7): 1313–1325.

Heuvelink, G. B. M. 1998. Uncertainty analysis in environmental modelling under a change of spatial scale. *Nutr. Cycl. Agroecosyst.* 50: 255–264.

Heuvelink, G. B. M., and E. J. Pebesma. 1999. Spatial aggregation and soil process modelling. *Geoderma* 89: 47–65.

Higdon, D. 1998. A process-convolution approach to modelling temperatures in the North Atlantic Ocean. *Environ. Ecol. Stat.* 5(2): 173–190.

Hill, J., C. Diemer, O. Stöver, and Th. Udelhoven. 1999. A local correlation approach for the fusion of remote sensing data with different spatial resolutions in forestry applications. International Archives of Photogrammetry and Remote Sensing, Vol. 32, Part 7-4-3 W6, Valladolid, Spain.

Hill, M. O. 1973. The intensity of spatial pattern in plant communities. *J. Ecol.* 61(1): 225–235.

Hobson, R. D. 1972. Surface roughness in topography: Quantitative approach. *Spatial Analysis in Geomorphology*, ed. R. J. Chorley, 221–245. New York: Harper and Row.

Hodgson, M. E. 1995. What cell size does the computed slope/aspect angle represent? *Photogramm. Eng. Remote Sens.* 60: 513–517.

Hodgson, M. E., J. R. Jensen, L. Schmidt, S. Schill, and B. Davis. 2003. An evaluation of LIDAR- and IFSAR-derived digital elevation models in leaf-on conditions with USGS Level 1 and Level 2 DEMs. *Remote Sens. Environ.* 84: 295–308.

Holben, B. N. 1986. Characteristics of maximum-value composite images from temporal AVHRR data. *Int. J. Remote Sens.* 7(11): 1417–1434.

Holdridge, L. R. 1947. Determination of world plant formations from simple climatic data. *Science* 105(2727): 367–368.

Holm, R. G., R. D. Jackson, B. Yuan, M. S. Moran, P. N. Slater, and S. F. Bigger. 1989. Surface reflectance factor retrieval from Thematic Mapper data. *Remote Sens. Environ.* 27: 47–57.

Holmes, K. W., O. A. Chadwick, and P. C. Kyriakidis. 2000. Error in a USGS 30-meter digital elevation model and its impact on terrain modeling. *J. Hydrol.* 233: 154–173.

Hong, S., J. M. H. Hendrickx, and B. Borchers. 2011. Down-scaling of SEBAL derived evapotranspiration maps from MODIS (250 m) to Landsat (30 m) scale. *Int. J. Remote Sens.* 32(21): 6457–6477.

Horn, B. K. P. 1981. Hill shading and the reflectance map. *Proc. IEEE* 69(1): 14–47.

Hosseini, R., F. Sinz, and M. Bethge. 2010. Lower bounds on the redundancy of natural images. *Vis. Res.* 50(22): 2213–2222.

Huang, H.-C., N. Cressie, and J. Gabrosek. 2002. Fast, resolution-consistent spatial prediction of global processes from satellite data. *J. Comput. Graph. Stat.* 11(1): 63–88.

Huck, F. O., C. L. Fales, R. Alter-Gartenberg, S. K. Park, and Z. Rahman. 1999. Information-theoretical assessment of sampled imaging systems. *Opt. Eng.* 38(5): 742–762.

Huck, F. O., C. L. Fales, and Z. Rahman. 1996. An information theory of visual communication. *Philos. Trans.: Math. Phys. Eng. Sci.* 354(1716): 2193–2248.

Huck, F. O., N. Halyo, and S. K. Park. 1981. Information efficiency of line-scan imaging mechanisms. *Appl. Opt.* 20: 1990–2007.

Hudson, G., and H. Wackernagel. 1994. Mapping temperature using kriging with external drift: Theory and an example from Scotland. *Int. J. Climatol.* 14: 77–91.

Hufkens, K., J. Bogaert, Q. H. Dong, L. Lu, C. L. Huang, M. G. Ma, T. Che, X. Li, F. Veroustraete, and R. Ceulemans. 2008. Impacts and uncertainties of upscaling of remote-sensing data validation for a semi-arid woodland. *J. Arid Environ.* 72(8): 1490–1505.

Hughes, M. L., P. F. McDowell, and W. A. Marcus. 2006. Accuracy assessment of georectified aerial photographs: Implications for measuring lateral channel movement in a GIS. *Geomorphology* 74(1–4): 1–16.

Hunsaker, C. T., M. F. Goodchild, M. A. Friedl, and T. J. Case (eds.). 2001. *Spatial Uncertainty in Ecology: Implications for Remote Sensing and GIS Applications.* New York: Springer.

Hunt, B. R. 1973. The application of constrained least squares estimation to image restoration by digital computer. *IEEE Trans. Comput.* C-22(9): 805–812.

Hunter, G., and M. Goodchild. 1997. Modeling the uncertainty of slope and aspect estimates derived from spatial databases. *Geogr. Anal.* 29(1): 35–49.

Hyppänen, H. 1996. Spatial autocorrelation and optimal spatial resolution of optical remote sensing data in boreal forest environment. *Int. J. Remote Sens.* 17(17): 3441–3452.

Irons, J. R., B. L. Markham, R. F. Nelson, D. L. Toll, D. L. Williams, R. S. Latty, and M. L. Stauffer. 1985. The effects of spatial resolution on the classification of Thematic Mapper data. *Int. J. Remote Sens.* 6(8): 1385–1403.

Isaaks, E. H. 1990. The application of Monte Carlo methods to the analysis of spatially correlated data. PhD Thesis. Stanford: Stanford University.

Isaaks, E. H., and R. M. Srivastava. 1989. *An Introduction to Applied Geostatistics.* Oxford, UK: Oxford University Press.

Iverson, L. R., E. A. Cook, and R. L. Graham. 1989. A technique for extrapolating and validating forest cover across large regions: Calibrating AVHRR data with TM data. *Int. J. Remote Sens.* 10(11): 1805–1812.

Iwahashi, J., and R. J. Pike. 2007. Automated classifications of topography from DEMs by an unsupervised nested-means algorithm and a three-part geometric signature. *Geomorphology* 86(3–4): 409–440.

Jackett, C. J., P. J. Turner, J. L. Lovell, and R. N. Williams. 2011. Deconvolution of MODIS imagery using multiscale maximum entropy. *Remote Sens. Lett.* 2(3): 179–187.

Jackson, T. J. 2001. Multiple resolution analysis of L-band brightness temperature for soil moisture. *IEEE Trans. Geosci. Remote Sens.* 39(1): 151–164.

Jager, H. I., and A. W. King. 2004. Spatial uncertainty and ecological models. *Ecosystems* 7: 841–847.

Jain, A. K. 1989. *Fundamentals of Digital Image Processing*. Eaglewood, NJ: Prentice-Hall, Inc.

Jenny, H. 1941. *Factors of Soil Formation*. New York: McGraw-Hill.

Jenson, S. K., and J. O. Domingue. 1988. Extracting topographic structure from digital elevation data for geographic information system analysis. *Photogr. Eng. Remote Sens.* 54(11): 1593–1600.

Jerri, A. J. 1977. The Shannon sampling theorem and its various extensions and applications: A tutorial review. *Proc. IEEE* 65: 1565–1596.

Jetz, W., and C. Rahbek. 2002. Geographic range size and determinants of avian species richness. *Science* 297: 1548–1551.

Jetz, W., C. R. Rahbek, and J. W. Lichstein. 2005. Local and global approaches to spatial data analysis in ecology. *Global Ecol. Biogeogr.* 14: 97–98.

Jiang, B., and Z. Rahman. 2012. Information theoretic analysis of linear shift-invariant edge-detection operators. *Opt. Eng.* 51: 067013.

Jin, H., G. Mountrakis, and P. Li. 2012. A super-resolution mapping method using local indicator variograms. *Int. J. Remote Sens.* 33(24): 7747–7773.

Johannesson, G., N. Cressie, and H. S. Huang. 2007. Dynamic multi-resolution spatial models. *Environ. Ecol. Stat.* 14: 5–25.

Johnson, G. D. 2004. Small area mapping of prostate cancer incidence in New York State (USA) using fully Bayesian hierarchical modelling. *Int. J. Health Geogr.* 3(1): 29.

Johnson, G. D., W. L. Myers, G. P. Patil, and C. Taillie. 2001. Characterizing watershed delineated landscapes in Pennsylvania using conditional entropy profiles. *Land. Ecol.* 16: 597–610.

Johnston, R. J., D. Gregory, G. Pratt, and M. Watts. 2000. *The Dictionary of Human Geography*, 4th ed. Oxford: Blackwell Publishing.

Jones, A. R., and N. A. Brunsell. 2009. A scaling analysis of soil moisture-precipitation interactions in a regional climate model. *Theor. Appl. Climatol.* 98: 221–235.

Jones, N. L., J. R. Walker, and S. F. Carle. 2005. Hydrogeologic unit flow characterization using transition probability geostatistics. *Ground Water* 43: 285–289.

Jordan, R. L., B. L. Huneycutt, and M. Werner. 1995. The SIR-C/X-SAR Synthetic Aperture Radar system. *IEEE Trans. Geosci. Remote Sens.* 33(4): 829–839.

Journel, A. G. 1983. Nonparametric estimation of spatial distributions. *Math. Geol.* 15(3): 445–468.

Journel, A. G., and C. V. Deutsch. 1993. Entropy and spatial disorder. *Math. Geol.* 25(3): 329–355.

Journel, A. G., and C. J. Huijbregts. 1978. *Mining Geostatistics*. New York: Academic Press.

Ju, J. C., S. Gopal, and E. D. Kolaczyk. 2005. On the choice of spatial and categorical scale in remote sensing land cover classification. *Remote Sens. Environ.* 96(1): 62–77.

Justice, C. O., B. L. Markham, J. B. G. Townshend, and R. L. Kennard. 1989. Spatial degradation of satellite data. *Int. J. Remote Sens.* 10(9): 1539–1561.

Justice, C. O., J. Townshend, E. Vermote, E. Masuoka, R. Wolfe, N. Saleous, D. P. Roy, and J. T. Morisette. 2002. An overview of MODIS land data processing and product status. *Remote Sens. Environ.* 83: 3–15.

Kahneman, D. 2011. *Thinking Fast and Slow*. New York: Farrar, Straus and Giroux.

Kalmykov, A. I., Y. I. Sinitsyn, O. V. Sytnik, and V. N. Tsymbal. 1989. Information content of radar remote sensing systems from space. *Radiophys. Quantum Electron.* 32: 779–785.

Karlström, A., and V. Ceccato. 2002. A new information theoretical measure of global and local spatial association. *S. Rev. Reg. Res.* 22: 13–40.

Kauth, R. J., and G. S. Thomas. 1976. The Tasselled Cap—A graphic description of the spectral-temporal development of agricultural crops as seen by Landsat symposium on machine processing of remotely sensed data. In *Proceedings of the Symposium on Machine Processing of Remotely Sensed Data*, Purdue University, West Lafayette, Indiana, pp. 41–51.

Kavzoglu, T. 2004. Simulating Landsat ETM+ imagery using DAIS 7915 hyperspectral scanner data. *Int. J. Remote Sens.* 25(22): 5049–5067.

Keim, R. F., A. E. Skaugset, and D. S. Bateman. 1999. Digital terrain modeling of small stream channels with a total-station theodolite. *Adv. Water Resour.* 23: 41–48.

Kent, M., W. J. Gill, R. E. Weaver, and R. P. Armitage. 1997. Landscape and plant community boundaries in biogeography. *Prog. Phys. Geogr.* 21(3): 315–353.

Kerr, Y. 2007. Soil moisture from space: Where are we? *Hydrogeol. J.* 15(1): 117–120.

Kidner, D. B., and D. H. Smith. 2003. Advances in the data compression of digital elevation models. *Comput. Geosci.* 29(8): 985–1002.

Kienzle, S. 2004. The effect of DEM raster resolution on first order, second order and compound terrain derivatives. *Trans. GIS* 8: 83–111.

Kim, S., N. Bose, and H. Valenzuela. 1990. Recursive reconstruction of high resolution image from noisy undersampled multiframes. *IEEE Trans. Acoust. Speech Signal Process.* 38: 1013–1027.

Kim, D., D. M. Cairns, and J. Bartholdy. 2009. Spatial heterogeneity and domain of scale on the Skallingen salt marsh, Denmark. *Dan. J. Geogr.* 109(1): 95–104.

Kitanidis, P. K. 1995. Quasi-linear geostatistical theory for inversing. *Water Resour. Res.* 31(10): 2411–2419.

Kitanidis, P. K. 1996. On the geostatistical approach to the inverse problem. *Adv. Water Resour.* 19(6): 333–342.

Kitanidis, P. K., and E. G. Vomvoris. 1983. A geostatistical approach to the inverse problem in groundwater modeling (steady state) and one-dimensional simulations. *Water Resour. Res.* 19(3): 677–690.

Klinkenberg, B., and M. F. Goodchild. 1992. The fractal properties of topography: A comparison of methods. *Earth Surf. Proc. Landforms* 17(3): 217–234.

Köhl, M., S. Magnussen, and M. Marchetti. 2006. *Sampling Methods, Remote Sensing and GIS Multiresource Forest Inventory.* Heidelberg: Springer Verlag.

Krajewski, W. F. 1987. Cokriging of radar-rainfall and rain gage data. *J. Geophys. Res.* 92(D8): 9571–9580.

Krummel, J. R., R. H. Gardner, G. Sugihara, O'R. V. Neill, and P. R. Coleman. 1987. Landscape patterns in a disturbed environment. *Oikos* 48: 321–324.

Kullback, S., and R. A. Leibler. 1951. On information and sufficiency. *Ann. Math. Stat.* 22(1): 79–86.

Kumar, L., and A. K. Skidmore. 2000. Radiation-vegetation relationships in a Eucalyptus forest. *Photogramm. Eng. Remote Sens.* 66(2): 193–204.

Kundur, D., and D. Hatzinakos. 1996. Blind image deconvolution. *Signal Proces. Mag. IEEE* 13(3): 43–64.

Kyriakidis, P. C. 2004. A geostatistical framework for area-to-point spatial interpolation. *Geogr. Anal.* 36: 259–289.

Kyriakidis, P. C., and J. L. Dungan. 2001. A geostatistical approach for mapping thematic classification accuracy and evaluating the impact of inaccurate spatial data on ecological model predictions. *Environ. Ecol. Stat.* 8(4): 311–330.

Kyriakidis, P. C., and A. G. Journel. 1999. Geostatistical space-time models: A review. *Math. Geol.* 31: 651–684.

Kyriakidis, P. C., A. M. Shortridge, and M. F. Goodchild. 1999. Geostatistics for conflation and accuracy assessment of digital elevation models. *Int. J. Geogr. Inf. Sci.* 13(7): 677–708.

Kyriakidis, P. C., and E. H. Yoo. 2005. Geostatistical prediction and simulation of point values from areal data. *Geogr. Anal.* 37: 124–151.

Lagacherie, P., P. Andrieux, and R. Bouzigues. 1996. Fuzziness and uncertainty of soil boundaries: From reality to coding in GIS. In *Geographic Objects with Indeterminate Boundaries*, eds. P. A. Borrough and A. U. Frank, 257–286. London: Taylor & Francis.

Lagendijk, R. L., J. Biemond, and D. E. Boekee. 1990. Identification and restoration of noisy blurred images using the expectation-maximization algorithm. *IEEE Trans. Acoust. Speech Signal Process.* 38(7): 1180–1191.

LaGro, J. Jr. 1991. Assessing patch shape in landscape mosaics. *Photogramm. Eng. Remote Sens.* 57(3): 285–293.

Lam, N. S.-N. 1983. Spatial interpolation methods: A review. *Cartogr. Geogr. Inf. Sci.* 10(2): 129–150.

Lam, N. S.-N. 2004. Fractals and scale in environmental assessment and monitoring. In *Scale and Geographic Inquiry*, eds. E. Sheppard and R. B. McMaster, 23–40. Oxford: Blackwell Publishing.

Lam, N. S.-N., D. Catts, D. Quattrochi, D. Brown, and R. McMaster. 2004. Chapter 4: Scale. In *A Research Agenda for Geographic Information Science*, eds. R. B. McMaster and E. L. Usery, 93–128. Boca Raton, FL: CRC Press.

Lam, N. S. N., and D. A. Quattrochi. 1992. On the issues of scale, resolution, and fractal analysis in the mapping sciences. *Prof. Geogr.* 44(1): 88–98.

Lambin, E. F., and H. J. Geist (eds.). 2006. *Land Use and Land Cover Change: Local Processes, Global Impacts*. Berlin: Springer.

Lambin, E., and A. Strahler. 1994. Change-vector analysis in multitemporal space: A tool to detect and categorize land cover change processes using high temporal resolution satellite data. *Remote Sens. Environ.* 48(2): 231–244.

Landau, H. J. 1967. Necessary density conditions for sampling and interpolation of certain entire functions. *Acta Math.* 117(1): 37–52.

Lantuejoul, C. 2001. *Geostatistical Simulation. Models and Algorithms*. New York: Springer.

Lark, R. M. 2003. Two robust estimators of the cross-variogram for multivariate geostatistical analysis of soil properties. *Eur. J. Soil Sci.* 54: 187–202.

Lark, R. M., and R. Webster. 1999. Analysis and elucidation of soil variation using wavelets. *Eur. J. Soil Sci.* 50: 185–206.

Lavin, H., and M. Quick. 1974. The OTF in electro-optical imaging systems. *Proceedings of the Society of Photo-Optical Instrumentation Engineers*, vol. 46 (Image Assessment and Specification), pp. 279–286.

Lawford, G. J. 2006. Fourier series and the cartographic line. *Int. J. Geogr. Inf. Sci.* 20(1): 31–52.

Lawford, G. J., and I. Gordon. 2010. The effect of offset correlation on positional accuracy estimation for linear features. *Int. J. Geogr. Inf. Sci.* 24(1): 129–140.

Lawrence, P. J., and T. N. Chase. 2010. Investigating the climate impacts of global land cover change in the community climate system model. *Int. J. Climatol.* 30(13): 2066–2087.

Le, Q. B., S. J. Park, and P. L. G. Vlek. 2010. Land Use Dynamic Simulator (LUDAS): A multi-agent system model for simulating spatio-temporal dynamics of coupled human-landscape system: 2. Scenario-based application for impact assessment of land-use policies. *Ecol. Inf.* 5: 203–221.

Le Coz, M., F. Delclaux, P. Genthon, and G. Favreau. 2009. Assessment of Digital Elevation Model (DEM) aggregation methods for hydrological modeling: Lake Chad basin, Africa. *Comput. Geosci.* 35(8): 1661–1670.

Le Hégarat-Mascle, S., D. Vidal-Madjar, O. Taconet, and M. Zribi. 1997. Application of Shannon information theory to a comparison between L- and C-band SIR-C polarimetric data versus incidence angle. *Remote Sens. Environ.* 60: 121–130.

Lea, C., and A. C. Curtis. 2010. Thematic Accuracy Assessment Procedures: National Park Service Vegetation Inventory, version 2.0. Natural Resource Report NPS/2010/NRR—2010/204. Fort Collins, CO: National Park Service.

Lee, J. S. 1980. Digital image enhancement and noise filtering by use of local statistics. *IEEE Trans. Pattern Anal. Mach. Intell.* 2: 165–168.

Lee, J., P. Snyder, and P. Fisher. 1992. Modeling the effect of data errors on feature extraction from digital elevation models. *Photogramm Eng. Remote Sens.* 58(10): 1461–1467.

Legendre, P., M. R. T. Dale, M.-J. Fortin, J. Gurevitch, M. Hohn, and D. Myers. 2002. The consequences of spatial structure for the design and analysis of ecological field surveys. *Ecography* 25: 601–615.

Lehmann, T. M., C. Gonner, and K. Spitzer. 1999. Survey: Interpolation methods in medical image processing. *IEEE Trans. Med. Imag.* 18: 1049–1075.

Lempel, A., and J. Ziv. 1986. Compression of two-dimensional data. *IEEE Trans. Inf. Theory* 32(1): 2–8.

Leung, W. Y. V., P. J. Bones, and R. G. Lane. 2001. Statistical interpolation of sampled images. *Opt. Eng.* 40: 547–553.

Levin, A., Y. Weiss, F. Durand, and W. T. Freeman. 2011. Efficient marginal likelihood optimization in blind deconvolution. *Proc. IEEE CVPR* 2657–2664.

Levin, S. A. 1992. The problem of pattern and scale in ecology. *Ecology* 73(6): 1943–1967.

Li, D., J. Zhang, and H. Wu. 2012. Spatial data quality and beyond. *Int. J. Geogr. Inf. Sci.* 26(12): 2277–2290.

Li, J., and A. D. Heap. 2011. A review of comparative studies of spatial interpolation methods in environmental sciences: Performance and impact factors. *Ecol. Inf.* 6(3–4): 228–241.

Li, H., B. S. Manjunath, and S. K. Mitra. 1995. Multisensor image fusion using the wavelet transform. *Graph. Models Image Process.* 57(3): 235–245.

Li, H., and J. F. Reynolds. 1994. A simulation experiment to quantify spatial heterogeneity in categorical maps. *Ecology* 75(8): 2446–2455.

Li, S. T., J. T. Kwok, and Y. Wang. 2002. Using the discrete wavelet frame transform to merge Landsat TM and SPOT panchromatic images. *Inf. Fusion* 3: 17–23.

Li, Z., and P. Huang. 2002. Quantitative measures for spatial information of maps. *Int. J. Geogr. Inf. Sci.* 16(7): 699–709.

Liang, S. 2003. *Quantitative Remote Sensing of Land Surfaces*. New York: Wiley.

Liang, S., H. L. Fang, M. Z. Chen, C. J. Shuey, C. Walthall, C. Daughtry, J. Morisette, C. Schaaf, and A. Strahler. 2002. Validating MODIS land surface reflectance and albedo products: Methods and preliminary results. *Remote Sens. Environ.* 83(1–2): 149–162.

Liang, Y., J. Li, and K. Guo. 2012. Lossless compression of hyperspectral images using hybrid context prediction. *Opt. Express* 20(7): 8199–8206.

Lin, P. D., and C. S. Liu. 2011. Geometrical MTF computation method based on the irradiance model. *Appl. Phys. B: Lasers Opt.* 102(1): 243–249.

Lin, Z., and H. Y. Shum. 2004. Fundamental limits of reconstruction-based superresolution algorithms under local translation. *IEEE Trans. Pattern Anal. Mach. Intell.* 26(1): 83–97.

Lindsay, R. W., D. B. Percival, and D. A. Rothrock. 1996. The discrete wavelet transform and the scale analysis of the surface properties of sea ice. *IEEE Trans. Geosci. Remote Sens.* 34(3): 771–787.

Linfoot, E. H. 1955. Information theory and optical images. *J. Opt. Soc. Am.* 45(10): 808–818.

Ling, F., Y. Du, F. Xiao, H. Xue, and S. Wu. 2010. Super-resolution land-cover mapping using multiple sub-pixel shifted remotely sensed images. *Int. J. Remote Sens.* 31(19): 5023–5040.

Liu, J., J. Zhang, and S. Fang. 2005. Uncertainty characterization in remotely sensed land cover information. In *Proc. SPIE 6045, MIPPR 2005: Geospatial Information, Data Mining, and Applications*, 60453D (December 2, 2005); doi: 10.1117/12.651871.

Liu, Y. B., T. Hiyama, and Y. Yamaguchi. 2006. Scaling of land surface temperature using satellite data: A case examination on ASTER and MODIS products over a heterogeneous terrain area. *Remote Sens. Environ.* 105: 115–128.

Liu, Y., and A. G. Journel. 2009. A package for geostatistical integration of coarse and fine scale data. *Comput. Geosci.* 35(3): 527–547.

Lo, C. P., and L. J. Watson. 1998. The influence of geographic sampling methods on vegetation map accuracy evaluation in a swampy environment. *Photogramm. Eng. Remote Sens.* 64(12): 1189–1200.

Lobl, E. 2001. Joint advanced microwave scanning radiometer (AMSR) science team meeting. *Earth Obs.* 13(3): 3–9.

Loew, A., and W. Mauser. 2007. Generation of geometrically and radiometrically terrain corrected SAR image products. *Remote Sens. Environ.* 106: 337–349.

Longley, P. A., and M. Batty (eds.). 1996. *Spatial Analysis: Modelling in a GIS Environment.* Cambridge: GeoInformation International.

Longley, P. A., M. F. Goodchild, D. J. Maguire, and D. W. Rhind. 2011. *Geographical Information Systems and Science*, 3rd ed. Hoboken: Wiley.

Lovejoy, S. 1982. Area-perimeter relation for rain and cloud areas. *Science* 216: 185–187.

Lu, D., M. Batistella, E. Moran, S. Hetrick, D. Alves, and E. Brondizio. 2011. Fractional forest cover mapping in the Brazilian Amazon with a combination of MODIS and TM images. *Int. J. Remote Sens.* 32(22): 7131–7149.

Lu, D., and Q. Weng. 2007. A survey of image classification methods and techniques for improving classification performance. *Int. J. Remote Sens.* 28(5): 823–870.

Lukosz, W. 1966. Optical systems with resolving powers exceeding the classical limit. *J. Opt. Soc. Am.* 56(11): 1463–1471.

Lunetta, R. S., and C. D. Elvidge. 1998. *Remote Sensing Change Detection: Environmental Monitoring Methods and Applications.* Chelsea: Ann Arbor Press.

Luoto, M., and J. Hjort. 2008. Downscaling of coarse-grained geomorphological data. *Earth Surf. Proc. Land.* 33: 75–89.

Lyon, S. W., and P. A. Troch. 2010. Development and application of a catchment similarity index for subsurface flow. *Water Resour. Res.* 46(3): W03511.

Ma, T., C. Zhou, Y. Xie, B. Qin, and Y. Ou. 2009. A discrete square global grid system based on the parallels plane projection. *Int. J. Geogr. Inf. Sci.* 23(10): 1297–1313.

MacEachren, A. M., and J. V. Davidson. 1987. Sampling and isometric mapping of continuous geographic surfaces. *Am. Cartogr.* 14: 299–320.

Machuca-Mory, D. F. 2010. *Geostatistics with Location-Dependent Statistics.* Edmonton, Canada: University of Alberta.

Majumdar, A., and A. E. Gelfand. 2007. Multivariate spatial modeling using convolved covariance functions. *Math. Geol.* 39: 225–245.

Malila, W. A. 1985. Comparison of the information contents of Landsat TM and MSS data. *Photogramm. Eng. Remote Sens.* 51(9): 1449–1457.

Maling, D. H. 1968. How long is a piece of string? *Cartogr. J.* 5: 147–156.

Mallat, S. G. 1989. A theory for multiresolution signal decomposition: The wavelet representation. *IEEE Trans. Pattern Anal. Mach. Intell.* 11(7): 674–693.

Mandelbrot, B. B. 1967. How long is the coast of Britain? Statistical self-similarity and fractional dimension. *Science* 156: 636–638.

Mantel, N. 1967. The detection of disease clustering and a generalized regression approach. *Cancer Res.* 27: 209–220

Marceau, D. J. 1999. The scale issue in social and natural sciences. *Can. J. Remote Sens.* 25: 347–356.

Marceau, D. J., and G. J. Hay. 1999. Remote sensing contributions to the scale issue. *Can. J. Remote Sens.* 25: 357–366.

Marchand, B. 1972. Information theory and geography. *Geogr. Anal.* 4: 234–257.

Mark, D. M., and F. Csillag. 1989. The nature of boundaries on 'area-class' maps. *Cartographica* 26(1): 65–77.

Markham, B. L. 1985. The Landsat sensors' spatial responses. *IEEE Trans. Geosci. Remote Sens.* GE-23(6): 864–875.

Marks, R. J. II. 1991. *Introduction to Shannon Sampling and Interpolation Theory.* New York: Springer.

Martin, D. 1996. An assessment of surface and zonal models of population. *Int. J. Geogr. Inf. Syst.* 10: 973–989.

Marvasti, F. 1996. Interpolation of low-pass signals at half the Nyquist rate. *IEEE Signal Proc. Lett.* 3: 42–43.

Marvasti, F. A. 2001. *Nonuniform Sampling: Theory and Practice.* New York: Kluwer Academic Publishers/Plenum Publishers.

Masek, J. G., E. F. Vermote, N. E. Saleous, R. Wolfe, F. G. Hall, K. F. Huemmrich, G. Feng, J. Kutler, and T.-K. Lim. 2006. A Landsat surface reflectance data set for North America, 1990–2000. *Geosci. Remote Sens. Lett.* 3: 68–72.

Mason, D. C., M. O'Conaill, and I. McKendrick. 1994. Variable resolution block kriging using a hierarchical spatial data structure. *Int. J. Geogr. Inf. Syst.* 8: 429–449.

Massaro, D., F. Savazzi, C. Di Dio, D. Freedberg, V. Gallese et al. 2012. When art moves the eyes: A behavioral and eye-tracking study. *PLoS ONE* 7(5): e37285.

Mather, P. M., and M. Koch. 2011. *Computer Processing of Remotely Sensed Images: An Introduction,* 4th ed. New York: Wiley.

Matheron, G. 1971. *The Theory of Regionalized Variables and its Applications.* Les Cahiers du Centre de Morphologie Mathématique de Fontainebleau, No. 5. Fontainebleau: Ecole Supérieure des Mines de Paris.

Mayaux, P., F. Achard, and J. Malingreau. 1998. Global tropical forest area measurements derived from coarse resolution satellite imagery: A comparison with other approaches. *Environ. Conserv.* 25(1): 37–52.

Mayaux, P., H. Eva, J. Gallego, A. H. Strahler, M. Herold, S. Agrawal, S. Naumov, E. E. De Miranda, C. M. Di Bella, C. Ordoyne, Y. Kopin, and P. Roy. 2006. Validation of the Global Land Cover 2000 Map. *IEEE Trans. Geosci. Remote Sens.* 44(7): 1728–1739.

Mayaux, P., and E. F. Lambin. 1995. Estimation of tropical forest area from coarse spatial resolution data: A two-step correction function for proportional errors due to spatial aggregation. *Remote Sens. Environ.* 53(1): 1–15.

McBratney, A. B. 1998. Some considerations on methods for spatially aggregating and disaggregating soil information. *Nutrient Cycl. Agroecosyst.* 50: 51–62.

McBratney, A. B., M. L. Mendonca Santos, and B. Minasny. 2003. On digital soil mapping. *Geoderma* 117: 3–52.

McBratney, A. B., and R. Webster. 1986. Choosing functions for semi-variograms of soil properties and fitting them to sampling estimates. *J. Soil Sci.* 37: 617–639.

McCoy, S. W., J. W. Kean, J. A. Coe, D. M. Staley, T. A. Wasklewicz, and G. E. Tucker. 2010. Evolution of a natural debris flow: In situ measurements of flow dynamics, video imagery, and terrestrial laser scanning. *Geology* 38: 735–738.

McFarland, M. J., R. L. Miller, and C. M. U. Neale. 1990. Land surface temperature derived from the SSM/I passive microwave brightness temperatures. *IEEE Trans. Geosci. Remote Sens.* 28(5): 839–845.

McGarigal, K., and B. J. Marks. 1995. FRAGSTATS: Spatial Pattern Analysis Program for Quantifying Landscape Structure. General Technical Report PNW-GTR-351: Portland, OR: USDA Forest Service, Pacific Northwest Research Station.

McLennan, J. A. 2007. *The Decision of Stationarity.* PhD Dissertation. University of Alberta.

McMaster, K. J. 2002. Effects of digital elevation model resolution on derived stream network positions. *Water Resour. Res.* 38: 1042.

McMaster, R. B., and E. L. Usery (eds.). 2004. *A Research Agenda for Geographic Information Science.* Boca Raton, FL: CRC Press.

McPherson, J. M., W. Jetz, and D. J. Rogers. 2006. Using coarse-grained occurrence data to predict species distributions at finer spatial resolutions—Possibilities and limitations. *Ecol. Model.* 192(3–4): 499–522.

McRoberts, R. E. 2006. A model-based approach to estimating forest area. *Remote Sens. Environ.* 103: 56–66.

Meisel, J. E., and M. G. Turner. 1998. Scale detection in real and artificial landscapes using semivariance analysis. *Landscape Ecol.* 13: 347–362.

Memon, N. D., K. Sayood, and S. S. Magliveras. 1994. Lossless compression of multispectral image data. *IEEE Trans. Geosci. Remote Sens.* 32(2): 282–289.

Mendicino, G., and A. Sole. 1997. The information content theory for the estimation of the topographic index distribution used in TOPMODEL. *Hydrol. Proc.* 11: 1099–1114.

Merlin, O., A. A. Bitar, J. P. Walker, and Y. Kerr. 2009. A sequential model for disaggregating near-surface soil moisture observations using multi-resolution thermal sensors. *Remote Sens. Environ.* 113: 2275–2284.

Metternicht, G. I., and J. A. Zinck. 1998. Evaluating the information content of JERS-1 SAR and Landsat TM data for discrimination of soil erosion features. *ISPRS J. Photogramm. Remote Sens.* 53(3): 143–153.

Michalak, A. M. 2008. Technical note: Adapting a fixed-lag Kalman smoother to a geostatistical atmospheric inversion framework. *Atm. Chem. Phys.* 8: 6789–6799.

Michalak, A. M., L. Bruhwiler, and P. P. Tans. 2004. A geostatistical approach to surface flux estimation of atmospheric trace gases. *J. Geophys. Res.* 109: D14109.

Mikhail, E. M., J. S. Bethel, and J. C. McGlone. 2001. *Introduction to Modern Photogrammetry.* New York: Wiley.

Miller, H. J., and S. L. Shaw. 2001. *Geographic Information Systems for Transportation: Principles and Applications.* Oxford, UK: Oxford University Press.

Mills, P. 2011. Efficient statistical classification of satellite measurements. *Int. J. Remote Sens.* 32(21): 6109–6132.

Milne, A. E., K. A. Haskard, C. P. Webster, I. A. Truan, K. W. T. Goulding, and R. M. Lark. 2011. Wavelet analysis of the correlations between soil properties and potential nitrous oxide emission at farm and landscape scales. *Eur. J. Soil Sci.* 62(3): 467–478.

Milzow, C., and W. Kinzelbach. 2010. Accounting for subgrid scale topographic variations in flood propagation modeling using MODFLOW. *Water Resour. Res.* 46, W10521.

Ming, D., J. Yang, L. Li, and Z. Song. 2011. Modified ALV for selecting the optimal spatial resolution and its scale effect on image classification accuracy. *Math. Comput. Model.* 54: 1061–1068.

Mitasova, H., and J. Hofierka. 1993. Interpolation by regularized spline with tension: II. Application to terrain modeling and surface geometry analysis. *Math. Geol.* 25(6): 657–669.

Modis, K., and K. Papaodysseus. 2006. Theoretical estimation of the critical sampling size for homogeneous ore bodies with small nugget effect. *Math. Geol.* 38(4): 489–501.

Modis, K., G. Papantonopoulos, K. Komnitsas, and K. Papaodysseus. 2008. Mapping optimization based on sampling size in earth related and environmental phenomena. *Stoch. Environ. Res. Risk Assess.* 22(1): 83–93.

Moellering, H., and W. Tobler. 1972. Geographical variances. *Geogr. Anal.* 4: 34–50.

Molnar, D. K., and P. Y. Julien. 2000. Grid-size effects on surface runoff modeling. *J. Hydrol. Eng.* 5: 8–16.

Montgomery, M. R. 2008. The urban transformation of the developing world. *Science* 319: 761–764.

Moody, A. 1998. Using landscape spatial relationships to improve estimates of land-cover area from coarse resolution remote sensing. *Remote Sens. Environ.* 64(2): 202–220.

Moody, A., and C. E. Woodcock. 1994. Scale-dependent errors in the estimation of landcover proportions-implications for global land-cover datasets. *Photogramm. Eng. Remote Sens.* 60(5): 585–594.

Moody, A., and C. E. Woodcock. 1995. The influence of scale and the spatial characteristics of landscapes on land-cover mapping using remote sensing. *Land. Ecol.* 10(6): 363–379.

Moody, A., and C. E. Woodcock. 1996. Calibration-based models for correction of area estimates derived from coarse resolution land-cover data. *Remote Sens. Environ.* 58(3): 225–241.

Moon, Z. K., and F. L. Farmer 2001. Population density surface: A new approach to an old problem. *Soc. Nat. Resour.: Int. J.* 14(1): 39–51.

Moore, R. K. 1979. Tradeoff between picture element dimensions and noncoherent averaging in side-looking airborne radar. *IEEE Trans. Aerospace Electron. Syst.* AES-15(5): 697–708.

Moore, I. D., P. E. Gessler, G. A. Nielsen, and G. A. Peterson. 1993. Soil attribute prediction using terrain analysis. *Soil Sci. Soc. Am. J.* 57(2): 443–452.

Moore, I. D., R. B. Grayson, and A. R. Ladson. 1991. Digital terrain modelling: A review of hydrological, geomorphological, and biological applications. *Hydrol. Proc.* 5(1): 3–30.

Moran, P. A. P. 1950. Notes on continuous stochastic phenomena. *Biometrika* 37(1): 17–23.

Morisette, J. T., F. Baret, and J. L. Privette et al. 2006. Validation of global moderate resolution LAI products: A framework proposed within the CEOS Land Product Validation subgroup. *IEEE Trans. Geosci. Remote Sens.* 44(7): 1804–1817.

Morisette, J. T., J. L. Privette, and C. O. Justice. 2002. A framework for the validation of MODIS land products. *Remote Sens. Environ.* 83(1–2): 77–96.

Myers, D. E. 1982. Matrix formulation of co-kriging. *Math. Geol.* 14(3): 249–257.

Myers, D. E. 1988. Vector conditional simulation. In *Geostatistic*, ed. M. Armstrong. Dordrecht: Kluwer Academic Publishers.

Myers, D. E. 1989. To be or not to be ... stationary? That is the question. *J. Math. Geol.* 21(3): 347–362.

Myers, D. E. 1997. Statistical models for multiple-scaled analysis. In: *Scale in Remote Sensing and GIS*, eds. D. A. Quattrochi and M. F. Goodchild, 273–294. Boca Raton, FL: Lewis Publishers.

Naraghi, M., W. Stromberg, and M. Daily. 1983. Geometric rectification of radar imagery using digital elevation models. *Photogramm. Eng. Remote Sens.* 49: 195–199.

Narayanan, R. M., M. K. Desetty, and S. E. Reichenbach. 2002. Effects of spatial resolution on information content characterisation in remotely sensed imagery based on classification accuracy. *Int. J. Remote Sens.* 23: 537–553.

Narayanan, R. M., T. S. Sankaravadivelu, and S. E. Reichenbach. 2000. Dependence of image information content on gray-scale resolution. *GeoCarto Int.* 15(4): 17–30.

Neal, J. C., P. M. Atkinson, and C. W. Hutton. 2012. Adaptive space–time sampling with wireless sensor nodes for flood forecasting. *J. Hydrol.* 414–415, 136–147.

Neeser, F. D., and J. L. Massey. 1993. Proper complex random processes with applications to information theory. *IEEE Trans. Inf. Theory* 39(4): 1293–1302.

Nelder, J. A., and R. W. M. Wedderburn. 1972. Generalized linear models. *J. R. Stat. Soc.: Ser. A* 135(3): 370–384.

Nelson, M. D., R. E. Mcroberts, G. R. Holden, and M. E. Bauer. 2009. Effects of satellite image spatial aggregation and resolution on estimates of forest land area. *Int. J. Remote Sens.* 30(8): 1913–1940.

Nelson, R. 1989. Regression and ratio estimators to integrate AVHRR and MSS data. *Remote Sens. Environ.* 30(3): 201–216.

Niedda, M. 2004. Upscaling hydraulic conductivity by means of entropy of terrain curvature representation. *Water Resour. Res.* 40: W04206.

Nikolakopoulos, K. G., E. K. Kamaratakis, and N. Chrysoulakis. 2006. Comparison for two regions in Crete, Greece. *Int. J. Remote Sens.* 27(21): 4819–4838.

Norton, C. L., G. C. Brock, and R. Welch. 1977. Optical and modulation transfer functions. *Photogramm. Eng. Remote Sens.* 43: 613–636.

Nunez, J., X. Otazu, O. Fors, A. Prades, V. Pala, and R. Arbiol. 1999. Multiresolution-based image fusion with additive wavelet decomposition. *IEEE Trans. Geosci. Remote Sens.* 37(3): 1204–1211.

Nusser, S. M., and E. E. Klaas. 2003. Survey methods for assessing land cover map accuracy. *Environ. Ecol. Stat.* 10(3): 309–331.

Nyquist, H. 1928. Certain topics in telegraph transmission theory. *Trans. Am. Inst. Electr. Eng.* 47(2): 617–644.

Oberthür, T., P. Goovaerts, and A. Dobermann. 1999. Mapping soil texture classes using field texturing, particle size distribution and local knowledge by both conventional and geo-statistical methods. *Eur. J. Soil Sci.* 50: 457–479.

Oliver, C. J. 1991. Review article: Information from SAR images. *J. Phys. D: Appl. Phys.* 24: 1493–1514.

Oliver, C. J., and S. Quegan. 2004. *Understanding Synthetic Aperture Radar Images.* SciTech Publishing.

Olivera, F., M. S. Lear, J. S. Famiglietti, and K. Asante. 2002. Extracting low-scale river network from high-scale digital elevation models. *Water Resour. Res.* 38(11): 12–31.

Ollier, S., D. Chessel, P. Couteron, R. Pélissier, and J. Thioulouse. 2003. Comparing and classifying one-dimensional spatial patterns: An application to laser altimeter profile. *Remote Sens. Environ.* 85: 453–462.

O'Neill, R. V., and B. Rust. 1979. Aggregation error in ecological models. *Ecol. Model.* 7: 91–105.

O'Neill, R. V., J. R. Krummel, R. H. Gardner, G. Sugihara, B. Jackson, D. L. DeAngelis, B. T. Milne et al. 1988. Indices of landscape pattern. *Land. Ecol.* 1(3): 153–162.

Openshaw, S. 1983. *The Modifiable Areal Unit Problem. Concepts and Techniques in Modern Geography: CAT-MOG Series 38.* Norwich, UK: GeoBooks.

Openshaw, S., and P. J. Taylor. 1981. The modifiable areal unit problem. In *Quantitative Geography: A British View*, eds. N. Wrigley and R. J. Bennett. London: Routledge.

Ord, J. K., and A. Getis. 1995. Local spatial autocorrelation statistics: Distributional issues and an application. *Geogr. Anal.* 27: 286–306.

Ortiz, B. V., C. Perry, P. Goovaerts, G. Vellidis, and D. Sullivan. 2010. Geostatistical modeling of the spatial variability and risk areas of southern root-knot nematodes in relation to soil properties. *Geoderma* 156: 243–252.

O'Sullivan, J. A., R. E. Blahut, and D. L. Snyder. 1998. Information-theoretic image formation. *IEEE Trans. Inf. Theory* 44(6): 2094–2123.

Ouarda, T. B. M. J., C. Girard, G. S. Cavadias, and B. Bobée. 2001. Regional flood frequency estimation with canonical correlation analysis. *J. Hydrol.* 254(1–4): 157–173.

Ouedraogo, I., P. Savadogo, M. Tigabu, S. D. Dayamba, and P. C. Odén. 2011. Systematic and random transitions of land-cover types in Burkina Faso, West Africa. *Int. J. Remote Sens.* 32(18): 5229–5245.

Pajares, G., and J. M. de la Cruz. 2004. A wavelet-based image fusion tutorial. *Pattern Recogn.* 37(2004): 1855–1872.

Palubinskas, G., P. Reinartz, and B. Richard. 2010. Image acquisition geometry analysis for the fusion of optical and radar remote sensing data. *Int. J. Image Data Fusion* 1(3): 271–282.

Paola, C., and M. Leeder. 2011. Environmental dynamics: Simplicity v. complexity. *Nature* 469: 38–39.

Papritz, A., and H. Flühler. 1994. Temporal change of spatially autocorrelated soil properties: Optimal estimation by cokriging. *Geoderma* 62(1–3): 29–43.

Pardo-Igúzquiza, E. 1999. VARFIT: A Fortran-77 program for fitting variogram models by weighted least squares. *Comput. Geosci.* 25(3): 251–261.

Paredes, E. G., M. Bóo, M. Amor, J. D. Bruguera, and J. Döllner. 2012. Extended hybrid meshing algorithm for multiresolution terrain models. *Int. J. Geogr. Inf. Sci.* 26(5): 771–793.

Parenteau, M. P., and M. C. Sawada. 2011. The modifiable areal unit problem (MAUP) in the relationship between exposure to NO_2 and respiratory health. *Int. J. Health Geogr.* 10: 58.

Paris, J. F., and H. H. Kwong. 1988. Characterization of vegetation with combined Thematic Mapper (TM) and Shuttle Imaging Radar (SIR-B) image data. *Photogramm. Eng. Remote Sens.* 54(8): 1187–1193.

Park, S. C., M. K. Park, and M. G. Kang. 2003. Super-resolution image reconstruction: A technical overview. *IEEE Signal Proc. Mag.* 20(3): 21–36.

Park, S. K., and R. A. Schowengerdt. 1982. Image sampling, reconstruction, and the effect of sample-scene phasing. *Appl. Opt.* 21(17): 3142–3151.

Park, S. K., R. A. Schowengerdt, and M. A. Kaczynski. 1984. Modulation-transfer-function analysis for sampled image systems. *Appl. Opt.* 23(15): 2572–2582.

Parker, J. A., R. V. Kenyon, and D. E. Troxel. 1983. Comparison of interpolating methods for image resampling. *IEEE Trans. Med. Imag.* 2(1): 31–39.

Parzen, E. 1962. On estimation of a probability density function and mode. *Ann. Math. Stat.* 33(3): 1065–1076.

Patino, J. E., and J. C. Duque. 2013. A review of regional science applications of satellite remote sensing in urban settings. *Comput. Environ. Urban Syst.* 37(1): 1–17.

Peet, R. K. 1978. Latitudinal variation in southern Rocky Mountain forests. *J. Biogeogr.* 5: 275–289.

Peng, H., F. Long, and D. Chris. 2005. Feature selection based on mutual information: Criteria of max-dependency, max-relevance, and min-redundancy. *IEEE Trans. Pattern Anal. Mach. Intell.* 27: 1226–1238.

Pentland, A. P. 1984. Fractal-based description of natural scenes. *IEEE Trans. Pattern Anal. Mach. Intell.* 6: 661–674.

Percival, D. P. 1995. On estimation of the wavelet variance. *Biometrika* 82: 619–631.

Perron, J. T., J. W. Kirchner, and W. E. Dietrich. 2008. Spectral signatures of characteristic spatial scales and nonfractal structure in landscapes. *J. Geophys. Res.* 113: F04003. doi: 10.1029/2007JF000866.

Peters-Lidard, C. D., F. Pan, and E. F. Wood. 2001. A re-examination of modeled and measured soil moisture spatial variability and its implications for land surface modeling. *Adv. Water Resour.* 24: 1069–1083.

Petersen, D. P., and D. Middleton. 1962. Sampling and reconstruction of wave-number-limited functions in N-dimensional Euclidean spaces. *Inf. Contr.* 5(4): 279–323.

Petrov, Y., and Z. Li. 2003. Local correlations, information redundancy, and sufficient pixel depth in natural images. *J. Opt. Soc. Am. A* 20(1): 56–66.

Peucker, T. K., and D. H. Douglas. 1975. Detection of surface specific points by local parallel processing of discrete terrain elevation data. *Comput. Graph. Image Proc.* 4(4): 75–387.

Peuquet, D. J. 2001. Making space for time: Issues in space-time data representation. *Geoinformatica* 5(1): 11–32.

Pfeifer, M., M. Disney, T. Quaife, and R. Marchant. 2012. Terrestrial ecosystems from space: A review of earth observation products for macroecology applications. *Global Ecol. Biogeogr.* 21(6): 603–624.

Pierce, L., J. Kellndorfer, W. Walker, and O. Barros. 2006. Evaluation of the horizontal resolution of SRTM elevation data. *Photogramm. Eng. Remote Sens.* 72(11): 1235–1244.

Pierce, L. E., F. T. Ulaby, K. Sarabandi, and M. C. Dobson. 1994. Knowledge-based classification of polarimetric SAR images. *IEEE Trans. Geosci. Remote Sens.* 32(5): 1081–1086.

Pike, R. J. 1988. The geometric signature: Quantifying landslide-terrain types from digital elevation models. *Math. Geol.* 20: 491–511.

Pike, R. J. 2000. Geomorphometry-diversity in quantitative surface analysis. *Prog. Phys. Geogr.* 24(1): 1–20.

Pike, R. J., and W. J. Rozema. 1975. Spectral analysis of landforms. *Ann. Assoc. Am. Geogeogr.* 65: 499–516.

Plaza, A., Q. Du, J. Bioucas-Dias, X. Jia, and F. Kruse. 2011. Foreword to the special issue on spectral unmixing of remotely sensed data. *IEEE Trans. Geosci. Remote Sens.* 49(11): 4103–4110.

Plotnick, R. E., R. H. Gardner, W. W. Hargrove, K. Prestegaard, and M. Perlmutter. 1996. Lacunarity analysis: A general technique for the analysis of spatial patterns. *Phys. Rev. E* 53(5): 5461–5468.

Plotnick, R. E., R. H. Gardner, and R. V. O'Neill. 1993. Lacunarity indices as measures of landscape texture. *Land. Ecol.* 8(3): 201–211.

Poggio, L., A. Gimona, and I. Brown. 2012. Spatio-temporal MODIS EVI gap filling under cloud cover: An example in Scotland. *ISPRS J. Photogramm. Remote Sens.* 72: 56–72.

Pohl, C., and J. L. Van Genderen. 1998. Multisensor image fusion in remote sensing: Concepts, methods and applications. *Int. J. Remote Sens.* 19(5): 823–854.

Poli, D. 2007. A rigorous model for spaceborne linear array sensors. *Photogramm. Eng. Remote Sens.* 73(2): 187–196.

Poorter, H., and O. Nagel. 2000. The role of biomass allocation in the growth response of plants to different levels of light, CO_2, nutrients and water: A quantitative review. *Austr. J. Plant Physiol.* 27(12): 1191–1191.

Popper, K. 1963. *Conjectures and Refutations: The Growth of Scientific Knowledge*. London: Routledge.

Porcello, L., N. Massey, R. Innes, and J. Marks. 1976. Speckle reduction in synthetic aperture-radars. *J. Opt. Soc. Am.* 66: 1305–1311.

Porcu, E., and J. Mateu. 2007. Mixture-based modeling for space–time data. *Environmetrics* 18: 285–302.

Potter, L. C., E. Ertin, J. T. Parker, and M. Cetin. 2010. Sparsity and compressed sensing in Radar imaging. *Proc. IEEE* 98(6): 1006–1020.

Prasad, S. 2000. Information capacity of a seeing-limited imaging system. *Opt. Commun.* 177: 119–134.

Pratt, W. K. 1991. *Digital Image Processing*. New York: Wiley.

Preparata, F. P., and M. I. Shamos. 1985. *Computational Geometry: An Introduction*. New York: Springer.

Price, J. C. 1984. Comparison of the information content of data from the Landsat-4 thematic mapper and the multispectral scanner. *IEEE Trans. Geosci. Remote Sens.* GE-22(3): 272–281.

Price, J. C. 1994. How unique are spectral signatures. *Remote Sens. Environ.* 49: 181–186.

Price, J. C. 1997. Spectral band selection for visible-near infrared remote sensing: Spectral-spatial resolution tradeoffs. *IEEE Trans. Geosci. Remote Sens.* 35(5): 1277–1285.

Price, J. C. 1999. Combining multispectral data of differing spatial resolution. *IEEE Trans. Geosci. Remote Sens.* 37(3): 1199–1203.

Primot, J., and M. Chambon. 1997. Modulation transfer function assessment for sampled imaging systems: Effect of intensity variations in periodic thin-line targets. *Appl. Opt.* 36(29): 7307–7314.

Pultar, E., M. Cova, M. Yuan, and M. F. Goodchild. 2010. EDGIS: A dynamic GIS based on space time points. *Int. J. Geogr. Inf. Sci.* 24(3): 329–346.

Qi, D. S., and T. Hesketh. 2005. An analysis of upscaling techniques for reservoir simulation. *Petrol. Sci. Technol.* 23: 827–842.

Qian, Y. Y. 2003. A study on the housing security policy in China: Affordable and low-rent housing. *China's Real Estate* 2003(8): 57–60 (in Chinese).

Qu, G. H., D. L. Zhang, and P. F. Yan. 2002. Information measure for performance of image fusion. *Electr. Lett.* 38(7): 313–315.

Quarmby, N. A., J. R. G. Townshend, J. J. Settle, M. Milnes, T. L. Hindle, and N. Silleos. 1992. Linear mixture modeling applied to AVHRR data for crop area estimation. *Int. J. Remote Sens.* 13(3): 415–425.

Quattrochi, D. A., and M. F. Goodchild (eds.). 1997. *Scale in Remote Sensing and GIS*. New York: CRC Press.

Quinn, P., K. Beven, and R. Lamb. 1995. The ln(a/tan beta) index: How to calculate it and how to use it within the TOPMODEL framework. *Hydrol. Proc.* 9: 161–182.

Rabus, B., M. Eineder, A. Roth, and R. Bamler. 2003. The Shuttle Radar Topography Mission—A new class of digital elevation models acquired by spaceborne radar. *ISPRS J. Photogramm. Remote Sens.* 57: 241–262.

Raffy, M. 1992. Change of scale in models of remote sensing: A general method for spatialisation of models. *Remote Sens. Environ.* 40(2): 101–112.

Rahman, A. F., J. A. Gamon, D. A. Sims, and M. Schmidt. 2003. Optimum pixel size for hyperspectral studies of ecosystem function in southern California chaparral and grassland. *Remote Sens. Environ.* 84: 192–207.

Razlighi, Q. R., N. Kehtarnavaz, and A. Nosratinia. 2009. Computation of image spatial entropy using quadrilateral Markov random field. *IEEE Trans. Image Process.* 18(12): 2629–2639.

Rebillard, P., and D. Evans. 1983. Analysis of coregistered Landsat, Seasat and SIR-A images of varied terrain types. *Geophys. Res. Lett.* 10(4): 277–280.

Reeves, G., and M. Gastpar. 2012. The sampling rate-distortion tradeoff for sparsity pattern recovery in compressed sensing. *IEEE Trans. Inf. Theory* 58(10): 3065–3092.

Remmel, T. K., and F. Csillag. 2006. Mutual information spectra for comparing categorical maps. *Int. J. Remote Sens.* 27(7): 1425–1452.

Reulke, R., S. Becker, N. Haala, and U. Tempelmann. 2006. Determination and improvement of spatial resolution of the CCD-line-scanner system ADS40. *ISPRS J. Photogramm. Remote Sens.* 60: 81–90.

Richards, J. A. 1993. *Remote Sensing: Digital Image Analysis*, 2nd ed. New York: Springer.

Richards, J. A., and X. P. Jia. 2006. *Remote Sensing Digital Image Analysis: An Introduction*. Berlin: Springer.

Richards, T., J. Gallego, and F. Achard. 2000. Sampling for forest cover change assessment at the pan-tropical scale. *Int. J. Remote Sens.* 21(6–7): 1473–1490.

Riitters, K. H. 2005. Downscaling indicators of forest habitat structure from national assessments. *Ecol. Indicators* 5(4): 273–279.

Riitters, K. H., R. V. O'Neill, J. D. Wickham, and K. B. Jones. 1996. A note on contagion indices for landscape analysis. *Land. Ecol.* 11(4): 197–202.

Riley, S. J., S. D. DeGloria, and R. Elliot. 1999. A terrain ruggedness index that quantifies topographic heterogeneity. *Intermountain J. Sci.* 5(1–4): 23–27.

Ripley, B. D. 1988. *Statistical Inference for Spatial Processes*. Cambridge: Cambridge University Press.

Ritter, P. 1987. A vector-based slope and aspect generation algorithm. *Photogramm. Eng. Remote Sens.* 53(8): 1109–1111.

Robinove, C. J. 1981. The logic of multispectral classification and mapping of land. *Remote Sens. Environ.* 11: 231–244.

Rodgers, C. D., and B. J. Connor. 2003. Intercomparison of remote sounding instruments. *J. Geophys. Res.* 108: 4116.

Rojas, F., A. S. Robert, and S. F. Biggar. 2002. Early results on the characterization of the Terra MODIS spatial response. *Remote Sens. Environ.* 83: 50–61.

Romme, W. H. 1982. Fire and landscape diversity in sub-alpine forests of Yellowstone National Park. *Ecol. Monogr.* 52(2): 199–221.

Rosenfield, G. H., and K. Fitzpatrick-Lins. 1986. A coefficient of agreement as a measure of thematic classification accuracy. *Photogramm. Eng. Remote Sens.* 52: 223–227.

Rossi, M., and D. Posa. 1992. A non-parametric bivariate entropy estimator for spatial processes. *Math. Geol.* 24(5): 539–553.

Rossi, R. E., P. W. Borth, and J. J. Tollefson. 1993. Stochastic simulation for characterizing ecological spatial patterns and appraising risk. *Ecol. Appl.* 3(4): 719–735.

Roy, D. P., L. Boschetti, C. O. Justice, and J. Ju. 2008. The collection 5 MODIS burned area product—Global evaluation by comparison with the MODIS active fire product. *Remote Sens. Environ.* 112(9): 3690–3707.

Rue, H., S. Martino, and N. Chopin. 2009. Approximate Bayesian inference for latent Gaussian models using integrated nested Laplace approximations (with discussion). *J. R. Stat. Soc.: Ser B* 71: 319–392.

Ruiz-Arias, J. A., J. Tovar-Pescador, D. Pozo-Vázquez, and H. Alsamamra. 2009. A comparative study of DEM-based models to estimate solar radiation on mountainous terrains. *Int. J. Geogr. Inf. Sci.* 23(8): 1049–1076.

Rupprecht, F., J. Oldeland, and M. Finckh. 2011. Modelling potential distribution of the threatened tree species *Juniperus oxycedrus*: How to evaluate the predictions of different modelling approaches? *J. Veg. Sci.* 22(4): 647–659.

Saalfeld, A. 1988. Conflation: Automated map compilation. *Int. J. Geogr. Inf. Syst.* 2(3): 217–228.

Sadowski, F. A., and J. E. Sarno. 1976. Forest Classification Accuracy as Influenced by Multispectral Scanner Spatial Resolution. Report No. 109600-71-F. Ann Arbor, MI: Environmental Research Institute of Michigan.

Sahr, K., D. White, and A. J. Kimerling. 2003. Geodesic discrete global grid systems. *Cartogr. Geogr. Inf. Sci.* 30: 121–134.

Saito, H., and G. Grasselli. 2010. Geostatistical downscaling of fracture surface topography accounting for local roughness. *Acta Geotech.* 5: 127–138.

Samet, H. 1981. An algorithm for converting rasters to quadtrees. *IEEE Trans. Pattern Anal. Mach. Intell.* 3: 93–95.

Samet, H. 1984. The quadtree and related hierarchical data structures. *ACM Comput. Surv.* 16: 187–260.

Samet, H. 2006. *Foundations of Multidimensional and Metric Data Structures*. San Francisco: Morgan Kaufmann Publishers, Inc.

Sampson, P. D., D. Damien, and P. Guttorp. 2001. Advances in modelling and inference for environmental processes with nonstationary spatial covariance. In *GeoENV III: Geostatistics for Environmental Applications*, eds. P. Monestiez, D. Allard, and R. Frodievaux, 17–32. Dordrecht: Kluwer Academic Publishers.

Sampson, P. D., and P. Guttorp. 1992. Nonparametric estimation of nonstationary spatial covariance structure. *J. Am. Stat. Assoc.* 87(417): 108–119.

Sappington, J. M., K. M. Longshore, and D. B. Thomson. 2007. Quantifiying landscape ruggedness for animal habitat analysis: A case study using bighorn sheep in the Mojave desert. *J. Wildl. Manage.* 71(5): 1419–1426.

Saveliev, A. A., S. S. Mukharamova, and A. F. Zuur. 2007. Analysis and modelling of lattice data. In *Analysing Ecological Data*, eds. A. F. Zuur, E. N. Ieno, and G. M. Smith, 321–339. New York: Springer.

Scarlatos, L., and T. Pavlidis. 1992. Hierarchical triangulation using cartographic coherence. *Graph. Models Image Process.* 54: 147–161.

Schabenberger, O., and C. A. Gotway. 2005. *Statistical Methods for Spatial Data Analysis*, 1st ed. London: Chapman & Hall/CRC Press.

Schabenberger, O., and F. J. Pierce. 2002. *Contemporary Statistical Models for the Plant and Soil Sciences*. Boca Raton, FL: CRC Press.

Scheifinger, H., and H. Kromp-Kolb. 2000. Modelling global radiation in complex terrain: Comparing two statistical approaches. *Agric. Forest Meteorol.* 100: 127–136.

Scheiner, S. M. 1992. Measuring pattern diversity. *Ecology* 73(5): 1860–1867.

Schmidt, J., and R. Andrew. 2005. Multi-scale landform characterization. *Area* 37(3): 341–350.

Schmullius, C. C., and D. L. Evans. 1997. Review article synthetic aperture radar (SAR) frequency and polarization requirements for applications in ecology, geology, hydrology, and oceanography: A tabular status quo after SIR-C/X-SAR. *Int. J. Remote Sens.* 18(13): 2713–2722.

Schoorl, J. M., M. P. W. Sonneveld, and A. Veldkamp. 2000. Three-dimensional landscape process modeling, the effect of DEM resolution. *Earth Surf. Proc. Land.* 25: 1025–1034.

Schowengerdt, R. A. 2007. *Remote Sensing: Models and Methods for Image Processing*, 3rd ed. New York: Academic Press.

Schowengerdt, R. A., S. K. Park, and R. Gray. 1984. Topics in the two-dimensional sampling and reconstruction of images. *Int. J. Remote Sens.* 5(2): 333–347.

Schreier, P. J., and L. L. Scharf. 2010. *Statistical Signal Processing of Complex-Valued Data: The Theory of Improper and Noncircular Signals.* Cambridge, UK: Cambridge University Press.

Schrödinger, E. 1955. *What is Life? The Physical Aspect of the Living Cell.* Cambridge: The University Press.

Settle, J. J., and N. A. Drake. 1993. Linear mixing and the estimation of ground cover proportions. *Int. J. Remote Sens.* 14: 1159–1177.

Sèze, G., and W. B. Rossow. 1991. Effects of satellite data resolution on measuring the space/time variations of surfaces and clouds. *Int. J. Remote Sens.* 12(5): 921–952.

Shannon, C. E. 1948. A mathematical theory of communication. *Bell Syst. Tech. J.* 27: 379–423, 623–656.

Shannon, C. E. 1949. Communication in the presence of noise. *Proc. IRE* 37(1): 10–21.

Sharpnack, D. A., and G. Akin. 1969. An algorithm for computing slope and aspect from elevations. *Photogramm. Eng.* 35(3): 247–248.

Shary, P. A. 1995. Land surface in gravity points classification by complete system of curvatures. *Math. Geol.* 27: 373–390.

Shary, P. A., L. S. Sharaya, and A. V. Mitusov. 2002. Fundamental quantitative methods of land surface analysis. *Geoderma* 107: 1–32.

Shelberg, M. C., H. Moellering, and N. S. Lam. 1982. Measuring the fractal dimensions of empirical cartographic curves. *Proceedings: Autocarto Five*, 481–490.

Shen, H., L. Du, L. Zhang, and W. Gong. 2012. A blind restoration method for remote sensing images. *IEEE Geosci. Remote Sens. Lett.* 9(6): 1137–1141.

Sheppard, C., and K. Larkin. 2003. Information capacity and resolution in three-dimensional imaging. *Optik* 114: 548–550.

Short, N. M. 2005. The Remote Sensing Tutorial, published by the Federation of American Scientists. Available at http://www.fas.org/irp/imint/docs/rst/index.html (accessed February 28, 2013).

Simic, A., J. M. Chen, J. Liu, and F. Csillag. 2004. Spatial scaling of net primary productivity using subpixel information. *Remote Sens. Environ.* 93: 246–258.

Simone, G., A. Farina, F. C. Morabito, S. B. Serpico, and L. Bruzzone. 2002. Image fusion techniques for remote sensing applications. *Inf. Fusion* 3(2002): 3–15.

Singh, V. P. 1997. The use of entropy in hydrology and water resources. *Hydrol. Proc.* 11: 587–626.

Sivapalan, M., K. Beven, and E. F. Wood. 1987. On hydrologic similarity: 2. A scaled model for runoff prediction. *Water Resour. Res.* 23: 2266–2278.

Skidmore, A. K. 1989. A comparison of techniques for calculating gradient and aspect from a gridded digital elevation model. *Int. J. Geogr. Inf. Syst.* 3(4): 323–334.

Skøien, J., and G. Blöschl. 2006. Sampling scale effects in random fields and implications for environmental monitoring. *Environ. Monitor. Assess.* 114(1–3): 521–552.

Skole, D. L., and C. J. Tucker. 1993. Tropical deforestation and habitat fragmentation in the Amazon: Satellite data from 1978 to 1988. *Science* 260(5116): 1905–1910.

Slater, J. A., G. Garvey, C. Johnston, J. Haasen, B. Heady, G. Kroenung, and J. Little. 2006. The SRTM data "finishing" process and products. *Photogramm. Eng. Remote Sens.* 72(3): 237–247.

Slater, P. N. 1980. *Remote Sensing: Optics and Optical Systems.* Boston: Addison-Wesley Publishing Co., Inc.

Slatton, K. C., M. M. Crawford, and Evans, B. L. 2001. Fusing interferometric radar and laser altimeter data to estimate surface topography and vegetation heights. *IEEE Trans. Geosci. Remote Sens.* 39(11): 2470–2482.

Smith, B., and D. M. Mark. 2001. Geographical categories: An ontological investigation. *Int. J. Geogr. Inf. Sci.* 15(7): 591–612.

Smith, B., and D. Sandwell. 2003. Accuracy and resolution of Shuttle Radar Topography Mission data. *Geophys. Res. Lett.* 30(9): 1467.

Smith, B., and A. C. Varzi. 2000. Fiat and bona fide boundaries. *Philos. Phenomenol. Res.* 60: 401–420.

Smith, T., D. Andresen, L. Carver, R. Dolin, C. Fischer, J. Frew, M. F. Goodchild et al. 1996. A digital library for geographically referenced materials. *Computer* 29(5): 54–60 (corrected author list appears in 29(7): 14).

Snedecor, G. W., and W. G. Cochran. 1967. *Statistical Methods*, 6th ed. Ames: The Iowa State University Press.

So, Y., and W. F. Kuhfeld. 1995. Multinomial Logit Models. In *SUGI 20 Conference Proceedings.* Cary, NC: SAS Institute Inc.

Song, L., A. J. Kimerling, and K. Sahr. 2002. Developing an equal area global grid by small circle subdivision. In *Discrete Global Grids*, eds. M. F. Goodchild and A. J. Kimerling. Santa Barbara, CA: National Center for Geographic Information and Analysis (NCGIA), University of California.

Srygley, R. B., and A. L. R. Thomas. 2002. Unconventional lift-generating mechanisms in free-flying butterflies. *Nature* 420: 660–664.

Starck, J. L., E. Pantin, and F. Murtagh. 2002. Deconvolution in astronomy: A review. *Publ. Astron. Soc. Pac.* 114(800): 1051–1069.

Steele, B. M., J. C. Winne, and R. L. Redmond. 1998. Estimation and mapping of misclassification probabilities for thematic land cover maps. *Remote Sens. Environ.* 66: 192–202.

Stehman, S. V., and R. L. Czaplewski. 1998. Design and analysis for thematic map accuracy assessment: Fundamental principles. *Remote Sens. Environ.* 64: 331–344.

Stehman, S. V., M. C. Hansen, M. Broich, and P. V. Potapov. 2011. Adapting a global stratified random sample for regional estimation of forest cover change derived from satellite imagery. *Remote Sens. Environ.* 115(2): 650–658.

Stein, A. 2005. Use of single- and multi-source image fusion for statistical decision-making. *Int. J. Appl. Earth Obs. Geoinf.* 6(3–4): 229–239.

Stein, A., I. G. Staritsky, J. Bouma, A. C. Van Eijnsbergen, and A. K. Bregt. 1991. Simulation of moisture deficits and areal interpolation by universal cokriging. *Water Resour. Res.* 27(8): 1963–1973.

Steininger, M. K., F. Godoy, and G. Harper. 2009. Effects of systematic sampling on satellite estimates of deforestation rates. *Environ. Res. Lett.* 4(034015).

Stoy, P. C., M. Williams, M. Disney, A. Prieto-Blanco, B. Huntley, R. Baxter, and P. Lewis. 2009. Upscaling as ecological information transfer: A simple framework with application to arctic ecosystem carbon exchange. *Land. Ecol.* 24: 971–986.

Strahler, A. H., C. E. Woodcock, and J. A. Smith. 1986. On the nature of models in remote sensing. *Remote Sens. Environ.* 20(2): 121–139.

Strebelle, S. 2002. Conditional simulation of complex geological structures using multiple point statistics. *Math. Geol.* 34: 1–22.

Stroet, C. B. M. T., and J. J. J. C. Snepvangers. 2005. Mapping curvilinear structures with local anisotropy kriging. *Math. Geol.* 37(6): 635–649.

Strome, W. M. 1979. Communication to A. P. Colvocoresses, reference 1372-113-2-2, February 19, 1979, Canada Centre for Remote Sensing.

Su, W., J. Li, Y. Chen, Z. Liu, J. Zhang, T. M. Low, I. Suppiah, and S. A. M. Hashim. 2008. Textural and local spatial statistics for the object-oriented classification of urban areas using high resolution imagery. *Int. J. Remote Sens.* 29(11): 3105–3117.

Sukhov, V. I. 1967. Information capacity of a map entropy. *Geod. Aerophotogr.* 10(4): 212–215.

Summerell, G., V. Shoemark, S. Grant, and J. P. Walker. 2009. Using passive microwave response to soil moisture change for soil mapping: A case study for the Livingstone Creek Catchment. *IEEE Trans. Geosci. Remote Sens. Lett.* 6(4): 649–652.

Swain, P. H. 1978. Bayesian classification in a time-varying environment. *IEEE Trans. Syst. Man Cybern.* 8(12): 879–883.

Switzer, P., and A. A. Green. 1984. Min/Max Autocorrelation Factors for Multivariate Spatial Imagery. Technical Report no. 6: Stanford, CA: Department of Statistics, Stanford University.

Takagi, M. 1998. Accuracy of digital elevation model according to spatial resolution. *Int. Arch. Photogramm. Remote Sens.* 32(4): 613–617.

Takeyama, M. and H. Couclelis. 1997. Map dynamics: Integrating cellular automata and GIS through Geo-Algebra. *Int. J. Geogr. Inf. Sci.* 11: 73–91.

Tan, B., C. E. Woodcock, J. Hu, P. Zhang, M. Ozdogan, D. Huang, W. Yang et al. 2006. The impact of gridding artifacts on the local spatial properties of MODIS data: Implications for validation, compositing, and band-to-band registration across resolutions. *Remote Sens. Environ.* 105(2): 98–114.

Tarnavsky, E., S. Garrigues, and M. E. Brown. 2008. Multiscale geostatistical analysis of AVHRR, SPOT-VGT, and MODIS global NDVI products. *Remote Sens. Environ.* 112(2): 535–549.

Tate, N., and P. M. Atkinson. 2001. *Modeling Scale in Geographical Information Science.* Chichester: Wiley.

Ter Braak, C. J. F. 1986. Canonical correspondence analysis: A new eigenvector technique for multivariate direct gradient analysis. *Ecology* 67(5):1167–1179.

Ter Braak, C. J. F. 1987. The analysis of vegetation-environment relationships by canonical correspondence analysis. *Vegetation* 69(3): 69–77.

Tercan, A. E., and P. A. Dowd. 1995. Approximate local confidence intervals under change of support. *Math. Geol.* 27(1): 149–172.

Terrell, D. G., and D. W. Scott. 1992. Variable kernel density estimation. *Ann. Stat.* 20: 1236–1265.

Thomas, C., T. Ranchin, L. Wald, and J. Chanussot. 2008. Synthesis of multispectral images to high spatial resolution: A critical review of fusion methods based on remote sensing physics. *IEEE Trans. Geosci. Remote Sens.* 46(5): 1301–1312.

Thome, K., B. Markham, J. Barker, P. Slaterm, and S. Biggar. 1997. Radiometric calibration of Landsat. *Photogramm. Eng. Remote Sens.* 63(7): 853–858.

Tian, Y. H., Y. J. Wang, Y. Zhang, Y. Knyazikhin, J. Bogaert, and R. B. Myneni. 2002. Radiative transfer based scaling of LAI retrievals from reflectance data of different resolutions. *Remote Sens. Environ.* 84: 143–159.

Tobler, W. R. 1970. A computer movie simulating urban growth in the Detroit Region. *Econ. Geogr.* 46: 234–240.

Tobler, W. R. 1979. Smooth pycnophylactic interpolation for geographical regions. *J. Am. Stat. Assoc.* 74: 519–536. With discussion.

Tobler, W. R. 1988. Resolution, resampling, and all that. In *Building Database for Global Science*, ed. H. Mounsey and R. Tomlinson, 129–37. London: Taylor & Francis.

Tobler, W. R., and Z. Chen. 1986. A quadtree for global information storage. *Geogr. Anal.* 18(4): 360–371.

Toraldo di Francia, G. 1969. Degrees of freedom of an image. *J. Opt. Soc. Am.* 59: 799–804.

Torlegård, K., A. Östman, and R. Lindgren. 1986. A comparative test of photogrammetrically sampled digital elevation models. *Photogrammetria* 41: 1–16.

Toutin, T. 1995. Multi-sources data fusion with an integrated and unified geometric modeling. *EARSeL J.* 4(2): 118–129.

Toutin, T. 2004. Review article: Geometric processing of remote sensing images: Models, algorithms and methods. *Int. J. Remote Sens.* 25(10): 1893–1924.

Toutin, T. 2011. State-of-the-art of geometric correction of remote sensing data: A data fusion perspective. *Int. J. Image Data Fusion* 2(1): 3–35.

Townshend, J. R. G. 1981. The spatial resolving power of earth resources satellites. *Prog. Phys. Geogr.* 5: 32–55.

Townshend, J. R. G., and C. O. Justice. 1986. Analysis of the dynamics of African vegetation using the normalized difference vegetation index. *Int. J. Remote Sens.* 7(11): 1435–1445.

Townshend, J. R. G., and C. O. Justice. 1988. Selecting the spatial resolution of satellite sensors required for global monitoring of land transformations. *Int. J. Remote Sens.* 9(2): 187–236.

Tran, T. D. B., A. Puissant, D. Badariotti, and C. Weber. 2011. Optimizing spatial resolution of imagery for urban form detection—The cases of France and Vietnam. *Remote Sens.* 3: 2128–2147.

Tso, B., and P. Mather. 2009. *Classification Methods for Remotely Sensed Data*, 2nd ed. Boca Raton, FL: CRC Press.

Tu, T. M., S. C. Su, H. S. Shyu, and P. S. Huang. 2001. A new look at IHS-like image fusion methods. *Inf. Fusion* 2(3): 177–186.

Tucker, D. M., M. Liotti, G. F. Potts, G. S. Russell, and M. I. Posner. 2004. Spatiotemporal analysis of brain electrical fields. *Hum. Brain Mapp.* 1: 134–152.

Turner, M. G., R. V. O'Neill, R. H. Gardner, and B. T. Milne. 1989. Effects of changing spatial scale on the analysis of landscape pattern. *Land. Ecol.* 3: 153–162.

Tveite, H., and S. Langaas. 1999. An accuracy assessment method for geographical line data sets based on buffering. *Int. J. Geogr. Inf. Sci.* 13(1): 27–47.

US Bureau of the Budget. 1947. *United States National Map Accuracy Standards*. Washington, DC: Bureau of the Budget.

US Geological Survey (USGS). 1986. Standards for Digital Elevation Models, Open File Report 86-004.

US Geological Survey (USGS). 1993. *Digital Elevation Models Data User Guide 5*. Reston, VA: USGS, 50 pp.

US Geological Survey (USGS). 2004. *Shuttle Radar Topography Mission*, 3 Arc Second scene SRTM_u03_n008e004, Unfilled Unfinished 2.0: College Park, MD: Global Land Cover Facility, University of Maryland, February 2000.

Unser, M. 2000. Sampling-50 years after Shannon. *Proc. IEEE* 88(4): 569–587.

Unwin, A., and D. Unwin. 1998. Exploratory spatial data analysis with local statistics. *J. R. Stat. Soc.: Ser. D* 47: 415–421.

Valor, E., V. Caselles, C. Coll, F. Sanchez, E. Rubio, and F. Sospedra. 2000. Simulation of a medium-scale-surface-temperature instrument from Thematic Mapper data. *Int. J. Remote Sens.* 21(16): 3153–3159.

Van de Vyver, H., and E. Roulin. 2009. Scale-recursive estimation for merging precipitation data from radar and microwave cross-track scanners. *J. Geophys. Res.* 114: D08104.

Van der Meer, F. 1997. What does multisensor image fusion add in terms of information content for visual interpretation? *Int. J. Remote Sens.* 18(2): 445–452.

Van Gardingen, P. R., G. M. Foody, and P. J. Curran (eds.). 1997. *Scaling-Up: From Cell to Landscape*. Cambridge: Cambridge University Press.

Van Niel, K. P., S. W. Laffan, and B. G. Lees. 2004. Effect of error in the DEM on environmental variables for predictive vegetation modeling. *J. Veg. Sci.* 15(5): 747–756.

Van Oort, P. A. J., A. K. Bregt, S. De Bruin, A. J. W. De Wit, and A. Stein. 2004. Spatial variability in classification accuracy of agricultural crops in the Dutch national land-cover database. *Int. J. Geogr. Inf. Sci.* 18: 611–626.

van Westen, C. J., E. Castellanos, and S. L. Kuriakose. 2008. Spatial data for landslide susceptibility, hazard, and vulnerability assessment: An overview. *Eng. Geol.* 102(3–4): 112–131.

Vargas-Guzmán, J. A., and R. Dimitrakopoulos. 2003. Computational properties of min/max autocorrelation factors. *Comput. Geosci.* 29(6): 715–723.

Vargas-Guzmán, J. A., D. E. Myers, and A. W. Warrick. 2000. Derivatives of spatial variances of growing windows and the variogram. *Math. Geol.* 32(7): 851–871.

Ver Hoef, J. M., and N. Cressie 1993. Multivariable spatial prediction. *Math. Geol.* 25(2): 219–239.

Verghese, G. C., and T. Kailath. 1979. A further note on backwards Markovian models. *IEEE Trans. Inf. Theory* 25: 121–124.

Vieira, C. A., P. M. Mather, and P. Aplin. 2004. Assessing the positional and thematic accuracy of remotely sensed data. In *Proceedings XXth Congress of the International Society for Photogrammetry and Remote Sensing (ISPRS): Geo-imagery Bridging Continents*, Istanbul, Turkey, pp. 979–984.

Viscarra Rossel, R. A., and R. M. Lark. 2009. Improved analysis and modelling of soil diffuse reflectance spectra using wavelets. *Eur. J. Soil Sci.* 60: 453–464.

Vogelmann, J. E., S. M. Howard, L. Yang, C. R. Larson, B. K. Wylie, and N. Van Driel. 2001. Completion of the 1990's national land cover data set for the conterminous United States from Landsat Thematic Mapper data and ancillary data sources. *Photogramm. Eng. Remote Sens.* 67(6): 650–662.

Vogiatzis, M. 2008. Cadastral mapping of forestlands in Greece: Current and future challenges. *Photogramm. Eng. Remote Sens.* 74(1): 39–46.

Vucetic, S., T. Fiez, and Z. Obradovic. 2000. Examination of the influence of data aggregation and sampling density on spatial estimation. *Water Resour. Res.* 36(12): 3721–3730.

Wackernagel, H. 2003. *Multivariate Geostatistics: An Introduction with Applications*, 3rd ed. Berlin: Springer.

Wald, L., T. Ranchin, and M. Mangolini. 1997. Fusion of satellite images of different spatial resolutions: Assessing the quality of resulting images. *Photogramm. Eng. Remote Sens.* 63(6): 691–699.

Walpole, R. E., R. H. Myers, and S. L. Myers. 1998. *Probability and Statistics for Engineers and Scientists*, 6th ed. Upper Saddle River, NJ: Prentice Hall International, Inc.

Walsh, T. A., and T. E. Burk. 1993. Calibration of satellite classifications of land area. *Remote Sens. Environ.* 46(3): 281–290.

Wang, Z., and A. C. Bovik. 2002. A universal image quality index. *IEEE Signal Proc. Lett.* 9(3): 81–84.

Watson, G. S. 1966. The statistics of orientation data. *J. Geol.* 74: 786–797.

Webster, R., and M. A. Oliver. 1990. *Statistical Methods in Soil and Land Resource Survey*. Oxford, UK: Oxford University Press.

Wechsler, S. 2007. Uncertainties associated with digital elevation models for hydrologic applications: A review. *Hydrol. Earth Syst. Sci.* 11: 1481–1500.

Wechsler, S., and C. Kroll. 2006. Quantifying DEM uncertainty and its effects on topographic parameters. *Photogramm. Eng. Remote Sens.* 72: 108–1090.

Wedderburn, R. W. M. 1974. Quasi-Likelihood functions, generalized linear models, and the Gauss-Newton method. *Biometrika* 61(3): 439–447.

Weih, R. C. Jr. 2010. Assessing the vertical accuracy of Arkansas five-meter digital elevation model for different physiographic regions. *J. Arkansas Acad. Sci.* 64: 123–128.

Werneck, G. L. 2008. Georeferenced data in epidemiologic research. *Cien. Saude Colet.* 13(6): 1753–1766.

Western, A. W., G. Blöschl, and R. B. Grayson. 1998. How well do indicator variograms capture spatial connectivity of soil moisture? *Hydrol. Proc.* 12: 1851–1868.

White, D., J. A. Kimerling, and S. W. Overton. 1992. Cartographic and geometric components of a global sampling design for environmental monitoring. *Cartogr. Geogr. Inf. Sci.* 19(1): 5–22.

White, E. M., A. T. Morzillo, and R. J. Alig. 2009. Past and projected rural land conversion in the US at state, regional, and national levels. *Land. Urban Plann.* 89(1–2): 37–48.

Whitehouse, D. 2009. *Renaissance Genius: Galileo Galilei & His Legacy to Modern Science.* New York: Sterling.

Whittaker, R. H. 1967. Gradient analysis of vegetation. *Biol. Rev. Camb. Philos. Soc.* 42(2): 207–264.

Wiens, J. A. 1989. Spatial scaling in ecology. *Funct. Ecol.* 3: 385–397.

Wikle, C. K. 2003. Hierarchical models in environmental science. *Int. Stat. Rev.* 71: 181–199.

Williams, M., R. Bell, L. Spadavecchia, L. E. Street, and M. T. Van Wijk. 2008. Upscaling leaf area index in an Arctic landscape through multiscale observations. *Glob. Change Biol.* 14: 1517–1530.

Willsky, A. S. 2002. Multiresolution Markov models for signal and image processing. *Proc. IEEE* 90(8): 1396–1458.

Wilson, A. K. 1988. The effective resolution element of Landsat Thematic Mapper. *Int. J. Remote Sens.* 9(8): 1303–1314.

Wilson, J. P., and P. A. Burrough. 1999. Dynamic modeling, geostatistics, and fuzzy classification: New sneakers for a new geography? *Ann. Assoc. Am. Geogr.* 89: 736–746.

Wilson, J. P., and J. C. Gallant. 2000. *Digital Terrain Analysis: Principles and Applications.* New York: Wiley.

Wilson, M. F. J., B., O'Connel, C. Brown, J. C. Guinan, and A. J. Grehan. 2007. Multiscale terrain analysis of multibeam bathymetry data for habitat mapping on the continental slope. *Mar. Geod.* 30: 3–35.

Wood, E. F., M. Sivapalan, K. Beven, and L. Jenson. 1988. Effects of spatial variability and scale with implications to hydrologic modeling. *J. Hydrol.* 102(1–4): 29–47.

Woodcock, C. E., R. Allen, M. Anderson, A. Belward, R. Bindschadler et al. 2008. Free access to landsat data. *Science* 320(5879): 1011.

Woodcock, C. E., and A. H. Strahler. 1987. The factor of scale in remote sensing. *Remote Sens. Environ.* 21(3): 311–332.

Woodcock, C. E., A. H. Strahler, and D. L. B. Jupp. 1988. The use of variograms in remote sensing: I. Scene models and simulated images. *Remote Sens. Environ.* 25: 323–348.

Woolrich, M. W., B. D. Ripley, M. Brady, and S. M. Smith. 2001. Temporal autocorrelation in univariate linear modeling of FMRI data. *NeuroImage* 14: 1370–1386.

Wu, H., J. S. Kimball, N. Mantua, and J. Stanford. 2011. Automated upscaling of river networks for macroscale hydrological modeling. *Water Resour. Res.* 47, W03517.

Wu, H., and Z. L. Li. 2009. Scale issues in remote sensing: A review on analysis, processing and modeling. *Sensors* 9: 1768–1793.

Wu, J. 2004. Effects of changing scale on landscape pattern analysis: Scaling relations. *Land. Ecol.* 19: 125–138.

Wu, J., and H. Li. 2006. Concepts of scale and scaling. In *Scaling and Uncertainty Analysis in Ecology: Methods and Applications*, eds. J. Wu, K. B. Jones, H. Li, and O. L. Loucks, 3–15. Dordrecht, the Netherlands: Springer.

Wu, J., H. Li, K. B. Jones, and O. L. Loucks. 2006. Scaling with known uncertainty: A synthesis. In *Scaling and Uncertainty Analysis in Ecology: Methods and Applications*, eds. J. Wu, K. B. Jones, H. Li, and O. L. Loucks, 329–346. Dordrecht, the Netherlands: Springer.

Wu, J. G., D. E. Jelinski, M. Luck, and P. T. Tueller. 2000. Multiscale analysis of landscape heterogeneity: Scale variance and pattern metrics. *Geogr. Inf. Sci.* 6(1): 6–19.

Wulder, M., J. White, M. Gillis, N. Walsworth, M. Hansen, and P. Potapov. 2010. Multiscale satellite and spatial information and analysis framework in support of a large-area forest monitoring and inventory update. *Environ. Monitor. Assess.* 170(1–4): 417–433.

Xie, Y. 2008. Remote sensing imagery in vegetation mapping: A review. *J. Plant Ecol.* 1(1): 9–23.

Yang, W. L., and J. W. Merchant. 1997. Impacts of upscaling techniques on land cover representation in Nebraska. *GeoCarto Int.* 12(1): 27–39.

Yocky, D. A. 1996. Multiresolution wavelet decomposition image merger of Landsat Thematic Mapper and SPOT panchromatic data. *Photogramm. Eng. Remote Sens.* 62(9): 1067–1074.

Yuan, M. 1999. Representing geographic information to enhance GIS support for complex spatiotemporal queries. *Trans. GIS* 3: 137–160.

Zebker, H. A., and R. M. Goldstein. 1986. Topographic mapping from interferometric synthetic aperture radar observations. *J. Geophys. Res.* 91(B5): 4993–4999.

Zebker, H. A., C. L. Werner, P. A. Rosen, and S. Hensley. 1994. Accuracy of topographic maps derived from ERS-1 interferometric radar. *IEEE Trans. Geosci. Remote Sens.* 32(4): 823–836.

Zeiler, M. 1999. *Modeling Our World: The ESRI Guide to Geodatabase Design.* Redlands: ESRI Press.

Zevenbergen, L. W., and C. R. Thorne. 1987. Quantitative analysis of land surface topography. *Earth Surf. Proc. Land.* 12: 47–56.

Zhan, X., R. DeFries, J. R. G. Townshend, C. DiMiceli, M. Hansen, C. Huang, and R. Sohlberg. 2000. The 250 m global land cover change product from the Moderate Resolution Imaging Spectroradiometer of NASA's Earth Observing System. *Int. J. Remote Sens.* 21(6 and 7): 1433–1460.

Zhang, C., and F. Qiu. 2011. A point-based intelligent approach to areal interpolation. *Prof. Geogr.* 63: 262–276.

Zhang, J. 2008. *Scale and Uncertainty in Spatial Information and Fusion.* Wuhan: Wuhan University Press (in Chinese).

Zhang, J. 2010. *Geographic Information Systems and Science.* Wuhan: Wuhan University Press (in Chinese).

Zhang, J., and M. F. Goodchild. 2002. *Uncertainty in Geographical Information.* New York: Taylor & Francis.

Zhang, J., P. Kyriadis, and R. E. J. Kelly. 2009. Geostatistical approaches to conflation of continental snow data. *Int. J. Remote Sens.* 30(20): 5441–5451.

Zhang, J. P. et al. 2007a. Geostatistical framework for conflation of heterogeneous geospatial data. In *Proceedings of SPIE*, vol. 6790: 679038-1.

Zhang, J. et al. 2007b. Anomaly detection in MODIS land products via time series analysis. *Geo-Spat. Inf. Sci.* 10: 44–50.

Zhang, J. X., and N. Yao. 2008a. The geostatistical framework for spatial prediction. *Geo-Spat. Inf. Sci.* 11(3): 180–185.

Zhang, J. X., and N. Yao. 2008b. Indicator and multivariate geostatistics for spatial prediction. *Geo-Spat. Inf. Sci.* 11(4): 243–246.

Zhang, J. X., and J. You. 2011. Discriminant models for uncertainty characterization in area class change categorization. *Geo-Spat. Inf. Sci.* 14(4): 255–261.

Zhang, J. X., J. P. Zhang, Y. W. Tang, and J. You. 2009. Spatio-temporal contextual approaches to mapping land cover change based on remote sensing imagery. In *Proceeding of SPIE: International Symposium on Spatial Analysis, Spatial-Temporal Data Modeling, and Data Mining*, vol. 7492, 74923Y, October 13, 2009, Wuhan, China. eds. Y. L. Liu and X. Tang. International Society for Optical Engineering.

Zhang, J. X., J. P. Zhang, and N. Yao. 2009a. Geostatistics for spatial uncertainty characterization. *Geo-Spat. Inf. Sci.* 12(1): 7–12.

Zhang, J. X., J. P. Zhang, and N. Yao. 2009b. Uncertainty characterization in remotely sensed land cover information. *Geo-Spat. Inf. Sci.* 12(3): 165–171.

Zhang, W., and D. R. Montgomery. 1994. Digital elevation model grid size, landscape representation, and hydrologic simulations. *Water Resour. Res.* 30: 1019–1028.

Zhang, X., N. Drake, and Wainwright, J. 2002. Scaling land surface parameters for global-scale soil erosion estimation. *Water Resour. Res.* 38(9): 1180.

Zhang, X., N. A. Drake, J. Wainwright, and M. Mulligan. 1999. Comparison of slope estimates from low resolution DEMs: Scaling issues and a fractal method for their solution. *Earth Surf. Proc. Land.* 24: 763–779.

Zhao, Y., and M. M. Wall. 2004. Investigating the use of the variogram for lattice data. *J. Comput. Graph. Stat.* 13: 719–738.

Zhou, J., D. L. Civco, and J. A. Silander. 1998. A wavelet transform method to merge Landsat TM and SPOT panchromatic data. *Int. J. Remote Sens.* 19(4): 743–757.

Zhou, Q. M., and X. J. Liu. 2006. *Digital Terrain Analysis.* Beijing: Science Press (in Chinese).

Zhukov, B., D. Oertel, F. Lanzl, and G. Reinhackel. 1999. Unmixing-based multisensor multi-resolution image fusion. *IEEE Trans. Geosci. Remote Sens.* 37: 1212–1226.

Zimmerman, D. L., and D. M. Holland. 2005. Complementary co-kriging: Spatial prediction using data combined from several environmental monitoring networks. *Environmetrics* 16: 219–234.

Index

A

Accuracy assessment; *see also* Errors
    in area class, 271–273
    in interval/ratio fields, 268–270
    positional, 273–274
Additive noise, images with, 242–246
Additive wavelet method, image fusion, 162
Aerial photography, geometric characteristics, 51
Airborne Visible and Infrared Imaging
        Spectrometer (AVIRIS), 276–277
Amplified signal, 49
Approximation error, 31
AR, *see* Autoregressive (AR) process
Area-based weighting, 10
Area-class mapping
    accuracy in, 271–273
    discriminant-space model, 204–209
    downscaling, 220–223
    GLM, 210–212
    methods for, 209–213
    spatial scales and patterns, 213–218
    upscaling, 219–220
Area-to-point kriging
    deconvolution in, 153
    downscaling by, 120–124
Area-to-point simulation, 291–293; *see also*
        Geostatistical simulation
Autoregressive (AR) process, 140
AVIRIS, *see* Airborne Visible and Infrared
        Imaging Spectrometer (AVIRIS)
Azimuthal direction, 56, 57
Azimuth resolution, 57–58

B

Backward geocoding, 62
BIOCLIM, 202
Block kriging, 108; *see also* Kriging
    upscaling by, 118–120
Block simulation, 293–296; *see also*
        Geostatistical simulation
    drawbacks, 293–294
    joint direct, 295
Bona fide geo-objects, 205
Brightness temperatures, 63

C

Canonical correspondence analysis (CCA), 207, 209
Cartographic scale (map scale), 2
CCA, *see* Canonical correspondence analysis (CCA)
Cell-counting method, terrain analysis, 189
Change-of-support problem, 9–10
Chirp signal, 59
Choropleth map, 94
Circular Map Accuracy Standard (CMAS),
        273–274
Cluster sampling, 39
CMAS, *see* Circular Map Accuracy Standard
        (CMAS)
Cokriging, 149–154; *see also* Kriging
    cross-scale, 154
    on matrix formulation, 152
    variants, 153
    vs. kriging, 152
Communication channel, 235
Complexity science, 307
Conditional entropy, 232
Conditional mutual information, 236
Confidence interval
    defined, 275
    hypothesis testing and, 274–278
Conflation, multiscale data, *see* Multiscale data
        conflation
Continuous fields, 18
Contours, 22, 23
Convolution, upscaling and, 130, 132
Correlograms, 8, 13, 92, 99
Covariance functions, 8
Cross-scale cokriging, 154
Crowdsourcing, 305–306
Curvature, terrain analysis, 180–185

D

Data modeling, 2–3
Data support, 2
    sampling framework and, 36, 43–44
Decoding, 235
Deconvolution/de-regularization
    in area-to-point kriging, 153
    downscaling and, 80–82
    gridded data downscaling, 137–141

variograms, 80–82
Wiener deconvolution, 137–138
Degradation–reclassification method, 219–220
Delaunay triangulation, 23
DEMs, *see* Digital elevation models (DEMs)
Depression angle, 56
Dielectric constant, 58
Digital elevation models (DEMs), 25, 276
terrain analysis, 170, 172–173, 187, 193–195
Digital numbers (DNs), 236
Direct approach, downscaling, 222
Discrete objects, 18
Discrete wavelet transform (DWT), 161–162
Discretization, 23–25
Discriminant-space model, 205
area-class mapping, 204–209
Distortions, radar images, 61
Divider method, terrain analysis, 189
DNs, *see* Digital numbers (DNs)
Dominance, 214
Downscaling, 9–12, 108; *see also* Upscaling
area-class information, 220–223
direct approach, 222
iterative approach, 222–223
by area-to-point kriging, 120–124
deconvolution, 130, 137–141
gridded data, 137–145; *see also* Gridded data
scaling
terrain variables, 196–198
resolution *vs.* accuracy, 197–198
DWT, *see* Discrete wavelet transform (DWT)
Dyadic tree-based models, 163–166

E

Edge refinement, 30–31
Effective Radiometric Resolution Element
(ERRE), 45, 46–47
Elementary sampling units (ESU), 278, 279
Encoding, 235
Entropy, 230–237, 303
defined, 230
ERRE, *see* Effective Radiometric Resolution
Element (ERRE)
Errors
gross, 269
imaging process, 242–246
propagation
analytical approaches, 282–287
defined, 267
random, 269
ESU, *see* Elementary sampling units (ESU)
EV, *see* Romme's relative evenness index (EV)
Exponential model, variogram, 74
Extent, 24
External drift method, 114
External orientation, elements for, 51

F

Far range, 56
Feature space, 206
First expansion, 30, 31
Forward geocoding, 62
Fractal methods, 10
Fractal models, 5
Fusion of subtree estimates, 165

G

GAC, *see* Global area coverage (GAC) data
Galilei, Galileo, 299
Gamma index, 100, 101
Gaussian model, variogram, 74
Geary's c index, 96, 97, 99, 101
Generalized linear model (GLM), 210–212
Geo-atom, 19, 20–21
Geo-dipoles, spatial interactions and, 31
Geo-fields, 21–25
discretization, 23–25
objects and, 27
Geographic information science (GIScience),
1, 18
Geographic location, 18
Geographic representations
geo-atom, 19, 20–21
geo-fields, 21–25
geo-objects, 25–27
hierarchical data structures, 28–31
overview, 17–20
scale dependence of representations, 32–33
temporal dimension, 32
Geographic scale (or observational scale), 1; *see
also* Scale
Geographic/spatial information, *see* Spatial/
geographic information
Geographic variance method, 5
Geometric characteristics, images, 50–55
Geometric measure, 257
Geo-objects, 25–27
dimensionality of, 25–26
Geoportals, 306
Geospatial measurements, 13, 299
bias, 269
overview, 35–36
precision, 269
spatial sampling framework, 36–47
Geostatistical data
variograms and, 91, 92
vs. lattice data, 91, 93
Geostatistical inverse modeling, 109, 124–126
applications of, 126
Geostatistical models, 13
geostatistical inverse modeling, 109, 124–126
indicator variogram modeling, 75–77

kriging, 109–115; *see also* Kriging
   overview, 67–68
   random fields, 68–70
   variograms, 68–70; *see also* Variograms
   covariance functions and, 70–75
Geostatistical simulation, 287–296
   area-to-point simulation, 291–293
   over a point support, 287–290
   over blocks, 293–296
GIFOV, *see* Ground instantaneous field of view
   (GIFOV)
GIScience, *see* Geographic information science
   (GIScience)
GLM, *see* Generalized linear model (GLM)
Global area coverage (GAC) data, 131
Global *vs.* local statistics, 100
Grain, 2
Gridded data scaling, 129–146
   downscaling, 137–145
      deconvolution, 137–141
      subpixel mapping, 144–145
      super-resolution mapping, 141–144
   upscaling, 131–137
      mechanistic approaches, 134–137
      statistical approaches, 131–134
Grids, 22, 23
Gross errors, 269
Ground instantaneous field of view (GIFOV),
   47
Ground range, 56
Ground sampling interval (GSI), 47
GSI, *see* Ground sampling interval (GSI)

H

Hierarchical data structures, 28–31
   quadtrees, 28–29
   TIN model, 29–31
Hill's blocked quadrat variance method, 217
Hill's paired quadrat variance, 217, 218
Hue, 154
Hypothesis testing, confidence interval and,
   274–278

I

IFOV, *see* Instantaneous field of view (IFOV)
IHS, *see* Intensity–hue–saturation (IHS)
Image coarsening, 130
Image decomposition, multiresolution, 159, 160
Image fusion
   conventional approaches, 154–155
   multiresolution representations based on
      wavelets, 155–160
   wavelet-based, 161–163
Image processing, 227, 229
Image reconstruction, multiresolution, 159, 160

Image resolution, information content and, 250–255
Images, informational characteristics of, 238–242
Incidence angle, 56
Indicator auto/cross-covariances, 75, 76
Indicator auto/cross-variograms, 75, 76
Indicator kriging, 116–118
Indicator variable, defined, 75
Indicator variogram modeling, 75–77
Information content, 227–264
   image resolution and, 250–255
   information theory, 230–238
   in map data, 255–261
   in remotely sensed images, 238–250
Information representation, 2–3
Information theory, 230–238
Inner orientation, elements for, 51
INSAR, *see* Interferometric synthetic aperture
   radar (INSAR)
Instantaneous field of view (IFOV), 36, 44, 46
Intensity, 154
Intensity–hue–saturation (IHS) image, 148
   RGB color space to, 154–155
   wavelet decomposition and, 162
Interferometric synthetic aperture radar
   (INSAR), 166
Interval/ratio fields, accuracy in, 268–270
Intrinsic scale, 1; *see also* Scale
Irregular lattice, 93–94; *see also* Regular lattice
Irregular points, 22, 23
Iterative approach, downscaling, 222–223

J

Janssen, Zacharias, 299
Joint entropy, 232

K

Karhunen–Loeve transform, *see* Principal
   component analysis (PCA) methods
Kernel density estimation methods
   area-class mapping, 209–210
k-nearest neighbor (k-NN)
   area-class mapping, 209
k-NN, *see* k-nearest neighbor (k-NN)
Kriging, 14, 95, 108, 109–115
   area-to-point, 120–124
   block, 108, 118–120
   indicator, 116–118
   with locally varying means, 149
   ordinary, 112, 113, 115
   simple, 109–110
   universal, 113, 115
   variance, 110–115
   vs. cokriging, 152
   weights, 110
Kullback–Leibler distance, 233

L

LAC, *see* Large area coverage (LAC) data
Lacunarity analysis, 216–217
Lagrange multiplier, 139
LAI, *see* Leaf-area index (LAI)
Large area coverage (LAC) data, 131
Lattice data, 13, 93–95
    local models, 99–103
    spatial autocorrelation and, 95–99
    vs. geostatistical data, 91, 93
Leaf-area index (LAI), 10
Least squares minimization problem, 138
Linear model of regionalization, 74
Lippershey, Hans, 299
Local models, 99–103
Local variance (LV) method, 5–6
    terrain analysis, 188–189
Local variogram models
    terrain analysis, 187–188
Local *vs.* global statistics, 100
Look angle, 56, 57
Luminance, 154

M

MAE, *see* Mean absolute error (MAE)
MAF-based point-support simulation, 296
Magnitude, spatial variation, 300
Map data, information content in, 255–261
MAPE, *see* Mean absolute percentage error
    (MAPE)
Map scale (cartographic scale), 2; *see also* Scale
MAUP, *see* Modifiable areal unit problem
    (MAUP)
Maximum fusion scheme, 163
Maximum-likelihood classification, 210
Maximum-likelihood (ML) estimation method,
    140
MBG, *see* Model-based geostatistics (MBG)
ME, *see* Mean error (ME)
Mean absolute error (MAE), 269; *see also* Errors
Mean absolute percentage error (MAPE), 269
Mean error (ME), 269; *see also* Errors
Measurement scale, 2
    statistics for determining, 82–86
Mechanistic approaches
    gridded data upscaling, 134–137
Microwave remote sensing
    PM sensors, 62–64
    resolution and, 55–58
    signals, imaging and distortions, 58–62
Mixture modeling, 221
Model-based geostatistics (MBG), 304
Moderate resolution-imaging spectroradiometer
    (MODIS), 17–18, 135, 266–267
    data on NASA Terra satellite, 229

Modifiable areal unit problem (MAUP), 94–95
MODIS, *see* Moderate resolution-imaging
    spectroradiometer (MODIS)
Modulation transfer function (MTF)
    equivalent bandwidth and, 45–46
Moran's *I* index, 97–98, 99, 101
MPG, *see* Multiple-point geostatistics (MPG)
MS, *see* Multispectral image (MS)
MTF, *see* Modulation transfer function (MTF)
Multinomial logistic regression, 211
Multiple-point geostatistics (MPG), 304–305
Multiplicative noise, information with, 247–250
Multiresolution analysis, image fusion and,
    155–160
Multiscale data conflation, 147–168
    dyadic tree-based models, 163–166
    multivariate geostatistics, 151–154
    overview, 147–149
Multispectral image (MS)
    PAN image and, fusion of, 163
Multivariable variogram modeling, 153; *see also*
    Variograms
Multivariate geostatistics, 151–154
Mutual information, 230, 233–235; *see also*
    Information content

N

NAPP, *see* National Aerial Photography Program
    (NAPP)
National Aerial Photography Program (NAPP),
    17
Near range, 56
Networks, 306–307
Noise, additive, 242–246
NSIDC, *see* US National Snow and Ice Data
    Center (NSIDC)
Nugget, variograms, 8
Nyquist rate, 40, 43
Nyquist–Shannon sampling theorem, 40, 43

O

Object field model, 31
Observational scale, 1
Octree, 28
OGC, *see* Open Geospatial Consortium (OGC)
    Simple Feature Specification
Open Geospatial Consortium (OGC) Simple
    Feature Specification, 26–27
Operational scale (process/intrinsic scale), 1
Optical remote sensing, 47–55
    resolution and, 47–55
Ordinary cokriging estimator, 151
Ordinary kriging, 112, 113, 115
Orography-based methods
    upscaling, terrain variables, 195

P

PAN, *see* Panchromatic image (PAN)
Panchromatic image (PAN)
    MS image and, fusion of, 163
Passive microwave (PM), 62–64
Patch-based indices, 214
PCA, *see* Principal component analysis (PCA)
    methods
Phase space, 205
Pixel-based contagion index, 214–215
PM, *see* Passive microwave (PM)
Point spread function (PSF), 40, 130
Point-support variogram models, 153
Polygons, 22, 23, 26
Polyline, 26
Positional accuracy, 273–274; *see also* Accuracy
    assessment
Power model, variogram, 74
Prdgram, terrain analysis, 191
Precision, of measurements, 269
Principal component analysis (PCA) methods,
    154–155, 206
Process scale, 1, 2
PSF, *see* Point spread function (PSF)
Punctual (point) support, 44

Q

Quadtrees, 28–29

R

Radar system, 55–58
    distortion of images, 61
Radiometric resolution, 49–50
Random errors, 269; *see also* Errors
    stochastic model and, 269
Random fields, 68–70
Random function (RF), 68–69, 71, 74, 86, 299
Range, variograms, 8
Range (or look) direction, 56, 57
Raster, 22
Rayleigh criterion, 45
RCTG, *see* Relative contagion index (RCTG)
Reality phenomena, 299, 300, 302
Realizations, 287
Red–green–blue (RGB) color space, 148
    to IHS, 148, 154–155
Reflection coefficient reflect, 59
Regionalized variable, 69
Regression methods, 10
Regularization, variograms, 77–80
Regular lattice, 93, 94; *see also* Irregular lattice
Regular points, 22, 23
Relative contagion index (RCTG), 215
Relative entropy, 230, 233, 236

Relative patchiness index (RPI), 215
Relative root mean squared error (RRMSE), 269;
    see also Errors
Remotely sensed images
    information content in, 238–250
Representative elementary area (REA), 189–190
Resolution, 2, 33
    image, information content and, 250–255
    microwave remote sensing and, 55–58
    optical remote sensing and, 47–55
    super, information capacity and, 253–255
    vs. accuracy, downscaling, 197–198
RF, *see* Random function (RF)
RMSE, *see* Root mean squared error (RMSE)
Romme's relative evenness index (EV), 214
Root mean squared error (RMSE), 269, 274; *see
    also* Errors
Roughness, 58–59
RPI, *see* Relative patchiness index (RPI)
RRMSE, *see* Relative root mean squared error
    (RRMSE)
Ruggedness index, 192–193

S

Sample size, 37, 40
Sampling density, 37, 40
Sampling interval, 40
Sampling scheme, 37, 39–40
Saturation, 154
Scalar geo-field, 21
Scale
    area-class mapping, *see* Area-class mapping
    downscaling, 9–12; *see also* Downscaling
    issue of, 1–4
    models of, 4–9
    spatial variation, 300
    terrain analysis, *see* Terrain analysis
    upscaling, 9–12; *see also* Upscaling
Scale dependence of representations, 32–33
Scale sensitivity analysis, 4
Scale variance analysis
    in area-class mapping, 218
Scaling, defined, 107
Scanner-based imaging, geometric
    characteristics, 53–54
SD, *see* Standard deviation (SD)
Second-order stationarity, 70
Self-information, 235
Self-similarity, complex curves, 5
Semivariogram, 72, 73
Sequential indicator simulation (SIS), 290
Shannon's sampling theorem, 22
Sill, variograms, 8
Simple kriging, 109–110
Simple random sampling, 37–38
Single mean value, 44

SIS, *see* Sequential indicator simulation (SIS)
Slant range, 56, 57
Slope and aspect, terrain analysis, 176–180
Space photography, geometric characteristics, 51
Space–time data, 300, 303
Spatial autocorrelation, 8, 92
    Geary's c index, 96, 97, 99
    lattice data and, 95–99
    Moran's I index, 97–98, 99
Spatial data modeling, 12
Spatial degradation, 130
Spatial dependence, 108
Spatial/geographic information, 17, 22
    capacity and super resolution, 253–255
    content, 227–264; *see also* Information
      content
    in images with additive noise, 242–246
    with multiplicative noise, 247–250
    mutual information, 233–235
    uncertainty, 265–298
Spatial heterogeneity, 9, 216
    scale modeling and, 11
Spatially binary variables, 31
Spatially unary variables, 31
Spatial patterns, area-class mapping, 213–218
Spatial resolution, in remotely sensed images,
    44–47
Spatial responsivity, 47–50
Spatial sampling
    data support and, 36, 43–44
    design, 36–43
    framework, 36–47, 300, 301, 302
Spatial scales, in area-class mapping, 213–218
Spatial statistics, 8–9
Spectral analysis, 190
Spectral response, 47–50
Spectral responsivity specres λ, 48–50
Spherical model, variogram, 73
Standard deviation (SD), 269; *see also* Errors
Stationary stochastic process, 237
Statistical approaches
    gridded data upscaling, 131–134
Stochastic model, random errors and, 269
Stochastic simulation, 267
Stratified sampling, 38–39
Subpixel mapping
    gridded data, 144–145
Super-resolution mapping
    gridded data, 141–144
Systematic sampling, 38

T

Temporal dimension, 32
Terrain analysis
    curvature, 180–185
    digital elevation data, 171–176

overview, 169–171
secondary properties, 186–187
slope and aspect, 176–180
topographic wetness index, 186–187
topography, 187–193; *see also* Topography,
    models of scale in
variables, methods for scaling
    downscaling, 196–198
    upscaling, 193–196
Terrestrial Hydrology Model with
    Biogeochemistry (THMB), 194–195
TFL, *see* Tobler's first law of geography (TFL)
Thematic measure, 257
Thiessen polygons (Voronoi/Dirichlet polygons),
    23
THMB, *see* Terrestrial Hydrology Model with
    Biogeochemistry (THMB)
Thresholding, area-class mapping, 209
TINs, *see* Triangulated irregular networks (TINs)
Tobler's first law of geography (TFL), 21
Topographic wetness index, 186–187
Topography, models of scale in, 187–193
    cell-counting method, 189
    divider method, 189
    local variance method, 188–189
    local variogram models, 187–188
    Prdgram, 191
    REA, 189–190
    ruggedness index, 192–193
    spectral analysis, 190
    VRM, 193
    wavelet analysis, 191–192
Topologic measure, 257
Transition probability-based models, 75–77
Tree-based models, data conflation, 163–166
TREES, *see* TRopical Ecosystem Environment
    observation by Satellite (TREES)
    project
Triangulated irregular networks (TINs), 22, 23,
    29–31
Triangulation refinement, 30–31
TRopical Ecosystem Environment observation by
    Satellite (TREES) project, 201

U

Uncertainty, spatial information, 265–298
    accuracy and, *see* Accuracy assessment
    geostatistical simulation, 287–296
    overview, 265–268
Univariate block simulation, 294
Universal kriging, 113, 115
Unrestrained deconvolution, 138
Upscaling, 9–12, 107; *see also* Downscaling
    area-class information, 219–220
    by block kriging, 118–120
    convolution and, 130, 132

gridded data, 131–137; *see also* Gridded data scaling
image coarsening and, 130
terrain variables, 193–196
orography-based methods, 195
statistical methods, 194
US Federal Geographic Data Committee, 18
US National Snow and Ice Data Center (NSIDC), 63

V

Validation, geostatistical approaches to, 278–282
Variograms, 8–9, 13
deconvolution/de-regularization, 80–82
measurement scale and, 82–86
multivariable, 153
regularization, 77–80
Vector, 22
Vector maps, 227
Vector ruggedness measure (VRM), 193
Voronoi regions, 257
VRM, *see* Vector ruggedness measure (VRM)

W

Wavelet analysis, 191–192
Wavelet-based image fusion, 161–163
Wavelet decomposition, IHS and, 162
Wavelet transform, 155–160
Wavelet transform method, 7
Whittaker–Shannon interpolation formula, 42
Wiener deconvolution, 137–138

Printed and bound by CPI Group (UK) Ltd, Croydon, CR0 4YY

18/10/2024

01776264-0012